Target Sites for Herbicide Action

TOPICS IN APPLIED CHEMISTRY

Series Editors: Alan R. Katritzky, FRS
Kenan Professor of Chemistry
University of Florida, Gainesville, Florida

Gebran J. Sabongi
Laboratory Manager, Encapsulation Technology Center
3M, St. Paul, Minnesota

BIOCATALYSTS FOR INDUSTRY
Edited by Jonathan S. Dordick

CHEMICAL TRIGGERING
Reactions of Potential Utility in Industrial Processes
Gebran J. Sabongi

THE CHEMISTRY AND APPLICATION OF DYES
Edited by David R. Waring and Geoffrey Hallas

HIGH-TECHNOLOGY APPLICATIONS OF ORGANIC COLORANTS
Peter Gregory

INFRARED ABSORBING DYES
Edited by Masaru Matsuoka

STRUCTURAL ADHESIVES
Edited by S. R. Hartshorn

TARGET SITES FOR HERBICIDE ACTION
Edited by Ralph C. Kirkwood

A Continuation Order Plan is available for this series. A continuation order will bring delivery of each new volume immediately upon publication. Volumes are billed only upon actual shipment. For further information please contact the publisher.

Target Sites for Herbicide Action

Edited by
Ralph C. Kirkwood
University of Strathclyde
Glasgow, United Kingdom

Plenum Press • New York and London

Library of Congress Cataloging-in-Publication Data

Target sites for herbicide action / edited by Ralph C. Kirkwood.
p. cm. -- (Topics in applied chemistry)
Includes bibliographical references and index.
ISBN 0-306-43846-1
1. Weeds--Physiology. 2. Plants, Effect of herbicides on.
3. Herbicides. 4. Weeds--Control. I. Kirkwood, R. C. (Ralph C.)
II. Series.
SB611.T36 1991
632'.954--dc20 91-22530
CIP

ISBN 0-306-43846-1

A Division of Plenum Publishing Corporation
233 Spring Street, New York, N.Y. 10013

Printed in the United States of America

Contributors

Peter M. Bramley • Department of Biochemistry, Royal Holloway and Bedford New College, University of London, Egham, Surrey TW20 0EX, United Kingdom

Richard H. Bromilow • Department of Insecticides and Fungicides, AFRC Institute of Arable Crops Research, Rothamsted Experimental Station, Harpenden, Herts AL5 2JQ, United Kingdom

Keith Chamberlain • Department of Insecticides and Fungicides, AFRC Institute of Arable Crops Research, Rothamsted Experimental Station, Harpenden, Herts AL5 2JQ, United Kingdom

John R. Coggins • Department of Biochemistry, University of Glasgow, Glasgow G12 8QQ, United Kingdom

Alan D. Dodge • School of Biological Sciences, University of Bath, Bath BA2 7AY, United Kingdom

John L. Harwood • Department of Biochemistry, University of Wales College of Cardiff, Cardiff CF1 1ST, United Kingdom

Kriton K. Hatzios • Department of Plant Pathology, Physiology and Weed Science, Virginia Polytechnic Institute and State University, Blacksburg, Virginia 24061-0330

Ralph C. Kirkwood • Department of Bioscience and Biotechnology, University of Strathclyde, Glasgow G4 0NR, United Kingdom

Christopher R. Leake • Environmental Sciences Department, Schering Agrochemicals Ltd., Chesterford Park Research Station, Saffron Walden, Essex CB10 1XL, United Kingdom; *present address:* NOR-AM Research Center, Pikeville, North Carolina 27863

David M. Mousdale • Department of Biochemistry, University of Glasgow, Glasgow G12 8QQ, United Kingdom

W. John Owen • DowElanco Ltd., Letcombe Laboratory, Letcombe Regis, Wantage, Oxon OX12 9JT, United Kingdom

Ken E. Pallett • Department of Plant Biochemistry, Rhone-Poulenc Agriculture, Ongar, Essex CM5 0HW, United Kingdom

Preface

The development of chemicals to selectively control the growth of weeds has been a fascinating success story which has unfolded largely during the last four decades. The dramatic growth of herbicide research that followed the wartime discoveries of the auxin-type herbicides (phenoxyalkanoic acids) resulted in a whole range of compounds and mixtures which are used to eliminate broad- or narrow-leaved weeds from agricultural, horticultural, or forestry crops.

Today, the safe use of this armament of compounds requires our understanding of their mode of action, metabolism, and environmental persistence. The most recently developed herbicides are highly effective inhibitors of specific enzyme systems, and formulation may be an important factor determining their efficient delivery at specific target sites.

In this book, the major target sites of herbicide action are discussed in Chapters 1–5, with particular reference to photosynthesis; amino acid, lipid, and carotenoid synthesis; and other primary target sites. The effects of synergists or antagonists as modifiers of herbicide action are described in Chapter 6.

The importance of efficient target site delivery as a fundamental factor in herbicide activity and selectivity is generally recognized. Delivery of a potentially lethal dose of active ingredient may depend on a whole range of factors including the efficiency of application, retention, absorption, translocation, immobilization, and detoxification. These aspects are considered in the remaining chapters, with particular reference to the pathways and mechanisms involved in the uptake, translocation, and metabolism of soil- and foliage-applied herbicides.

Most, if not all, of the compounds mentioned in this volume have arisen through the systematic screening of novel chemicals and the identification of active lead compounds. The subsequent development of more active analogues and suitable formulation/delivery systems illustrates the ingenuity and expertise of the R & D chemists/biologists involved in the launch of a new herbicide. While this approach has been eminently successful, there is a view that, ideally, novel compounds should be tailored to fit specific enzyme receptor sites. This aim will be achieved only with improved understanding of the molecular architecture of the site(s) and the molecular requirements for transport into and within the plant. The approach taken in this book should serve to illuminate the problems and possible solutions.

It is hoped that this volume will be of interest and value to those concerned with scientific agriculture and horticulture, including growers, government advisors, representatives of the agrochemical industry, and researchers in herbicide activity, mode of action, or environmental fate. It should also be useful to those undergraduates and postgraduates concerned with plant biochemistry and physiology, since the study of herbicide action has led to a better understanding of the mechanisms these chemicals inhibit.

Ralph C. Kirkwood

Glasgow, United Kingdom

Contents

2. Amino Acid Synthesis

David M. Mousdale and John R. Coggins

3. Lipid Synthesis

John L. Harwood

4. Carotenoid Biosynthesis

Peter M. Bramley

5. Other Primary Target Sites for Herbicides

Ken E. Pallett

6. Modifiers of Herbicide Action at Target Sites

Kriton K. Hatzios

7. Fate of Soil-Applied Herbicides: Factors Influencing Delivery of Active Ingredients to Target Sites

Christopher R. Leake

Target Sites for Herbicide Action

1

Photosynthesis

ALAN D. DODGE

1.1. INTRODUCTION

Photosynthesis is essential for all plant life on earth, for by this process alone are plants able to convert solar energy into chemical energy, thus providing the raw materials for growth and development. Crop plants produce flowers, fruits, seeds, tubers, petioles, roots, and leaves for our energy and dietary needs and our enjoyment, while weeds produce fruits, seeds, tubers, and rhizomes, making the crop production more difficult! In practical terms, agricultural cultivation is a highly managed ecosystem in which plants have been bred for productivity rather than survival and therefore must be artificially protected from the competing plants (weeds) by herbicide treatment or physical removal of the latter.

The use of herbicides that inhibit or interact with the photosynthetic machinery is seen, superficially, to be advantageous because of the likelihood of fewer problems of animal toxicity. However, the identical nature of the process in both crop and weed species, apart from C_4 plants, means that herbicide selectivity must be achieved by differential uptake, movement, or metabolism. Following the discovery of the herbicidal action of *N*′-(4-chlorophenyl)-*N*,*N*-dimethylurea (subsequently known as CMU or monuron) by Bucha and Todd in 1951,[1] Wessels and Van der Veen[2] and Cooke[3] showed that this compound was a potent inhibitor of photosynthetic electron transport. This, the first so-called "Hill reaction inhibitor," was followed subsequently by numerous other phenylureas, triazines, uracils,

ALAN D. DODGE • School of Biological Sciences, University of Bath, Bath BA2 7AY, United Kingdom.

phenylcarbamates, triazinones, hydroxybenzonitriles, phenols, and other compounds. The years between 1955 and 1975 saw the peak of discovery of photosynthetic electron transport inhibitors, including, for example, at least 15 new phenylureas and a similar number of triazines. Although there have been few novel compounds released as electron transport inhibitor herbicides during the last 15 years, there have been major changes in our understanding of the site of action of this class of herbicides and of the events that follow and lead to plant death.

The other major class of herbicides involved with photosynthesis are compounds that act as divertors of electron flow at photosystem I (PSI). After the initial discovery of diquat and paraquat in the late 1950s,[4,5] other related compounds have had limited use. One of these, morfamquat, has now been withdrawn. Thirty years later, there are no new chemicals or herbicide families to rival paraquat and diquat at this particular target site. We have, however, an increased understanding of the nature of the damaging radical species that are generated as a result of the action of these compounds. With these herbicides, as also with the electron transport inhibitors, toxic oxygen species have a major role.

The inhibition of photosynthetic carbon metabolism has not proved to be a fruitful ground for the discovery and development of herbicidal compounds. Consideration of the action and chemistry of enzyme inhibitors, however, could prove to be a profitable exercise for the identification of new herbicidal target sites.

1.2. THE GENERAL STRUCTURE OF THE CHLOROPLAST AND THE ELECTRON TRANSPORT SYSTEM

The complete photosynthetic apparatus for capturing light energy and incorporating carbon dioxide into organic molecules occurs within the chloroplast. In higher plants, the chloroplasts are generally discoid in shape, with a length of 5–10 mm and a width of around 1 mm. In palisade mesophyll cells of leaves, there may be more than 100 chloroplasts per cell, while in the spongy mesophyll cells, about half this number. Transmission electron microscopy has revealed that a chloroplast is bounded by an envelope of two distinct membranes, and these surround a generally clear stroma in which there is a highly organized membrane system. The most conspicuous features of this membrane system are the hollow disks, called thylakoids, which are arranged in stacks called grana. Membrane interconnections between the stacks are called stroma lamellae. The thylakoid system thus provides an extensive membrane with an area approximately 300 to 800 times per unit leaf area,[6] for light capture and associated electron transport.

Figure 1.1 shows in outline a portion of a thylakoid membrane. It is visualized as a lipid bilayer in which various protein components are

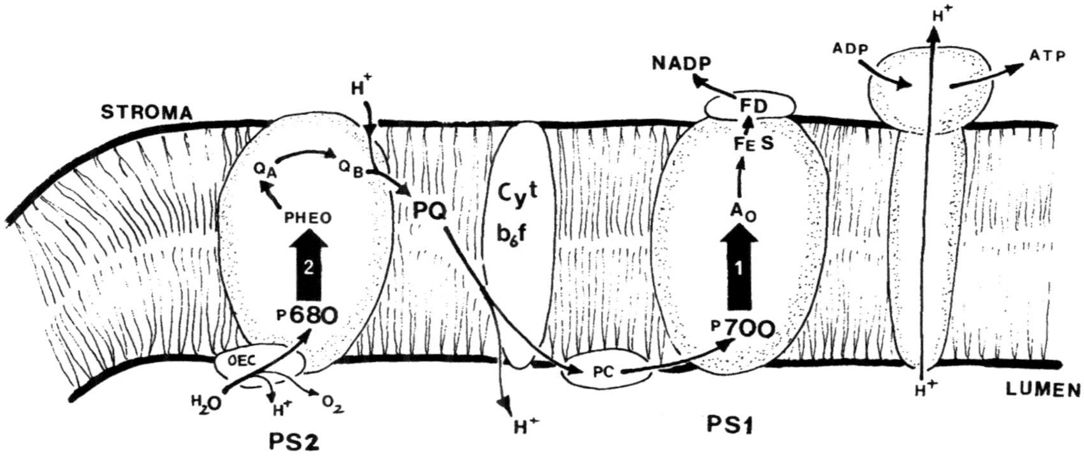

Figure 1.1. Simplified scheme showing electron transport in a portion of a chloroplast thylakoid membrane. Electrons flow from water via an oxygen-evolving complex (OEC) to photosystem II (PS2), pheophytin (PHEO), plastquinone (PQ), plastocyanin (PC) to photosystem I (PSI). A_0, Chlorophyll; FeS, iron–sulfur centres; FD, ferredoxin. Phosphorylation is catalyzed by proton transport through a transmembrane proton channel (CF_0) to the ATP-synthetase complex (CF_1).

embedded. The lipid bilayer is composed predominantly of the two glycolipids monogalactosyldiacylglycerol and digalactosyldiacylglycerol, with the 18:3 fatty acid α-linolenate representing around 90% of the fatty acid content. The presence of a high level of unsaturated fatty acid gives the membrane a high degree of fluidity, thus allowing the lateral movement of pigment proteins for maximizing energy capture and utilization. Two photosystems, PSII and PSI, composed of pigment protein complexes and associated proteins, are responsible for absorbing light energy and mediating the utilization of this energy to drive electrons from water through a series of carriers including plastoquinone, cytochrome *f*, plastocyanin, ferredoxin, and ferredoxin:$NADP^+$ oxidoreductase to $NADP^+$. As a consequence of this electron flow, there is a vectorial movement of protons across the thylakoid from the stroma to the lumen. The electrochemical potential gradient thus generated is utilized, via the CF_0-CF_1 complex, to catalyze the phosphorylation of ADP to ATP.

1.3. THE ORGANIZATION OF PHOTOSYSTEM II

Although it has been known for over 30 years that electron transport inhibitor herbicides interact with PSII, our concept of the structural organization of this photosystem, and the exact site of interaction, has advanced greatly during the last few years.

PSII consists of two distinct functional areas—a reaction center and a group of pigment proteins that act as the light-harvesting complex. The reaction center contains only about 1-2% of the pigment of the complete photosystem. It consists of two proteins, known as D1 and D2, of 32 and 30 kDa size, respectively, plus cytochrome *b*-559 and two other proteins of unknown function.[7] Associated with the D1 and D2 proteins is P680, a specialized chlorophyll molecule that not only acts as the energy trap for the light-absorbing pigment of PSII, but also as a redox component, acting as the primary electron donor; molecules of pheophytin and nonheme iron are also associated. Furthermore, D1 and D2 act as the apoproteins for specialized forms of plastoquinone, Q_A for D2 and Q_B for D1 (Figure 1.2). Electron flow from water to plastoquinone is thus initiated because excitation energy channeled to P680 results in a charge separation between $P680^+$ and $pheophytin^-$. Electrons are donated to $P680^+$ immediately from tyrosine residues in the D1 protein,[8] which in turn receives electrons from water via a manganese cluster and associated proteins. This results in the evolution of oxygen and the release of protons into the thylakoid lumen. Recombination of electrons in the $P680^+$, pheophytin couple is prevented by the passage of electrons to Q_A. Q_A acts as a one-electron carrier, passing electrons via nonheme iron to Q_B. The relationship of Q_B to the D1 protein is not identical

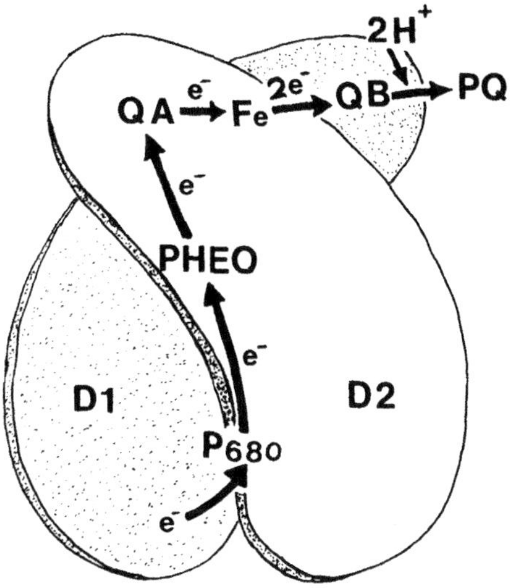

Figure 1.2. Simplified scheme showing the two polypeptides of the PSII reaction center, D1 and D2, together with associated quinones Q_A and Q_B.

to that of Q_A to D2. Reduction of Q_B is a two-electron process, and this is followed by protonation. Q_B remains tightly bound until the receipt of a second electron and protonation, and then this plastoquinol links with the mobile plastoquinone pool within the thylakoid membrane.

The other major functional area of PSII is that of the antenna proteins. Current ideas suggest that there are several hundred chlorophyll molecules associated with each reaction center, and they can be separated into distinct apoproteins with molecular weights of 29, 25–28, 27 or 31, and 21 kDa.[7] As these proteins do not appear to be associated with herbicide binding, they will not be discussed further.

1.3.1. Detailed Structure of the D1 Protein and the Herbicide Binding Site

Extensive work during the last decade has clearly shown that the D1 protein of the PSII reaction center is also the "herbicide binding" protein.[9] This protein, formerly known as the 32-kDa or Q_B protein, consists of 353 amino acid residues; according to data derived from hydrophobicity plots, it is folded in such a way that there are five transmembrane segments (Figure 1.3).[10] This protein appears also to have another unique feature in that it is rapidly degraded and synthesized during normal photosynthetic function.[11] It has been shown that turnover is light related and proceeds from zero in darkness to a turnover of 50–80 times that of other photosynthetic membrane proteins under high light.[12] The region of the protein between the fourth and fifth helices is thought to be the site of Q_B interaction and is in close proximity to the nonheme iron electron donor that is linked to

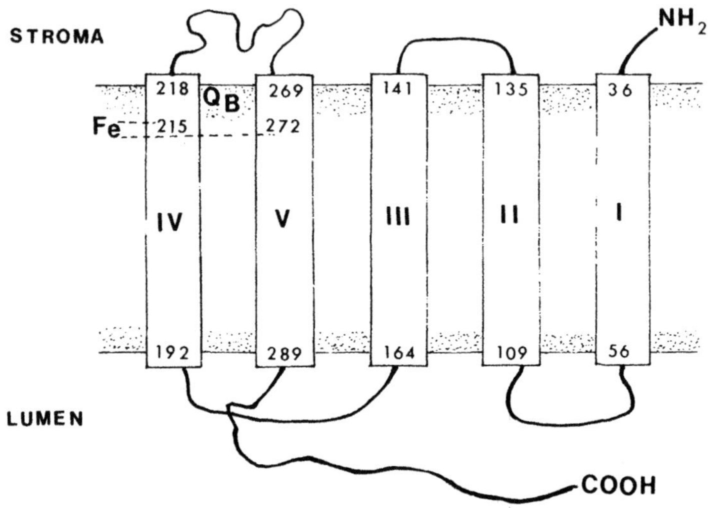

Figure 1.3. Diagrammatic scheme showing the possible folding of the D_1 polypeptide through a thylakoid membrane in five transmembrane segments. Numbers refer to the amino acid residues. Based on predictions from hydropathy index plots by Trebst.[10] See also Marder and Barber.[7]

histidine residues at amino acids 215 and 272. Considerable evidence accumulated from displacement studies, the use of azido probes with mutations, and X-ray crystallography of the related photosynthetic bacterial L protein has given an increasingly clearer picture of the relationship between electron transport inhibitors and the D1 protein.

The diversity of chemical structures that interact, apparently at a similar site, has given rise to a considerable amount of quantitative structure–activity relationship (QSAR) speculation.[13-19] The high-affinity noncovalent binding was thought to be due in part to the presence of a common structural element, an sp^2 hybrid carbon next to an electron-withdrawing group such as O or N and linked to a lipophilic carrier. The work of Tischer and Strotmann[20] showed that herbicides from various families could displace each other from the chloroplast, thus implying a degree of functional overlap.

The appearance of weeds that showed considerable resistance to the triazine herbicides[21] eventually led to detailed biochemical studies that identified resistance mechanisms at the level of a D1 protein amino acid change. Hirschberg and McIntosh[22] showed that in *Amaranthus hybridus* this was due to a replacement of serine 264 by glycine. Remarkably, this particular amino acid change did not greatly affect the inhibitory properties of the phenylurea diuron.[23] Subsequent work has shown, however, that changes at amino acids 219 and 275 will give resistance to diuron but have little effect with atrazine.[24] Studies on mutations of the D1 protein have been extensively reviewed,[25,26] and the most surprising feature of this work is that the amino acid sequence pattern is so highly conserved throughout the plant kingdom.

The Nobel Prize winning X-ray crystallography work of Deisenhofer, Huber, Michel, and colleagues[27–29] has revealed the detailed structure of the photosynthetic reaction center of the purple bacteria *Rhodopseudomonas viridis*. There is little effect of any of the PSII inhibitors on bacterial electron transport apart from the triazine herbicide terbutryn.[30] Figure 1.4 shows the possible orientation of this herbicide with the bacterial L reaction center protein. It shows the possibility of hydrogen bridges between the nitrogen of the peptide bond of isoleucine 224 and the hydroxyl group of serine 223. It is likely that atrazine would bind similarly to the D1 protein of higher plants, for which, by analogy, amino acid changes and also azido triazine labeling[31] support the idea of a similar binding niche.[24,26]

In addition to attempting to identify the exact sites of interaction of herbicides with specific amino acids, other workers have stressed the importance of an association with thylakoid lipids,[32] and also the specific structure of plastoquinone.[18] It is well known that PSII herbicides compete with Q_B for its binding site and thus, by displacing the quinone, inhibit electron

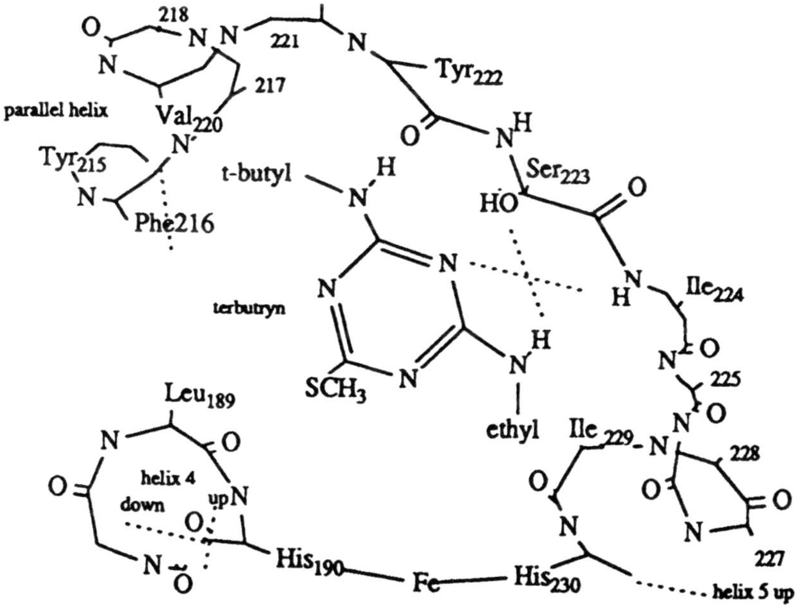

Figure 1.4. Scheme showing the binding of terbutryn to the L subunit of *Rhodopseudomonas viridis*. From Trebst[26] based on Refs. 27–29.

transport.[33] Vermaas *et al.*[34] showed that an analogue of plastoquinone, azidoquinone, would replace plastoquinone, and this would inhibit also the binding of atrazine and ioxynil. Furthermore, Oettmeier and Soll[35] showed that a short-chain plastoquinone analogue competed with diuron for a binding site in the chloroplast thylakoid. On the basis of this and other information,[36] Gardner[18] has proposed a model for herbicide binding that takes account of certain structural relationships between plastoquinone and various electron transport inhibitors (Figure 1.5). For example, in the case of diuron, the carbonyl group overlaps the carbonyl C-4 of plastoquinone, and the substituted phenyl ring occupies the same region as the isoprenoid tail adjacent to C-5.

Thus far, no mention has been made of the inhibitory binding characteristics of certain phenol-type herbicides such as the hydroxybenzonitriles, DNOC, and dinoseb. There are clear chemical differences between these

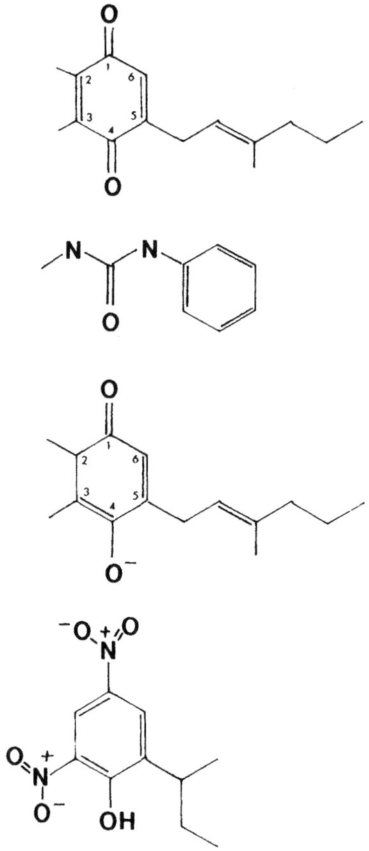

A

B

C

D

Figure 1.5. Stereochemical models for the PSII herbicides diuron (B) and dinoseb (D) in relation to plastoquinone (A) or the plastoquinone semiquinone (C). Based on the work of Gardner.[18]

compounds and those that interact at the phenylurea–triazine site. This has been noted in different inhibitory characteristics[37] and the possibility of different sites of action.[38] Detailed work with these herbicides is described by Van Rensen,[39] together with QSAR studies by Trebst and Draber.[17] It has been postulated that interaction with the D1 protein is likely, being related to steric characteristics rather than lipophilicity or electronic parameters. Furthermore, Gardner[18] has developed an analogy between dinoseb and plastoquinone (Figure 1.5). In this instance, however, he postulated an analogy between the herbicide and the semiquinone of plastoquinone, with the phenol hydroxyl group of dinoseb in the region of the C-4 semiquinone and the nitro group *para* to the hydroxyl group related to the C-1 carbonyl.

1.3.2. The Consequences of Electron Transport Inhibition

The displacement of Q_B and the failure to pass electrons to the plastoquinone pool will, in the first instance, lead to an inhibition of $NADP^+$ reduction and hence, indirectly, CO_2 incorporation.[40] Although early workers suggested that photosynthetic inhibition would lead to plant starvation, the appearance of phytotoxic symptoms such as chlorophyll bleaching was enhanced if plants were illuminated.[41–43] This has given rise to the concept that when light energy absorbed by photosynthetic pigments is not utilized, it will overtax the normal chloroplast protective systems and lead to chloroplast damage and bleaching.[40] Duysens and Sweers[44] first demonstrated that fluorescence emission was enhanced if algae were treated with diuron. Since that time, numerous investigators have used this technique, not only to study inhibition of electron flow, but also to monitor the rate of herbicide penetration.[45,46]

The absorption of light energy by a photosynthetic pigment such as chlorophyll will, in the first instance, promote the generation of the singlet state with a lifetime of around 10^{-6}–10^{-8} s. In addition to the emission of fluorescence, which indicates energy wastage, excitation energy is passed by resonance transfer to excite the acceptor P680. If $P680^+$ accumulates because the displacement of Q_B as pheophytin is reduced, the strongly oxidizing potential could instigate damage to adjacent pigments and polypeptides. It is possible that cytochrome *b*-559 of the PSII reaction center could have a role in promoting electron cycling as a means of reducing $P680^+$ and thus preventing damage.[47] It is also possible that P680 could be dissociated from the light-harvesting antennae of PSII.[48] The conversion of singlet chlorophyll by intersystem crossing to the triplet state could be equally damaging. Triplet chlorophyll can directly interact in a damaging way with pigments, proteins, and lipids (type I reaction). Furthermore, it can interact with triplet or ground-state oxygen ($^3O_2$) to generate the potentially damaging singlet oxygen ($^1O_2$). This is a type II reaction. Investigations with leaves[40] and isolated chloroplasts[49] showed that photodynamic damage was enhanced by the presence of oxygen. Recent studies with isolated PSII reaction centers[50,51] have shown that they are particularly susceptible to photodynamic damage, particularly in the presence of oxygen. A role for singlet oxygen has been proposed.[24]

Singlet oxygen is a very reactive form of oxygen and will interact with membrane fatty acids to produce lipid peroxides, with protein amino acids (especially histidine and methionine), and also with nucleic acid bases. During normal photosynthesis, the close proximity of excited photosensitizer pigments (chlorophylls) and oxygen means that a great deal of protection is built into the chloroplast (see Section 1.5 below and Chapter 5 of this volume). Within the photosystems, carotenoid pigments such as β-carotene and lutein are able to quench both triplet chlorophyll and $^1O_2$. It is not surprising therefore that carotenoid-less mutants are rapidly bleached.[52] The $^1O_2$ and free-radical scavenger α-tocopherol is present in the thylakoids in a ratio of one molecule to 24 chlorophylls.[53] Furthermore, ascorbate and glutathione, which are both present in the stroma in millimolar amounts, also quench $^1O_2$.[54] Because of the reactivity of $^1O_2$ and the difficulty of identifying it in complex biochemical systems, current evidence for its generation in herbicide-treated leaves is circumstantial.[40] A model system developed by Percival and Dodge[49] was analogous to an inhibited chloroplast. In these experiments with isolated thylakoids, photodynamic damage was inhibited in the absence of oxygen or by the presence of a water-soluble carotenoid, crocetin. It was enhanced, however, by the use of a deuterium buffer that prolongs the lifetime of $^1O_2$ in aqueous conditions by 2–10 μs. Attempts have been made to develop herbicidal chemicals that promote the photogeneration of singlet oxygen in either ultraviolet[55] or visible light.[56–58] One paradoxical feature of PSII herbicide-induced damage is that binding of the herbicide to the D1 protein actually inhibits D1 turnover.[11] Damage to the protein has been suggested to be catalyzed by the Q_B semiquinone radical, which could induce damage directly or indirectly by reducing oxygen to superoxide.[59] Recently, Matoo *et al.*[59] have shown that a free-radical scavenger will inhibit D1 protein breakdown. In normal circumstances, D1 turnover and repair appear to be essential for the maintenance of a fully functional photosynthetic system. In the long term, however, herbicide-inhibited turnover is bound to have an adverse effect.[24]

1.3.3. Modifying the D1 Protein and Engineering Herbicide Resistance

Considerable work during the last few years has identified the genetic basis of PSII inhibitor herbicide resistance and opened up the possibility

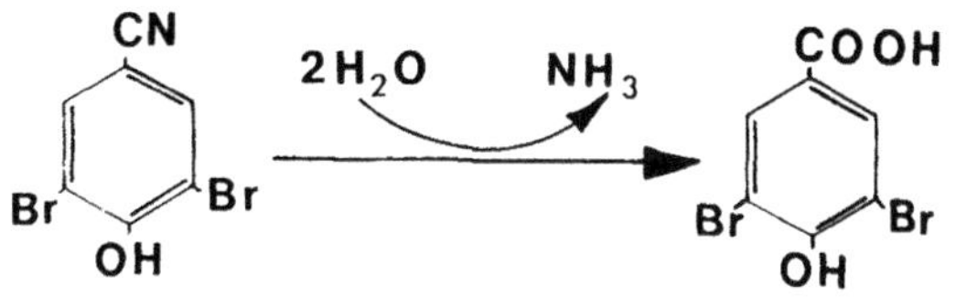

Figure 1.6. The conversion of bromoxynil to 3,5-dibromo-4-hydroxybenzoic acid by a nitrilase enzyme.

of transferring this resistance to normally tolerant plants. The D1 protein is encoded by a chloroplast gene, *psbA*,[60] and the nucleotide sequence has been established for tolerant and resistant plants.[61] Recent remarkable work by Cheung *et al.*[62] has shown that it is possible to transfer the *psbA* gene of *Amaranthus* into a nuclear gene in tobacco. Furthermore, these authors demonstrated that the product of the tobacco nuclear gene could be transported into the chloroplast via a transit peptide, the nuclear-encoded small subunit of the enzyme ribulose-bisphosphate carboxylase, and installed into the chloroplast thylakoid. The net result of this experiment was that tolerance of tobacco to atrazine was increased.

Another method of increasing resistance to a PSII herbicide is to promote the breakdown of the molecule within the plant before it reaches the active site. This has been achieved in an effective way in the case of bromoxynil resistance in tobacco.[63,64] Bromoxynil is broken down by a nitrilase enzyme to ammonia and a herbicidally inactive benzoic acid moiety (Figure 1.6). A strain of *Klebsiella pneumoniae* has been isolated which uses bromoxynil as a primary nitrogen source[65] by virtue of its nitrilase activity. The gene has been characterized[66] and transferred into tobacco using conventional gene transfer techniques. The transgenic tobacco thus produced was able to tolerate more than 20 times the normally lethal dose of bromoxynil.[64]

1.4. THE ORGANIZATION OF PHOTOSYSTEM I

Although the detailed structure of PSI has not received as much attention as that of PSII, it has been known for many years that PSI functions as a light-driven plastocyanin:ferredoxin oxidoreductase. Electron flow between these two proteins is mediated by a number of redox components that appear to be linked to a PSI core complex consisting of seven polypeptides. Approximately half of the PSI chlorophyll is linked to this core complex, and the remainder to a number of other polypeptides referred to as the light-harvesting complex.[7]

The sequence of redox components from plastocyanin to ferredoxin is probably as follows:

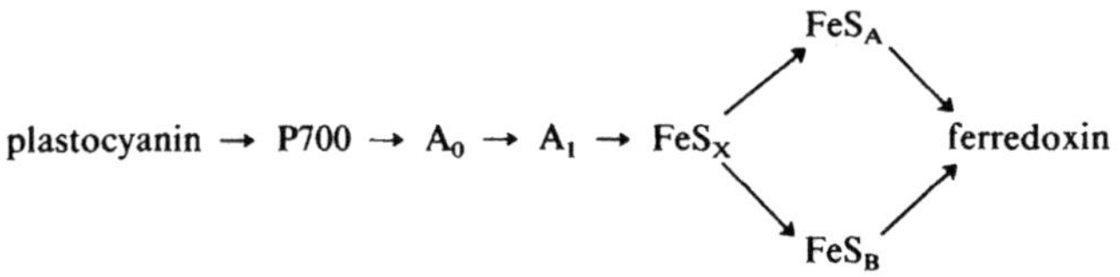

(1.1)

P700, the primary electron donor of PSI, and A_0, the electron acceptor, are chlorophylls, P700 having a redox potential of around −490 mV. A_1 is possibly a quinone, and FeS_X, FeS_A, and FeS_B are iron–sulfur centers with potentials of −705, −530, and −580 mV, respectively. FeS_A and FeS_B probably operate in parallel[67] and denote electrons to ferredoxin (redox potential, −420 mV) (Figure 1.7).

1.4.1. Herbicides Interacting with Photosystem I

It has been known since 1965[68] that the bipyridinium herbicides paraquat and diquat interact in the vicinity of ferredoxin and divert electrons from $NADP^+$. The immediate consequence is that carbon dioxide incorporation rapidly ceases.[69] Warden and Bolton[70] suggested that these herbicides divert electrons from $P430^-$, now known to be equivalent to the iron–sulfur centers A and B. The redox potentials of paraquat, −446 mV, and diquat, −349 mV, are appropriate for such an interaction. Recent work[71] has indicated that electron donation to paraquat is primarily from FeS_B although paraquat may also accept electrons from FeS_A, but with a lower affinity. It is also possible that paraquat could be reduced directly via ferredoxin.[72]

The criteria determining the herbicidal effectiveness of the bipyridiniums are related to redox potential, molecular size, and the ability to be reduced in a reversible manner by a one-electron transfer to a radical cation.[73] The dicationic nature of the molecule is important, not only for electron reception and donation, but probably also for facilitating binding

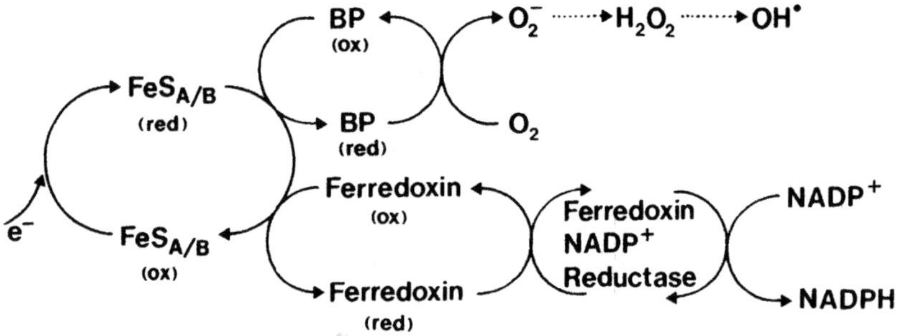

Figure 1.7. Electron transport at photosystem I through the iron–sulfur centers FeS_A and FeS_B to either ferredoxin or a bipyridinium herbicide (BP).

to negatively charged thylakoid membranes. Certain amphiphilic viologens with a similar redox potential to that of paraquat will interact with thylakoid membranes but do so via a hydrophobic relationship rather than an electrostatic one.[74] Electron donation to oxygen via these viologens was less than to paraquat, but this observation may have been associated with electron transport inhibition. Cuendet and Gratzel[74] also showed that Mg^{2+} could displace the dicationic paraquat from the chloroplast membrane. Subsequently, Bowyer *et al.*[75] demonstrated that Mg^{2+} would inhibit paraquat-mediated oxygen uptake by thylakoids. By contrast, it was shown that dioxanthiadiaza-2,5-pentalene, a heptapentalene derivative, with a redox potential of −375 mV, catalyzed oxygen uptake and induced herbicidal effects similar to those of paraquat but was not affected by divalent cations. This may reflect the higher lipophilicity of the heptapentalene and the fact that the thylakoid interaction did not involve an electrostatic relationship.[75]

As indicated previously, the initial events of bipyridinium action involve not only the generation of a potentially stable cation, but also the reoxidation of this cation by molecular oxygen. Early experiments with these herbicides noted a requirement for oxygen in eliciting toxic symptoms,[76] and subsequently this was shown to involve the generation of toxic oxygen radicals. Detailed studies of the relationship of bipyridiniums with oxygen[77] and work with pulse radiolysis techniques[78] showed that reduced paraquat interacts with oxygen with a reaction rate constant of $7.7 \times 10^8\ M^{-1}\ s^{-1}$, giving rise to a micromolar concentration of superoxide. Although it is not possible to identify superoxide generation *in vivo*, spin trapping experiments with isolated chloroplasts using the trap 5,5-dimethyl-1-pyrolline 1-oxide have clearly demonstrated superoxide generation in the presence of paraquat.[79–81]

Although some initial work suggested that superoxide was the primary damaging agent in bipyridinium action,[78] subsequent investigations have indicated that hydroxyl radicals generated from superoxide are more likely to be the most important toxic agent.[82] It is also possible that hydroxyl radicals could be generated from reduced paraquat and hydrogen peroxide.[83]

Current ideas suggest that superoxide ($O_2^{\cdot -}$) and hydrogen peroxide, produced either via superoxide dismutase or by spontaneous protonation, interact in the presence of an iron catalyst in a Fenton type reaction and generate hydroxyl radicals (OH·) (Figure 1.8).

As these hydroxyl radicals induce cellular breakdown, iron is probably released from iron–sulfur centers, ferredoxin, and cytochromes as well as from the store of phytoferritin.[84] The generation of OH· radicals in paraquat-treated thylakoids has been verified by ESR studies[81] as well as by using a specific probe with whole leaf tissue. It was shown with paraquat-treated *Lemna minor* leaves that there was a major increase in OH· generation.[85] In a different experimental approach, Vaughn and Duke[86] used cytochemical

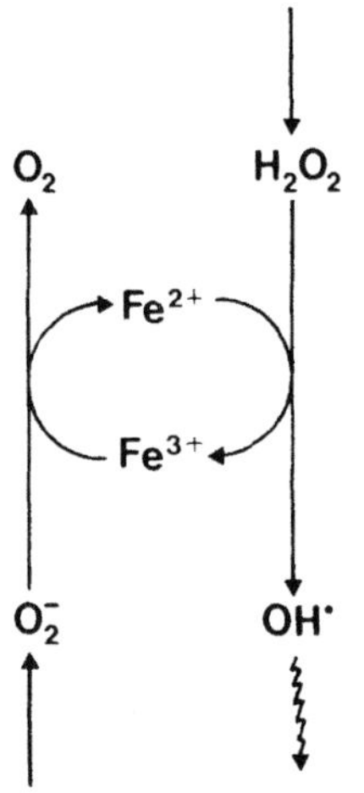

Figure 1.8. Scheme showing the formation of hydroxyl radicals (OH·) from superoxide $O_2^{\cdot -}$ and hydrogen peroxide.

and electron microscope techniques. Cerium perhydroxide, formed from interaction of peroxy and hydroxyl radicals with cerium chloride, was identified adjacent to thylakoid membranes.

In spite of considerable work during the last 30 years to identify other herbicidal PSI electron flow divertors, the bipyridiniums are still the only compounds of major commercial importance. Heptapentalenes have already been mentioned,[80] but these have yet to be commercialized. Likewise, older work with substituted quinones[87] did not yield herbicides. There is some evidence that some dinitrophenyl ether compounds are reduced via ferredoxin. In this instance, however, oxygen radicals are not of primary importance, rather the generation of a nitro radical[88,89] acts as a direct toxic agent.[90]

1.4.2. Radical-Induced Damage

It has already been mentioned that the major toxic agent of the PSI electron divertor herbicides is the OH· radical. Like singlet oxygen it promotes peroxidation of membrane fatty acids, and will react especially with deoxyribose sugar and thymine of DNA, and certain protein amino acids such as methionine and histidine. A short time after initial bipyridinium treatment major cellular disruption has occurred,[91] and the typical toxic symptoms of pigment bleaching are readily apparent. As the action of paraquat initially depends upon the maintenance of photosynthetic electron transport from PSII, experiments with chloroplasts isolated from treated flax leaves showed that damaging radicals inactivated this system within 20 hours.[69] Subsequent deleterious events in a general toxic cascade would also include the generation of $^1O_2$ from photosynthetically redundant chlorophyll as well as triplet chlorophyll-induced Type I reactions. Once cell integrity is abolished by damage to tonoplast and plasmalemma membranes, the release of vacuolar contents will further promote breakdown by osmotic changes, the release of toxic products, and the loss of cellular buffering capacity.[92]

1.4.3. Tolerance to Photosystem I Herbicides

Unlike tolerance of weeds to PSII herbicides, tolerance to paraquat or diquat is neither as widespread or as well understood. Furthermore, the level of tolerance is of a much lower level.[93]

Tolerant weeds are predominantly members of the Compositae and Gramineae and have been identified in the United Kingdom, Hungary, Egypt, Japan, and Australia. In principle, the most important reasons for tolerance are (i) that the herbicide does not reach the active site at PSI, being either sequestered or metabolized en route, (ii) that the herbicide reaches the active site but fails to intercept electrons, and (iii) that the generation of toxic radicals is promoted but the toxicity of these is prevented by enhanced scavenging systems. Work with *Conyza bonariensis*[94,95] and *Hordeum glaucum*[96] is strongly in favor of the first suggestion although a mechanism has yet to be identified for sequestration. On the other hand, Shaaltiel and Gressel[97] claimed that increased scavenging is of considerable importance. Although we are far from producing paraquat-tolerant plants by genetic engineering, a number of paraquat-tolerant biotypes of *Lolium perenne* have been identified as a result of a conventional plant breeding program.[98]

1.5. CHLOROPLAST ANTIOXIDATIVE PROTECTIVE SYSTEMS

The photosynthetic apparatus continues to function in an ever changing and potentially stressful environment because, among other factors, the chloroplast is endowed with a range of biochemical protective systems. It is evident that if these systems are inhibited or their production is prevented, cellular damage and possibly death will ensue. Conversely, if these biochemical systems could be rendered more efficient in crop plants, survival or possibly enhanced selectivity could be achieved.

A list of important chloroplast scavenging systems (in addition to peroxisome-located catalase) is presented in Table 1.1, together with the toxic species that they control. The function of the carotenoid pigments has been outlined already, and the inhibition of their production has severe toxic consequences (see Chapter 4 of this volume). The CuZn chloroplast enzyme superoxide dismutase is the first line of defense against chloroplast

Table 1.1. Some Toxic Species Generated within Plant Cells and Appropriate Scavenging Systems

Toxic species	Scavenging system
$O_2^{\cdot-}$ (Superoxide)	Superoxide dismutase(s)
	Ascorbate
H_2O_2	Ascorbate peroxidase
	Dehydroascorbate reductase
	Glutathione reductase
	Catalase
OH· and other free radicals	α-Tocopherol
	Ascorbate
	Glutathione
	Polyamines
	Flavonols
$^1O_2$ (singlet oxygen)	Carotenoids
	Ascorbate
	Glutathione
^{3}Chl (triplet chlorophyll)	Carotenoids
	Ascorbate
	Glutathione

superoxide[99]:

$$O_2^{\cdot-} + O_2^{\cdot-} + 2H^+ \rightarrow H_2O_2 + O_2 \qquad (1.2)$$

Tolerance to paraquat has been achieved by treating plants with a superoxide scavenger, copper penicillamine.[100,101] On the other hand, feeding plants with an inhibitor of superoxide dismutase, diethyl dithiocarbamate, actually enhanced the action of paraquat.[102,103]

The removal of superoxide dismutase-generated H_2O_2 within the chloroplast is accomplished via ascorbate and glutathione and the linked enzymes ascorbate peroxidase, dehydroascorbate reductase, and glutathione reductase. Generally, it is referred to as the "ascorbate-glutathione cycle."[104,105]

Some further aspects of the toxic effects of unscavenged H_2O_2 will be discussed below (Section 1.6.1). Experiments with isolated chloroplasts have shown that during illumination in the presence of paraquat, there is a rapid oxidation of both ascorbate[106] and glutathione.[107]

Schmidt and Kunert[108] treated *Ipomoea purpurea* cotyledons with either sodium ascorbate or L-galactano-1,4-lactone, a precursor of ascorbate, and protected the cotyledons against paraquat-induced damage. The action of the scavenger α-tocopherol appears to be related to ascorbate as this is needed to reduce the radical-induced tocopheroxyl to α-tocopherol.[109] Once the ratio of ascorbate to α-tocopherol falls below about 10:1, protective effectiveness is diminished, and radical damage may follow.[110] More recent work suggests that a number of compounds including polyamines[111] and flavonols[112] are effective free-radical scavengers. It is evident that a study of such chloroplast quenching systems could be a valuable area for ideas for the design and development of new herbicides.

1.6. PHOTOSYNTHETIC CARBON METABOLISM

The fundamental basis of photosynthetic carbon metabolism is the incorporation of carbon dioxide by ribulose-bisphosphate carboxylase (rubisco). This leads to the synthesis of three-carbon sugars which are either exported from the chloroplast or metabolized to regenerate the acceptor ribulose bisphosphate. Rubisco is a bifunctional enzyme in that, in parallel to carboxylation, it catalyzes an oxygenation reaction that leads to phosphoglycolate. This is the starting point for photorespiratory metabolism, which will be discussed below (Section 1.6.2). In C_4 plants, the conventional C_3 pattern of the photosynthetic carbon reduction Calvin cycle is confined to the bundle sheath cells. The surrounding mesophyll cells act as an ancillary carbon dioxide pump, fixing carbon dioxide via phosphoenolpyruvate carboxylase into C_4 acids. These are transported to the bundle sheath for decarboxylation.[113] In this way, photorespiration is limited because of the elevated carbon dioxide levels.

These complex areas of photosynthetic carbon metabolism could provide considerable potential for the development of specific enzyme inhibitors as herbicides, but to date little success has been achieved. Potential targets will be discussed in relationship to (a) C_3 metabolism, (b) photorespiration, and (c) C_4 metabolism.

1.6.1. Inhibitors of the Photosynthetic Carbon Reduction (Calvin) Cycle

Quantitatively, the most important enzyme of the photosynthetic carbon reduction cycle is rubisco, which represents around 50% of soluble leaf protein.[6] This has a major disadvantage in that inhibition of a large amount of leaf protein could require logistically difficult and probably prohibitively expensive herbicide levels.[114]

It has been suggested that during normal photosynthesis, only 40–60% of this enzyme is functioning at any time[115] and that the remainder could act as a possible trap for potential inhibitors. These could include xylulose 1,5-bisphosphate, which could arise by nonenzymatic epimerization of ribulose 1,5-bisphosphate.[116]

One approach to developing inhibitors of rubisco is to produce analogues of the natural intermediate of the carboxylase reaction, 2-carboxy-3-ketoribitol 1,5-bisphosphate (Figure 1.9B). Siegel and Lane[117] showed

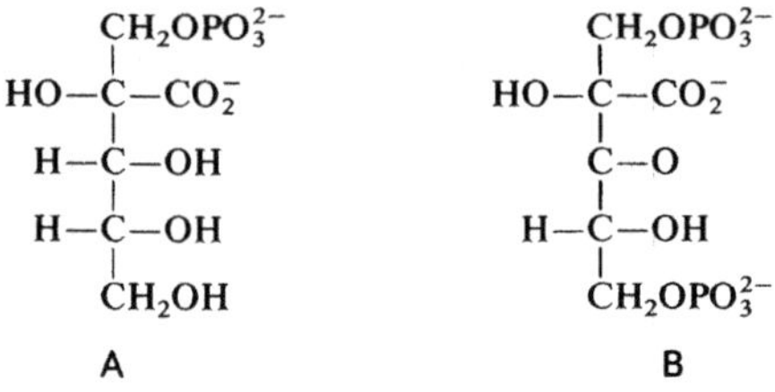

Figure 1.9. Structures of 2-carboxyarabinitol 1-phosphate (A) and 2-carboxy-3-ketoribitol 1,5-bisphosphate (B).

that competitive inhibition occurred between this intermediate and 2-carboxyribitol 1,5-bisphosphate. Furthermore, Pierce *et al.*[118] demonstrated inhibition with 2-C-(hydroxymethyl)-D-arabinitol 1,5-bisphosphate. Although both of these compounds effectively inhibited the isolated enzyme, with isolated chloroplasts they were ineffective because of their failure to pass through the membrane envelope. It is of considerable interest that recent work has shown that a related compound, 2-carboxyarabinitol 1-phosphate (Figure 1.9A), is a natural inhibitor of rubisco.[119] Whether this operates in a wide range of C_3 plants remains to be ascertained; apparently, it binds to rubisco at night to prevent carboxylation.

Heuer and Portis[120] have recently synthesized a number of phosphonate compounds as potential inhibitors of the carboxylase reaction. Experiments with isolated chloroplasts showed that phosphonoformate and phosphonopropionate entered the chloroplasts via the phosphate translocator in contrast to phosphonacetate, which did not. Furthermore, phosphonoformate was an effective inhibitor of 3-phosphoglycerate metabolism. As the authors noted, the amount of inhibitor that would have to be applied to whole plants would be considerable to ensure effective movement from cytoplasm to chloroplast stroma.[120]

In experiments with isolated chloroplasts, Stokes and Walker[121] found that DL-glyceraldehyde was an effective inhibitor of the carbon reduction cycle. This and subsequent studies[122,123] have shown that the conversion of ribulose 5-phosphate to ribulose 1,5-bisphosphate via the enzyme phosphoribulokinase is particularly sensitive to DL-glyceraldehyde in a noncompetitive manner.

Another class of potential inhibitors is the haloacetol phosphates of bromine, chlorine, and iodine. Hartman showed that there was irreversible binding of these compounds to the substrate active site of triose-phosphate isomerase because of structural similarities to dihydroxyacetone phosphate.[124,125] Subsequent work with isolated wheat chloroplasts[126] has indicated that iodoacetol phosphates enter the chloroplast by the phosphate translocator and primarily inhibit glyceraldehyde-3-phosphate dihydrogenase and, to a lesser extent, triose-phosphate isomerase. The inhibition of triose-phosphate isomerase by phosphoglycolate will be discussed further in Section 1.6.2.

Hydrogen peroxide generation has already been outlined in connection with bipyridinium action. Hydrogen peroxide and other peroxy compounds are being studied increasingly for use as algicides.[126,127] Although a primary aspect of peroxy damage is through free-radical processes, hydrogen peroxide itself is an effective inhibitor of some enzymes of the photosynthetic carbon reduction cycle; the most sensitive are fructose bisphosphatase and sedoheptulose bisphosphatase[128,129] and, to a lesser extent, glyceraldehyde-3-phosphate dehydrogenase and phosphoribulokinase.[130]

1.6.2. Inhibitors of Photorespiration

The oxygenase activity of rubisco results in the initial formation of phosphoglycolate. Accumulation of this compound within the chloroplast stroma results in inhibition of photosynthesis because phosphoglycolate, a transition state analogue of the substrate, is an effective inhibitor of triose-phosphate isomerase.[131-133] Mutants of *Hordeum vulgare*[131] and *Arabidopsis*[134] that lacked the enzyme phosphoglycolate phosphatase grew satisfactorily in 0.8–1% carbon dioxide, in which photorespiration would be limited, but became rapidly chlorotic under normal air. Kerr and co-workers[135,136] attempted to design herbicides which were inhibitors of phosphoglycolate phosphatase. Vanadyl glycolate and vanadyl sulfate were effective competitive inhibitors of the enzyme *in vitro* but failed to act in experiments with whole leaves. Ideas concerning the actual function of photorespiration have changed much during the last 20 or so years. Initially, photorespiration was seen as an essentially wasteful process and one that should be controlled to enhance crop growth. Present ideas suggest that the site chemistry of rubisco is such that, in addition to reacting with carbon dioxide, it will inevitably react with oxygen.[137] As a consequence, toxic phosphoglycolate must be removed and carbon salvaged. Because initial ideas centered on a wasteful role of photorespiration, a number of investigators attempted to find chemicals that would inhibit the process at a particular enzyme site. It now appears that the only effective way of controlling photorespiration is by raising the level of carbon dioxide, although, in the future, it might be possible to modify rubisco chemically or genetically. Work on inhibitors of glycolate metabolism, however, could provide a good starting point for herbicide design.

Glycolate which is generated within the chloroplast is exported and further metabolized within the peroxisomes. Here, the flavin mononucleotide (FMN)-linked enzyme glycolate oxidase catalyzes the conversion to glyoxylate with the generation of hydrogen peroxide:

$$\text{glycolate} + O_2 \xrightarrow{\text{glycolate oxidase}} \text{glyoxylate} + H_2O_2 \tag{1.3}$$

A number of chemicals have been shown to inhibit this enzyme and, in so doing, to cause the accumulation of glycolate and an inhibition of photosynthesis.[138] It is assumed that photosynthesis is limited by the prevention of photorespiratory carbon recycling. Zelitch first demonstrated the effectiveness of α-hydroxysulfonates, and, in particular, α-hydroxy-2-pyridinemethanesulfonic acid (α-HPMS).[139] This and other bisulfate addition compounds such as disodium sulfoglycolate are competitive inhibitors. On the basis of this work, Corbett and Wright[140] attempted to rationally design herbicides by producing a range of bisulfate addition compounds. Formaldehyde bisulfate was an effective inhibitor of glycolate oxidase, but this and a number of other compounds showed no useful herbicide response with whole plants.

Another type of inhibitor was devised by Kerr[141] on the basis of work by Walsh *et al.*[142] on an inhibitor of an FMN enzyme. This compound, L-2-hydroxy-3-butynoic acid (HBA), and its methyl ester caused the rapid accumulation of glycolate and an inhibition of glycolate oxidase in pea leaf disks.

Some discrimination in the use of glycolate oxidase inhibitors would be expected between C_3 and C_4 plants. In the latter, it is generally assumed that there is little or no photorespiratory carbon flow. Zelitch noted clear differences between glycolate accumulation in C_3 plants such as tobacco and sunflower and C_4 maize, with α-HPMS.[143] Jenkins *et al.* studied the effects of HBA on a C_3 barley[144] and a C_4 maize.[145] In both species, photosynthetic carbon dioxide incorporation was diminished, but much more so in barley. Glycolate accumulation accounted for 50–80% of ^{14}C incorporation in barley, in contrast to only 15% in maize.

During the operation of glycolate oxidase in normal photorespiration, hydrogen peroxide is formed and is removed by peroxisome-based catalase. The potential toxic action of hydrogen peroxide has already been discussed, and the herbicidal action of aminotriazole is assumed to be, in part, due to catalase inhibition.[146] Mutants of barley that are catalase deficient are rapidly bleached in normal air, in which photorespiration may operate, but will grow normally under CO_2-enriched air.[147] It is of interest that in this barley mutant and also in aminotriazole-treated plants, there was an enhancement in the level of leaf glutathione, probably due to increased pressure upon radical-scavenging systems.[148,149]

The further metabolism of glycine produced within the peroxisomes is via serine hydroxymethyltransferase within the mitochondria. *Arabidopsis* mutants that lacked this enzyme failed not only to convert glycine to serine, but also to carry out a glycine decarboxylation.[150] The photorespiratory

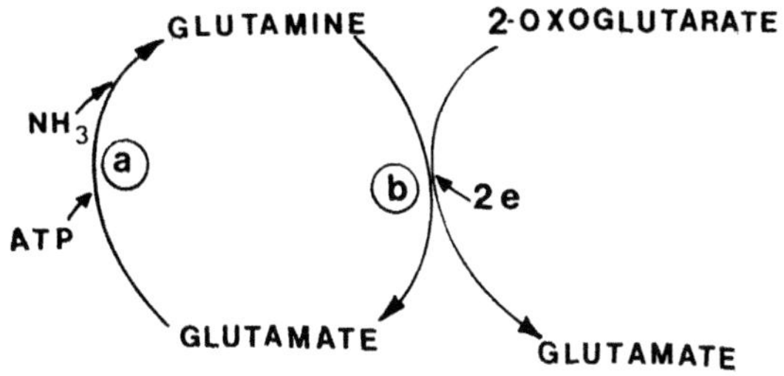

Figure 1.10. The glutamate synthase cycle. a, Glutamine synthetase; b, glutamate synthase.

inhibitor isonicotinyl hydrazide has been used for a number of years and shown to promote glycolate accumulation in *Chlorella*[151] and also to inhibit the conversion of glycine to serine.[152] One important feature of this conversion is the release of ammonia, which is recycled through the photorespiratory nitrogen cycle.[153] This involves the incorporation of ammonia through the glutamate synthase cycle (Figure 1.10). The herbicidal properties of inhibitors of ammonia incorporation have been clearly demonstrated for glufosinate (phosphinothricin).[154] The herbicidal compound bialaphos, produced by *Streptomyces hygroscopicus*,[155] is rapidly hydrolyzed by plants to produce the active compound glufosinate. Other structural analogues of glutamic acid, such as methionine sulfoximine, are excellent inhibitors of glutamine synthetase but have not been developed as herbicides.[156] The result of the use of these compounds is a rapid accumulation of ammonia,[154] an inhibition of photosynthesis,[156] and the appearance of chlorotic symptoms.[154] Ammonia not only uncouples chloroplast electron transport[157] but also promotes chloroplast photoinhibition.[158]

Table 1.2. The World's 10 "Worst Weeds"[a]

Weed species	Type of photosynthesis	Number of countries where known as a weed
Cyperus rotundus	C_4	91
Cynodon dactylon	C_4	90
Echinochloa crus-galli	C_4	65
Echinochloa colonum	C_4	67
Eleusine indica	C_4	64
Sorghum halepense	C_4	51
Imperata cylindrica	C_4	49
Eichornia crassipes	C_3	50
Portulaca oleracea	C_4	78
Chenopodium album	C_3	58

[a] Based on Holm *et al.*[159]

1.6.3. Inhibitors of C_4 Metabolism

Looked at worldwide, C_4 weeds are of considerable importance. Eight out of 10 of the "world's worst weeds," as classified by Holm *et al.*,[159] are C_4 plants (Table 1.2). Because of the unique biochemistry of these plants, the development of inhibitors, especially of C_4 enzymes, could be the basis for effective selective herbicides.[160] The mesophyll carboxylating enzyme is one primary target; another is pyruvate phosphodikinase, which catalyzes the unique conversion of pyruvate to phosphoenolpyruvate. Lawyer *et al.*[161] set out specifically to design "C_4 herbicides" by inhibiting the dikinase from maize leaves. Irreversible inhibition of the enzyme was obtained with 2-(bromomethyl)acrylic acid and 3-fluoropyruvate. This experimental approach proved to be unsuccessful, however, as no compound inhibited the enzyme totally, and no structural leads were given for synthesis of appropriate pyruvate analogues.

1.7. CONCLUSIONS

The development of herbicides that inhibit photosynthetic electron transport has proved to be outstandingly successful. Furthermore, because of the efficiency of these compounds and their use as tools to study photosynthesis, our knowledge of photosystem II in particular has been greatly enhanced by their use. Recent developments involving X-ray structural analysis of photosynthetic bacterial reaction centers as well as the ability to engineer herbicide resistance into crop plants have been outstanding scientific achievements.

The study of inhibitors of photosynthetic carbon metabolism has, by contrast, been remarkably unsuccessful. Attempts to design herbicidal inhibitors on rational grounds on the basis of specific enzyme inhibition have not afforded herbicidal compounds that are effective on whole plants. It is possible that in many instances the chloroplast envelope proves to be an insuperable barrier. Nevertheless, recent work identifying specific enzymes such as 5-enolpyruvylshikimate-3-phosphate (EPSP) synthase, acetyl-CoA carboxylase, and acetohydroxy acid synthase has demonstrated the effectiveness of enzyme inhibitors, although none were developed on "rational" grounds.

In looking for new leads for chemical synthesis, natural compounds might prove to be a good starting point, as has been shown in other areas of pesticide chemistry, for example, with the pyrethroids. An effective inhibitor of photosynthetic electron transport, grandinol (Figure 1.11), has recently been isolated from the leaves of *Eucalyptus grandis.*[162] A bipyridinium compound, orellanine, has been isolated from an aqueous

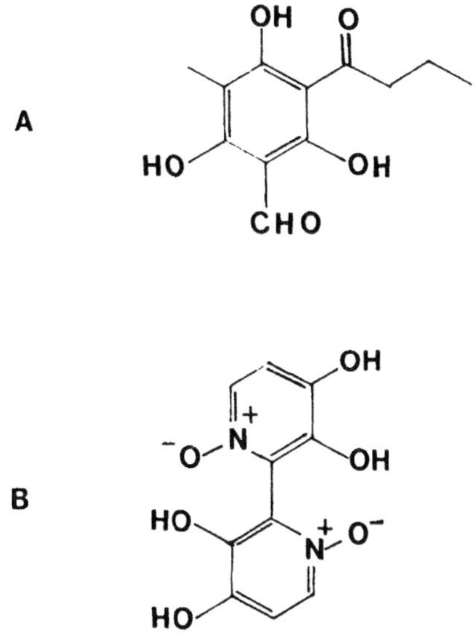

Figure 1.11. Structures of grandinol (A) and orellanine (B).

extract of the fungus *Cortinarius speciosissimus* (Figure 1.11) and shown to promote pigment breakdown in *Lemna minor* in a manner similar to that for diquat and paraquat.[163]

REFERENCES

1. H. C. Bucha and C. W. Todd, *Science 114,* 493 (1951).
2. J. S. C. Wessels and R. Van der Veen, *Biochim. Biophys. Acta 19,* 548 (1956).
3. A. R. Cooke, *Weeds 4,* 397 (1956).
4. R. C. Brian, R. F. Homer, J. Stubbs, and R. L. Jones, *Nature 181,* 446 (1958).
5. R. F. Homer, G. C. Mees, and T. E. Tomlinson, *J. Sci. Food Agric. 11,* 309 (1960).
6. D. W. Lawlor, *Photosynthesis: Metabolism, Control and Physiology,* Longman, Harlow, England (1987).
7. J. B. Marder and J. Barber, *Plant Cell Environ. 12,* 595 (1989).
8. R. J. Debus, B. A. Barry, G. T. Babcock, and L. McIntosh, *Proc. Natl. Acad. Sci. U.S.A. 85,* 427 (1988).
9. D. J. Kyle, *Photochem. Photobiol. 41,* 107 (1985).
10. A. Trebst, *Z. Naturforsch. C 41,* 240 (1986).
11. D. J. Kyle, I. Ohad, and C. J. Arntzen, *Proc. Natl. Acad. Sci. U.S.A. 81,* 4070 (1984).
12. A. K. Mattoo, H. Hoffman-Falk, J. B. Marder, and M. Edelman, *Proc. Natl. Acad. Sci. U.S.A. 81,* 1380 (1984).
13. C. Hansch, in: *Progress in Photosynthesis Research* (H. Metzer, ed.), Vol. 3, p. 1685, Metzner, Tübingen (1969).
14. K.-H. Buchel, *Pestic. Sci. 3,* 89 (1972).
15. A. Trebst and W. Draber, in: *Advances in Pesticide Science* (H. Geissbuhler, ed.) p. 223. Pergamon, Oxford (1979).

16. L. L. Shipman, *J. Theoret. Biol. 90,* 123 (1981).
17. A. Trebst and W. Draber, *Photosynth. Res. 10,* 381 (1986).
18. G. Gardner, *Photochem. Photobiol. 49,* 331 (1989).
19. P. N. Judson and G. P. Rowson, *Pestic. Sci. 27,* 327 (1989).
20. W. Tischer and H. Strotmann, *Biochim. Biophys. Acta 460,* 113 (1979).
21. G. E. Ryan, *Weed Sci. 18,* 614 (1970).
22. J. Hirschberg and L. McIntosh, *Science 222,* 1346 (1983).
23. K. Pfister, S. R. Radosevich, and C. J. Arntzen, *Plant Physiol. 64,* 995 (1979).
24. D. J. Chapman and J. Barber, *Proc. Brighton Crop Prot. Conf.—Weeds,* p. 1185 (1989).
25. L. Mets and A. Thiel, in: *Target Sites of Herbicide Action* (P. Böger and G. Sandmann, eds.), p. 1, CRC Press, Boca Raton, Florida (1989).
26. A. Trebst, in: *Herbicide Resistance in Weeds and Crops* (J. C. Caseley, G. W. Cussans, and R. K. Atkin, eds.), p. 145, Butterworth, Heinemann, Oxford, England (1991).
27. J. Deisenhofer, O. Epp, K. Mikki, R. Huber, and H. Michel, *Nature 318,* 618 (1985).
28. H. Michel and J. Deisenhofer, *Biochemistry 27,* 1 (1988).
29. J. Deisenhofer and H. Michel, *EMBO J. 8,* 2149 (1989).
30. J. J. Stein, A. Castelliri, J. P. Bogaiz, and C. A. Wraight, *J.. Cell. Biochem. 24,* 243 (1984).
31. P. K. Wolber, M. Eilmann, and K. E. Steinbeck, *Arch. Biochem. Biophys. 248,* 224 (1986).
32. D. J. Chapman, J. De-Felice, and J. Barber, *Planta 166,* 280 (1985).
33. C. A. Wraight, *Isr. J. Chem, 21,* 348 (1981).
34. W. F. J. Vermaas, C. J. Arntzen, L.-Q. Gru, and C.-A. Yu, *Biochim. Biophys. Acta 723,* 266 (1983).
35. W. Oettmeier and H.-J. Soll, *Biochim. Biophys. Acta 724,* 287 (1983).
36. D. A. Kleier, T. A. Andrea, J. K. J. Hegedus, G. M. Gardner, and B. Cohen, *Z. Naturforsch., C 42,* 733 (1987).
37. K. E. Pallett and A. D. Dodge, *Pestic. Sci.* 10, 216 (1979).
38. U. Johanningmeier, E. Neumann, and W. Oettmeier. *J. Bioenerg. Biomembr. 15,* 43 (1983).
39. J. J. S. Van Rensen, in: *Herbicides and Plant Metabolism* (A. D. Dodge, ed.), p. 21, Cambridge University Press, Cambridge (1989).
40. K. E. Pallett and A. D. Dodge, *J. Exp. Bot. 31,* 1051 (1980).
41. W. H. Minshall, *Weeds 5,* 29 (1957).
42. F. M. Ashton, *Weeds 13,* 164 (1965).
43. J. L. P. Van Oorschot and P. H. Van Leeuwen, *Weed Res. 14,* 81 (1974).
44. L. M. N. Duysens and H. E. Sweers, in: *Microalgae and Photosynthetic Bacteria* (edited by Japan Society of Plant Physiology), p. 353, University of Tokyo Press, Tokyo (1963).
45. M. Voss, G. Renger, C. Kotter, and P. Graber, *Weed Sci. 32,* 675 (1984).
46. D. Habash, M. P. Percival, and N. R. Baker, *Weed Res. 25,* 389 (1985).
47. L. K. Thompson and G. W. Brudvig, *Biochemistry 27,* 6653 (1988).
48. R. E. Cleland, A. Melis, and P. J. Neale, *Photosynth. Res 9,* 79 (1986).
49. M. P. Percival and A. D. Dodge, *Z. Naturforsch., C 39,* 482 (1984).
50. H. McTavish, R. Picorel, and M. Seibert, *Plant Physiol. 89,* 452 (1989).
51. W.-Z. He, A. Telfer, A. F. Drake, J. Hoadley, and J. Barber, *Proceedings of the VIIIth International Photosynthesis Congress,* Martinus Nijhoff, Dordrecht, The Netherlands.
52. I. C. Anderson and D. S. Robertson, *Plant Physiol. 35,* 531 (1960).
53. B. Halliwell, *Chloroplast Metabolism,* Clarendon Press, Oxford (1981).
54. K. J. Kunert and A. D. Dodge, in: *Target Sites of Herbicide Action* (P. Böger and G. Sandmann, eds.), p. 45, CRC Press, Boca Raton, Florida (1989).
55. A. Zweig and G. W. Nachtigall, *Photochem. Photobiol. 22,* 257 (1975).
56. J. P. Knox and A. D. Dodge, *Plant Sci. Lett. 37,* 3 (1984).
57. J. P. Knox and A. D. Dodge, *Planta 164,* 22 (1985).
58. J. P. Knox and A. D. Dodge, *Planta 164,* 30 (1985).

59. A. K. Mattoo, S. K. Sopery, B. M. Greenberg, F. E. Callahan, M. L. Ghirardi, and M. Edelman. *Herbicides Active in the Chloroplast*, p. L4, Monhein, Germany (1989).
60. G. Zurawski, H. J. Bohnert, P. R. Whitfield, and W. Bottomley. *Proc. Natl. Acad. Sci. U.S.A. 79*, 7699 (1982).
61. J. Hirschberg and L. McIntosh, *Science 222*, 1346 (1983).
62. A. Y. Cheung. L. Bogorad, M. Van Montagu, and J. Schell. *Proc. Natl. Acad. Sci. U.S.A. 85*, 391 (1988).
63. D. M. Stalker, K. E. McBride, and L. D. Malyj, *Science 242*, 419 (1988).
64. G. Freyssinet, B. Leroux, M. Lebrun, B. Pelissier, A. Sailland, and K. E. Pallett, *Proc. Brighton Crop Prot. Conf.—Weeds*, p. 1225 (1989).
65. K. E. McBride, J. W. Kenny, and D. M. Stalker, *Appl. Environ. Microbiol. 52*, 325 (1986).
66. D. M. Stalker, I. D. Malyj, and K. E. McBride, *J. Biol. Chem. 265*, 6310 (1988).
67. W. Haehnel, *Annu. Rev. Plant Physiol. 35*, 659 (1984).
68. G. Zweig, N. Shavit, and M. Avron, *Biochim. Biophys. Acta 109*, 332 (1965).
69. N. Harris and A. D. Dodge, *Planta 104*, 210 (1972).
70. J. T. Warden and J. R. Bolton, *Photochem. Photobiol. 20*, 251 (1974).
71. T. Fujii, E.-I. Yokoyama, K. Inoue, and H. Sakurai, *Biochim. Biophys. Acta 1015*, 41 (1990).
72. J. R. Bowyer, P. O'Neill, P. Camilleri, and C. M. Todd, *Biochim. Biophys Acta 932*, 124 (1988).
73. L. A. Summers, *The Bipyridinium Herbicides*, Academic Press, London (1980).
74. P. Cuendet and M. Gratzel, *Photobiochem. Photobiophys. 5*, 335 (1983).
75. J. R. Bowyer, P. Camilleri, and A. Stapleton, *FEBS Lett. 172*, 239 (1984).
76. G. C. Mees, *Ann. Appl. Biol. 48*, 601 (1960).
77. J. A. Farrington, M. Ebert, and E. J. Land. *J. Chem. Soc., Faraday Trans. 1, 74*, 665 (1978).
78. J. A. Farrington, M. Ebert, E. J. Land, and K. Fletcher, *Biochim. Biophys. Acta 314*, 372 (1973).
79. J. R. Harbour and J. R. Bolton, *Biochem. Biophys. Res. Commun. 64*, 803 (1975).
80. P. Camilleri, J. R. Bowyer, M. T. Clark, and P. O'Neill. *Biochim. Biophys. Acta 765*, 236 (1985).
81. A. D. Dodge, in: *Herbicides and Plant Metabolism* (A. D. Dodge, ed.), p. 37, Cambridge University Press, Cambridge (1989).
82. J. M. McCord, *Rev. Biochem. Toxicol. 1*, 109 (1979).
83. C. C. Winterbourn, *FEBS Lett. 128*, 339 (1981).
84. M. Saito, C. E. Thomas, and S. D. Aust. *J. Free Radicals Biol. Med. 1*, 179 (1985).
85. C. F. Babbs, J. A. Pham, and R. C. Coolbaugh, *Plant Physiol. 90*, 1267 (1989).
86. K. C. Vaughn and S. O. Duke, *Plant Cell Environ. 6*, 13 (1983).
87. J. R. Bowyer and P. Camilleri, in: *Herbicides* (D. M. Hutson and T. R. Roberts, eds.), p. 105, John Wiley, Chichester (1987).
88. D. J. Gillham and A. D. Dodge, *Pestic. Sci. 19*, 19 (1987).
89. D. J. Gillham and A. D. Dodge, *Pestic. Sci. 19*, 25 (1987).
90. W. M. Draper and J. E. Casida *J. Agric. Food Chem. 33*, 130 (1985).
91. N. Harris and A. D. Dodge, *Planta 104*, 201 (1972).
92. A. D. Dodge, *Endeavour 30*, 130 (1971).
93. A. D. Dodge, in: *Herbicide Resistance in Weeds and Crops* (J. C. Caseley, G. W. Cussans, and R. K. Atkin, eds.), p. 165, Butterworth, Heinemann, Oxford, England (1991).
94. E. P. Fuerst, H. Y. Nakatani, A. D. Dodge, D. Penner, and C. J. Arntzen, *Plant Physiol. 77*, 984 (1985).
95. K. C. Vaughn, M. A. Vaughan, and P. Camilleri, *Weed Sci. 37*, 5 (1989).
96. S. B. Powles and G. Cornic, *Aust. J. Plant Physiol. 14*, 81 (1987).
97. Y. Shaaltiel and J. Gressel, *Plant Physiol. 85*, 869 (1987).

98. J. S. Faulkner, in: *Symposium on Status, Biology and Control of Grass Weeds in Europe*, p. 349, European Weed Research Society (1975).
99. C. Jackson, J. Dench, A. C. Moore, B. Halliwell, C. Foyer, and D. O. Hall. *Eur. J. Biochem. 91*, 339 (1978).
100. R. J. Youngman, A. D. Dodge, E. Lengfelder, and E. F. Elstner, *Experientia 35*, 1295 (1979).
101. R. J. Youngman and A. D. Dodge. *Z. Naturforsch., C* 34, 1032 (1979).
102. D. J. Gillham, Aspects of Chloroplast Protection against Photo-Oxidative Damage, Ph.D. thesis, University of Bath, Bath (1986).
103. J. Gressel and Y. Shaaltiel, in: *Biotechnology for Crop Protection* (P. A. Hedin, J. J. Menn, and R. M. Hollingworth, eds.), p. 4, American Chemical Society, Washington, D.C. (1988).
104. C. H. Foyer and B. Halliwell, *Planta 133*, 21 (1976).
105. B. Halliwell, *Chem. Phys. Lipids 44* 327, (1987).
106. C. Foyer, J. Rowell, and D. A. Walker, *Planta 157*, 239 (1983).
107. M. W. Law, S. A. Charles, and B. Halliwell, *Biochem. J. 210*, 899 (1983).
108. A. Schmidt and K. J. Kunert, in: *Molecular Strategies for Crop Protection* (C. Arntzen and C. Ryan, eds.), p. 401, Alan R. Liss, New York (1987).
109. A. L. Tappel, *Geriatrics 23*, 97 (1968).
110. B. F. Finckh and K. J. Kunert, *J. Agric. Food Chem. 33*, 574 (1985).
111. G. Drolet, E. B. Dumbroff, R. L. Legge, and J. E. Thompson, *Phytochemistry 25*, 367 (1986).
112. V. Takahama, *Photochem. Photobiol. 38*, 363 (1983).
113. M. D. Hatch, *Biochim. Biophys. Acta 895*, 81 (1987).
114. C. C. Black, in: *Herbicide Physiology II* (S. O. Duke, ed.), p. 1, CRC Press, Boca Raton, Florida (1985).
115. J. T. Bahr and R. G. Jensen, *Arch. Biochem. Biophys 185*, 39 (1978).
116. C. Paech, J. Pierce, S. D. McCurry, and N. E. Tolbert, *Biochem. Biophys. Res. Commun. 83*, 1084 (1978).
117. M. I. Siegel and M. D. Lane, *J. Biol. Chem. 248*, 5486 (1973).
118. J. Pierce, N. E. Tolbert, and R. Barker, *Biochemistry 19*, 934 (1980).
119. S. Gutteridge, M. A. J. Parry, S. Burton, A. J. Keys, A. Feeney, J. C. Servaites, and J. Pierce, *Nature 324*, 274 (1986).
120. B. Heuer and A. R. Portis, *Plant Physiol. 84*, 649 (1987).
121. D. M. Stokes and D. A. Walker, *Biochem. J. 128*, 1147 (1972).
122. E. S. Bamberger and M. Avron, *Plant Physiol. 56*, 481 (1975).
123. A. R. Slabas and D. A. Walker, *Biochem. J. 153*, 613 (1976).
124. F. C. Hartman, *Biochem. Biophys. Res. Commun. 33*, 1084 (1968).
125. F. C. Hartman, *Biochemistry 9*, 1776 (1970).
126. G. Barroin and M. Feuillade, *Water Res. 20*, 619 (1986).
127. C. A. Clarke, S. J. L. Wright, and A. D. Dodge, VIth International Symposium Photosynthetic Prokaryotes, p. 225 (1988).
128. W. Kaiser, *Planta 145*, 377 (1979).
129. H. W. Heldt, C. J. Chon, R. M. Lilley, and A. Portis, in: *Photosynthesis* (D. O. Hall, J. Coombs, and T. W. Goodwin, eds.), p. 469, The Biochemical Society, London (1978).
130. K. Tanaka, T. Otsubo, and N. Kondo, *Plant Cell Physiol. 23*, 1009 (1982).
131. N. P. Hall, A. C. Kendall, P. J. Lea, J. C. Turner, and R. M. Wallsgrove, *Photosynth. Res. 11*, 89 (1987).
132. R. Wolfenden, *Nature 223*, 704 (1969).
133. L. E. Anderson, *Biochim. Biophys. Acta 235*, 237 (1971).
134. C. R. Somerville and W. L. Ogren. *Nature 280*, 833 (1979).
135. M. W. Kerr and D. P. Whitaker. *Ann. Proc. Phytochem. Soc. Eur. 26*, 45 (1985).
136. M. W. Kerr and C. F. Gear, *Biochem. Soc. Trans. 2*, 338 (1974).
137. G. Lorimer and T. J. Andrews, *Nature 243*, 359 (1973).

138. K. Asada, S. Kitoh, R. Deura, and Z. Kasai, *Plant Cell Physiol. 6,* 615 (1965).
139. I. Zelitch, *J. Biol. Chem. 233,* 1299 (1958).
140. J. R. Corbett and B. J. Wright, *Phytochemistry 10,* 2015 (1971).
141. M. W. Kerr, in: *Proceedings of the Third International Congress on Photosynthesis* (M. Avron, ed.), p. 1285, Elsevier, Amsterdam (1974).
142. C. T. Walsh, A. Schonbrunn, O. Lockridge, V. Massey, and R. M. Abeles, *J. Biol. Chem. 247,* 6004 (1972).
143. I. Zelitch, *Plant Physiol. 51,* 229 (1973).
144. C. L. D. Jenkins, L. J. Rogers, and M. W. Kerr, *Phytochemistry 21,* 1849 (1982).
145. C. L. D. Jenkins, L. J. Rogers, and M. W. Kerr, *Phytochemistry 21,* 1859 (1982).
146. E. Margoliash, A. Novogradsky, and A. Schejter, *Biochem. J. 74,* 339 (1960).
147. M. L. Parker and P. J. Lea, *Planta 159,* 572 (1983).
148. I. K. Smith, A. C. Kendall, A. J. Keys, J. C. Turner, and P. J. Lea, *Plant Sci. Lett. 37,* 29 (1984).
149. I. K. Smith, *Plant Physiol. 79,* 1044 (1985).
150. C. R. Somerville and W. L. Ogren, *Plant Physiol. 67,* 666 (1981).
151. C. P. Whittingham and G. G. Pritchard, *Proc. Roy. Soc. B 157,* 366 (1963).
152. T. Kisaki and N. E. Tolbert, *Plant Cell Physiol. 11,* 247 (1970).
153. A. J. Keys, I. F. Bird, M. J. Cornelius, P. J. Lea, R. M. Wallsgrove, and B. J. Miflin, *Nature 275,* 741 (1978).
154. H. Köcher, *Aspects Appl. Biol. 4,* 227 (1983).
155. Y. Kondo, T. Shomura, Y. Ogawa, T. Tsuroka, H. Watanabe, K. Totsukawa, T. Suzuki, C. Moriyama, J. Yoshida, S. Inouya, and T. Niida. *Scientific Reports of Meiji Seika Kaisha 13,* 34 (1973).
156. P. J. Lea and S. M. Ridley, in: *Herbicides and Plant Metabolism* (A. D. Dodge, ed.), p. 137, Cambridge University Press, Cambridge (1989).
157. D. W. Krogman, A. J. Jagendorf, and M. Avron, *Plant Physiol. 34,* 272 (1959).
158. G. H. Krause and H. Laasch, in: *Progress in Photosynthesis Research* (J. Biggins, ed.), Vol IV, p. 19, Martinus Nijhoff, Dordrecht (1987).
159. L. G. Holm, D. L. Plucknett, J. V. Pancho, and J. B. Herberger, *The World's Worst Weeds, Distribution and Biology,* Honolulu, University Press, Hawaii (1977).
160. C. L. D. Jenkins and M. D. Hatch, CSIRO, Division of Plant Industry, Biennial Report, p. 12 (1982).
161. A. L. Lawyer, S. R. Kelley, and J. I. Allen, in: *Progress in Photosynthesis Research* (J. Biggins, ed.), Vol. III, p. 831, Martinus Nijhoff, Dordrecht (1987).
162. S. Yoshida, T. Asami, T. Kawano, K. Yoneyama, W. D. Crow, D. M. Paton, and N. Takahashi, *Phytochemistry 27,* 1943 (1988).
163. K. Hoiland, *Trans. Br. Mycol. Soc. 81,* 633 (1983).

2

Amino Acid Synthesis

DAVID M. MOUSDALE AND JOHN R. COGGINS

2.1. AMINO ACID BIOSYNTHESIS IN PLANTS

Green plants share with most microbial species the capacity to synthesize all the major amino acids; the few exceptions are parasitic and depend on the nitrogen metabolism of the host plant. In addition, many nonprotein "uncommon" amino acids are produced, and frequently accumulated, in large amounts in a range of species; these are important in plant/plant, plant/insect, and plant/animal relationships.[1] The amino acids essential for mammalian diets are phenylalanine, tryptophan, lysine, leucine, isoleucine, valine, threonine, and methionine for adults, with the addition of arginine and histidine for immature animals (the third aromatic amino acid required for protein synthesis, tyrosine, can be formed metabolically, given an adequate dietary intake of phenylalanine). The biosynthetic routes for these 10 amino acids are particularly attractive targets for toxicologically safe herbicides and for pesticides aimed at plant pathogens; the potential for antibiotics may be more limited because medicinally important microorganisms tend to inhabit nutritionally rich habitats.

The general outline of the pathways of amino acid biosynthesis is shown in Figure 2.1. It should be stressed that the detailed enzymology of these pathways is far better understood in bacteria and some eucaryotic microorganisms than in higher plants; indeed, in most plant species, information on the relevant enzymology is fragmentary.[2] As will be seen later,

DAVID M. MOUSDALE AND JOHN R. COGGINS • Department of Biochemistry, University of Glasgow, Glasgow G12 8QQ, United Kingdom.

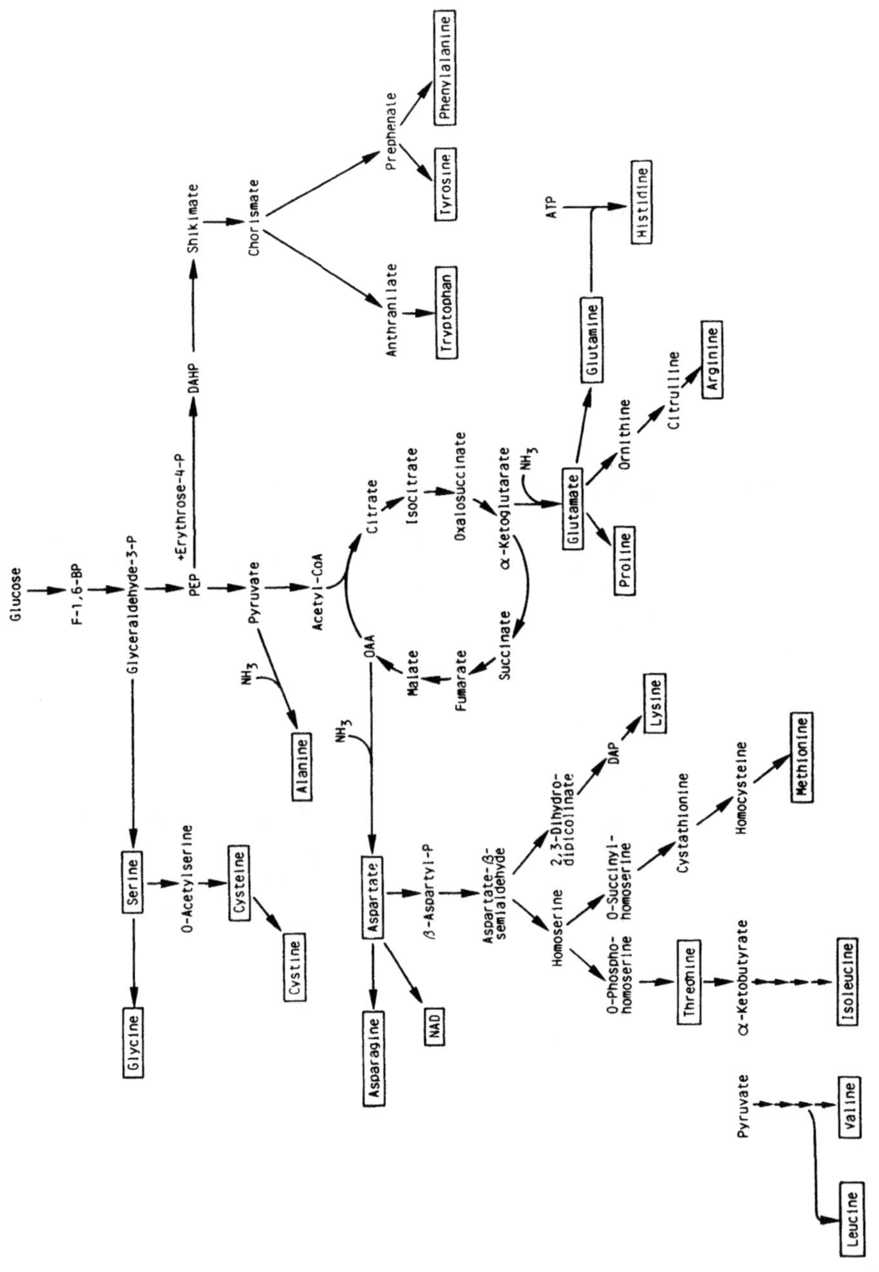

Figure 2.1. Outline of amino acid biosynthesis in plants and bacteria.

this has not prevented either the identification of the sites of action of herbicides acting on amino acid biosynthesis or the initial discovery of such herbicides (which to date has resulted invariably from random screening programs).

The general area of nitrogen metabolism in higher plants has obvious attractions for the rational development of novel herbicides. The availability of nitrogen in the edaphic environment is a major factor determining the success of any particular species.[3] The nitrogen economy is sufficiently critical that much of the nitrogen in deciduous species is recycled from the leaves during the autumn for over-winter storage.[4] Major crop species are, however, often supplied with a generous amount of nitrogen fertilizer, and the question of herbicide selectivity then becomes important for the post-emergence control of weeds in crop monocultures. In principle, the similarity of the biosynthetic pathways used by plants (both lower and higher) would suggest that any amino acid inhibitor herbicide would be of the broad-spectrum type. Differential uptake and metabolism of the herbicides by weed and crop species and the overproduction of the target enzyme or the genetic engineering of inhibitor-insensitive enzymes in selected crops have therefore received much attention. An important divergence between plants and bacterial model systems is that the location of amino acid synthesis inside chloroplasts (and some other types of plastid) is now well established.[5] The requirement for the putative herbicide to accumulate at this subcellular site offers further potential for herbicide selectivity as well as the generation of "plant-specific" chemicals.

In considering the commercially successful herbicides known to be inhibitors of amino acid biosynthesis, three aspects will be considered: the kinetic description of the inhibition, the molecular interactions involved, and the relevance of the proposed target site to the observed physiological effects. Because of the extensive body of research published on the shikimate pathway and the herbicide glyphosate, this example will be taken as a paradigm of the inhibitor–enzyme relationship. The other cases considered will be the branched-chain amino acid family, histidine biosynthesis, and glutamine synthetase.

2.2. GLYPHOSATE AND THE SHIKIMATE PATHWAY

2.2.1. The Shikimate Pathway

The common (prechorismate) shikimate pathway utilizes the same reactions in all the plant and microbial species investigated.[6,7] Figurc 2.2 shows the seven reactions involved. In green tissues, erythrose 4-phosphate and phosphoenolpyruvate (PEP) are the products of the Calvin cycle and

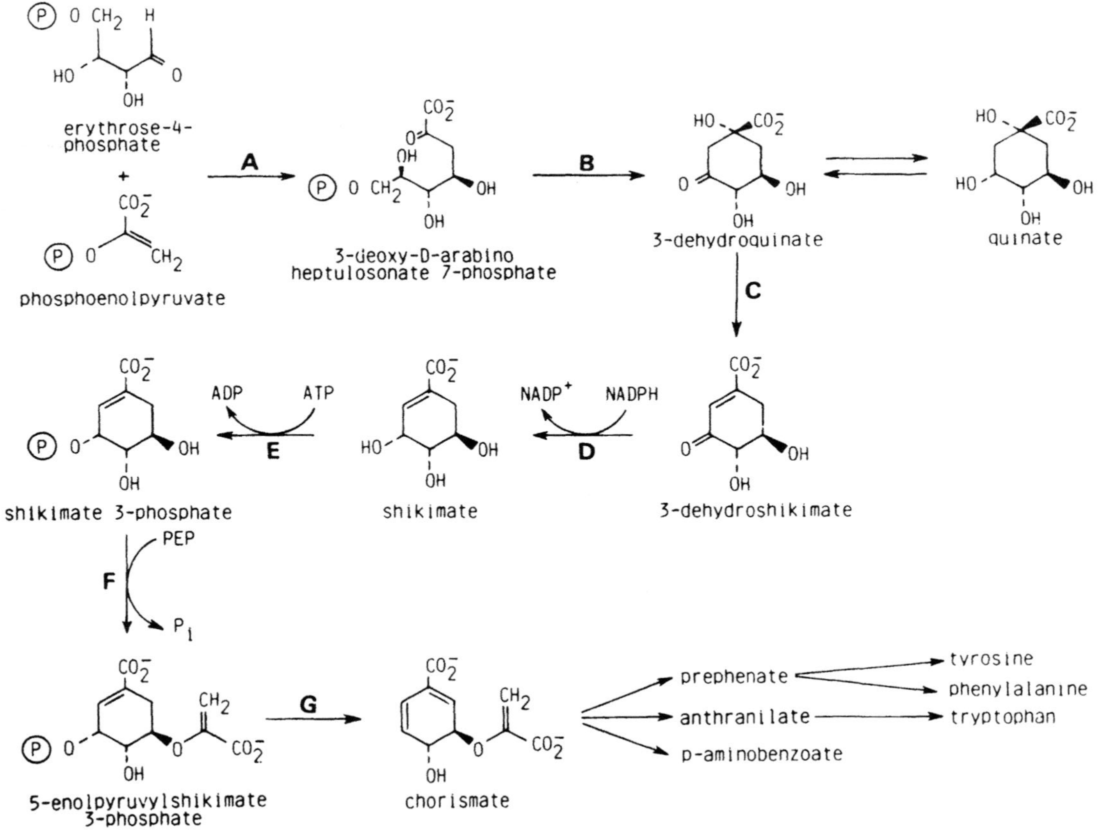

Figure 2.2. Enzymes and reactions of the shikimate pathway: A, DAHP synthase; B, dehydroquinate synthase; C, dehydroquinase; D, shikimate dehydrogenase; E, shikimate kinase; F, EPSP synthase; G, chorismate synthase.

glycolysis, respectively, and the successive reactions in this part of the pathway form the alicyclic precursors for aromatic compounds. Chorismic acid is the major branch point: from here, routes to phenylalanine (and tyrosine), tryptophan, *p*-aminobenzoic acid, and quinones diverge (Figure 2.3). Important alternative pathways in some plant species are the formation of quinic acid from dehydroquinate and the direct aromatization of dehydroshikimate to gallic and protocatechuic acids.[8-11] From this small number of primary end products, an enormous number of secondary products are synthesized; quantitatively, the major examples are lignin, lignin precursors, and flavonoids. Aromatic compounds can account for up to 35% of the dry weight of plants, and 20% or more of the carbon flux is processed via the shikimate pathway.[12]

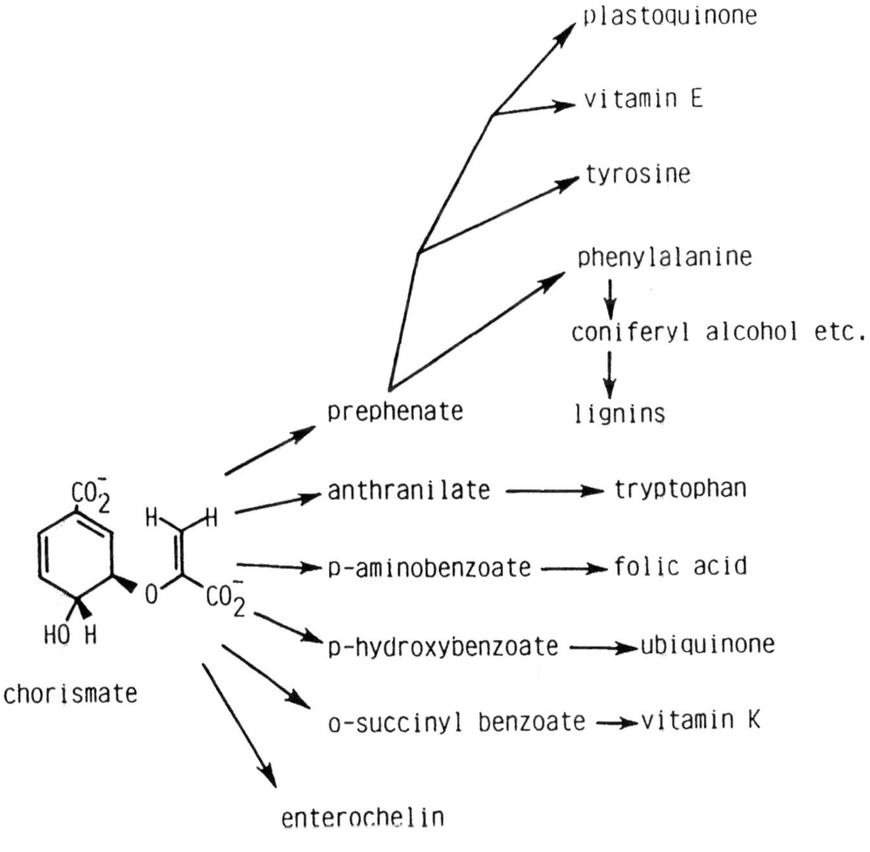

Figure 2.3. Major metabolic fates of chorismic acid in plants and bacteria.

2.2.2. 5-Enolpyruvylshikimate-3-Phosphate Synthase

5-Enolpyruvylshikimate-3-phosphate (EPSP) synthase (EC 2.5.1.19), also known as 3-phosphoshikimate 1-carboxyvinyltransferase, catalyzes the reversible transfer of the enolpyruvyl group from PEP to shikimate 3-phosphate (Figure 2.2). There is no known cofactor requirement. When purified to homogeneity, bacterial and plant EPSP synthases are monomeric monofunctional proteins of molecular weight 40,000–50,000.[13–18] Some fungi have a radically different molecular architecture in which EPSP synthase is part of a pentafunctional polypeptide which also contains the previous four enzymatic activities of the common shikimate pathway, the *arom* complex.[19–21]

2.2.3. Inhibition of EPSP Synthase by Glyphosate

Glyphosate (Roundup), a phosphonomethyl derivative of the amino acid glycine (Figure 2.4), is a potent inhibitor of all the wild-type microbial and plant EPSP synthases that have been isolated (Table 2.1). The inhibition markedly increases over the pH range 6–8.[22,23] While considerable differences in the inhibitor constants that have been measured are apparent (Table 2.1), the range of values is put into context by the report that glyphosate concentrations in the cytosol of treated plants can be in excess of 1 mM, with intrachloroplastic levels of $>50\ \mu M$.[24]

On the basis of steady-state kinetics, glyphosate behaves as a reversible and competitive inhibitor of the forward reaction with respect to PEP and as an uncompetitive inhibitor with respect to shikimate 3-phosphate; if the backward reaction is considered, then glyphosate is competitive with phosphate and uncompetitive with EPSP.[14,16,18,22,23,25–31] These kinetic patterns can be rationalized on the basis of a model in which shikimate 3-phosphate binds to the free enzyme, and PEP then interacts with the binary enzyme–substrate complex; in the reverse reaction, EPSP is the first substrate to bind.[27] Direct fluorometric monitoring of enzyme–substrate interactions has confirmed that both shikimate 3-phosphate and EPSP bind to the free enzyme.[32] Fast-reaction, pre-steady-state techniques have also shown that shikimate 3-phosphate binds ahead of PEP.[33] Rapid reaction/chemical quenching studies have confirmed the postulated route of the reaction via a tetrahedral intermediate.[34]

EPSP synthase appears to be unique in that, unlike all other PEP-utilizing enzymes studied, PEP is bound as the second substrate.[33] This has a clear mechanistic link in that generally glyphosate does not inhibit enzymes utilizing PEP as a substrate.[27] The only close analogy to the EPSP synthase-catalyzed reaction is found with UDP-*N*-acetylmuramic acid (UDP-NAM) synthase.[35,36] Like EPSP synthase, this enzyme also transfers the enolpyruvyl

Glyphosate

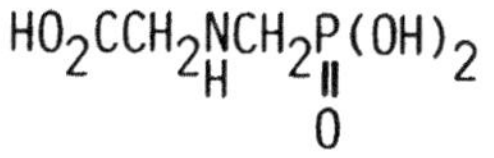

Amitrole

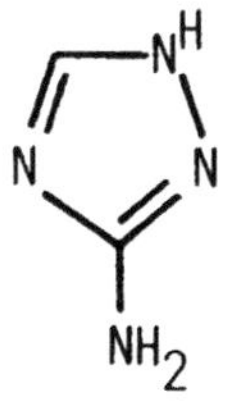

Phosphinothricin

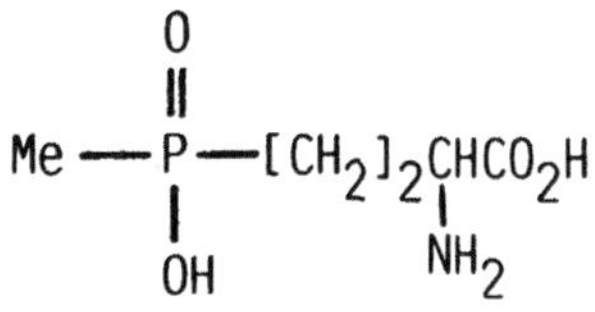

Imidazapyr

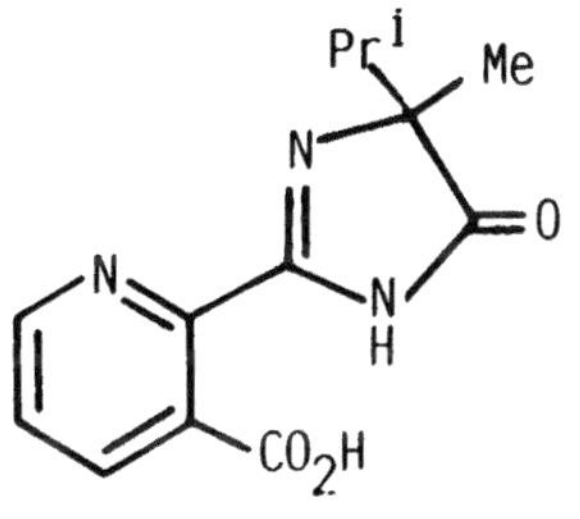

Sulfometuron methyl

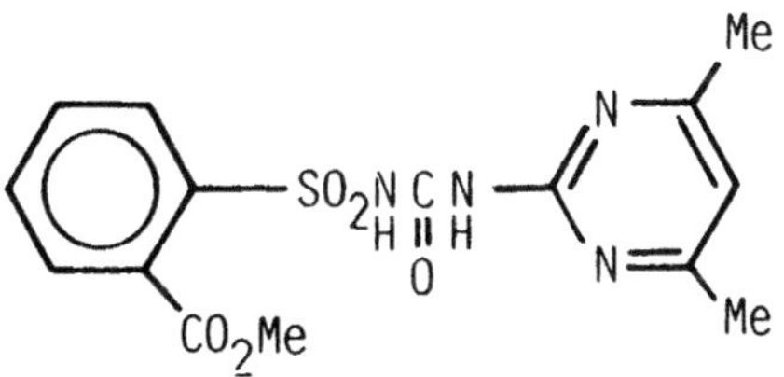

Figure 2.4. Structures of glyphosate, amitrole, phosphinothricin, and representative imidazolinone and sulfonylurea herbicides.

moiety to an alcohol as the second substrate.[37] Fosfomycin is a naturally occurring inhibitor of UDP-NAM synthase.[36,37] Fosfomycin does not, however, inhibit EPSP synthase, nor does glyphosate inhibit UDP-NAM synthase.[22,28,38] The action of the two inhibitors appears to be markedly different in that fosfomycin is a relatively weak inhibitor and can form a nonenzymatic complex with PEP; furthermore, PEP can bind directly to

Table 2.1. Inhibition of EPSP Synthases by Glyphosate

Species	K_i (μM)	K_mPEP (μM)	Reference
Escherichia coli	0.9	16	14
Escherichia coli	40	—	30
Escherichia coli	0.8	—	24
Klebsiella pneumoniae	1.0	17	22
Klebsiella pneumoniae	10	500	28
Neurospora crassa	1.1	2.7	27
Candida maltosa	12	—	29
Corydalis sempervirens	0.32	—	24
Nicotiana silvestris	1.25	—	23
Pisum sativum	0.08	5.2	16
Sorghum bicolor	0.16	8	18

UDP-NAM synthase and protect the enzyme against inactivation by cysteine-alkylating agents.[35]

The glyphosate-EPSP synthase interaction is also idiosyncratic in that glyphosate is an isolated ("single point") toxophore. Despite a considerable effort of synthetic organic chemistry, no glyphosate analogue shows greater inhibitory activity, and even very small changes in the structure result usually in abolition of the ability to inhibit EPSP synthase; Table 2.2 (with data from Ref. 38) shows for the *Neurospora crassa arom* complex enzyme how critical the exact molecular requirements for inhibition are. Glyphosine, the diphosphono analogue, has been marketed, not as a herbicide, but as a sugar-cane ripener (Polaris). The weak activity in enzyme screens (Table

Table 2.2. Structure-Function Relations of the Inhibition of *arom* EPSP Synthase by Glyphosate and Similar Compounds[a]

Compound	K_i/K_mPEP
$PO(OH)_2CH_2NHCH_2COOH$ (glyphosate)	0.3
$PO(OH)_2CH_2NHCH_3$	>200
$PO(OH)_2CH_2NH_2$	>30
$PO(OH)_2CH_2OCH_2COOH$	>60
$PO(OH)_2CH_2SCH_2COOH$	>180
$PO(OH)_2CH_2N(Me)CH_2COOH$	>90
$PO(OH)_2CH_2NHCH(Me)COOH$	>150
$PO(OH)_2CH_2NHCH_2CH_2COOH$	>50
$PO(OH)_2CH_2CH_2NHCH_2COOH$	>120
$PO(OH)_2CH_2CH(NH_2)COOH$	>200
$PO(OH)_2CH_2CH_2CH(NH_2)COOH$	>50
$[PO(OH)_2CH_2]_2NCH_2COOH$ (glyphosine)	>4

[a] Ref. 38.

2.2) is probably due to contaminating glyphosate since more recent investigations with analytically pure glyphosine show no detectable inhibition of pea seedling EPSP synthase (Mousdale and Coggins, unpublished data); the pea enzyme has a much lower K_i for glyphosate than does the *N. crassa* enzyme.[16,27] Similarly, the *N*-trichloromethyl analogue of glyphosate shows approximately one-third the inhibitory activity (on a molar basis) when tested on the pea enzyme in parallel with glyphosate (Mousdale and Coggins, unpublished data). Only one structural relative of glyhosate appears to have received serious consideration as a herbicide, namely, 3-amino-3-carboxypropyl methylphosphinate (HOE 00661); this compound was approximately an order of magnitude less potent than glyphosate in parallel trials.[39] With the successful crystallization of a bacterial EPSP synthase, the scope for definitive studies of the precise binding reactions of substrates and glyphosate (and its analogues) is greatly improved.[40] When three-dimensional modeling of enzyme–substrate–inhibitor interactions is achieved, the binding specificities can be visualized and protein engineering techniques employed to construct resistant enzymes.

Conventional protein modification studies with the *Escherichia coli* and petunia EPSP synthases have begun to identify the amino acid residues involved in substrate binding and possibly also in catalysis. A lysine and an arginine are essential and could be the cationic sites binding the anionic substrates; there is also evidence for the participation of a histidine residue and for a cysteine residue being important for the enzyme structure, even if it is not directly involved in catalysis.[22,41–45] Glyphosate has a positively charged imino group (pK_a 10.6) at physiological pH.[46] A glutamate residue may provide the binding site on the enzyme surface.[47] Evidence of the EPSP synthase–glyphosate interaction is clearly seen by the protection against chemical modifying agents afforded by the inhibitor.[22] A single-point mutation produced a glyphosate-tolerant *Salmonella typhimurium*; the replacement of proline 101 by a serine resulted in an EPSP synthase with an order-of-magnitude increase in the cellular glyphosate tolerance.[48] Mutant strains of two other bacterial species have been isolated.[49–51] Site-directed mutagenesis techniques have been used to generate a glyphosate-resistant petunia EPSP synthase.[52] The kinetic analysis of these enzymes shows that, compared with wild-type enzymes, the key change is that the K_m for PEP has been substantially increased.[33] The conclusion from these studies is that glyphosate has the same binding sites as PEP; molecular modeling does, moreover, show that in some conformations glyphosate would be very similar to PEP.

2.2.4. EPSP Synthase and Herbicide Selectivity

The effective inhibition of EPSP synthase by glyphosate is essential evidence that this enzyme is the target site for herbicidal action. Even more persuasive is the introduction of herbicidal resistance by either the increase in cellular levels of EPSP synthase or the insertion into sensitive cells of a glyphosate-resistant EPSP synthase. Overproduction of EPSP synthase in bacteria allows growth in the presence of glyphosate concentrations quite toxic to wild-type cultures, and, in yeast, heterologous overexpression of the *E. coli* EPSP synthase gene results in glyphosate resistance.[14,53,54]

In green plants, the shikimate pathway is localized in the chloroplast, and varying degrees of success in engineering glyphosate tolerance have been found using similar approaches. Nuclear-encoded EPSP synthase is synthesized initially as a pre-EPSP synthase containing a large transit peptide (approximately 15% of the length of the mature enzyme) which is essential for insertion of the precursor into the chloroplast, where a protease removes the targeting peptide.[55-57] Several avenues have been used in attempts to engineer glyphosate tolerance in higher plants. Glyphosate-insensitive EPSP synthase from *S. typhimurium* has been inserted into tobacco and tomato, and this has resulted in a severalfold increase in the amounts of applied glyphosate tolerated[58,59]; the chimeric genes are expressed but appear to be purely cytosolic.[60] This suggests that an extraplastidic shikimate pathway is functional in at least some of the cell types and organs represented in a growing plant.[61] Deliberate targeting of a resistant bacterial EPSP synthase into the chloroplasts using a plant transit peptide has produced transgenic plants with a high level of glyphosate tolerance.[52,60] Overexpression of the wild-type petunia and *Arabidopsis* EPSP synthases in several other plant species has increased the dose of glyphosate which can be tolerated.[55,62] Transfer and expression of an amplified carrot EPSP synthase gene in carrot somatic hybrid cells produced resistant cells with a five- to sixfold increase in EPSP synthase enzyme activity.[63] Isolation of plant cell cultures by selection on glyphosate-containing media correlates with the overproduction of EPSP synthase.[64-67]

No naturally resistant plant species has yet been characterized. Several species of the genus *Pseudomonas* were found to have EPSP synthase 50- to 100-fold less sensitive to glyphosate than the enzymes from bacteria or plants.[68] A cyanobacterium shows considerable resistance to glyphosate when using ammonia as a nitrogen source but not when fixing dinitrogen.[69] The few reports of apparent glyphosate resistance in plants can be accounted for by lack of uptake of the herbicide.[70,71] Glyphosate metabolism within treated plants generally is considered to be minimal though sporadic claims to the contrary have emerged.[72,73] Soil microorganisms certainly can break down glyphosate, some even being able to use it as a sole phosphorus source.[74,75] The utilization of modern molecular genetics to endow plants with the capability of metabolizing glyphosate appears promising because of their extreme sensitivity to modification of the active structure required for EPSP synthase inhibition (see Section 2.3).

Enzymatic sites for glyphosate action other than EPSP synthase have been suggested from *in vitro* studies.[76] All of these are probably of secondary importance and result from the chelating action of glyphosate with the metalloenzymes involved. The most extensively studied is the initial shikimate pathway enzyme 3-deoxyarabinoheptulosonate-7-phosphate (DAHP) synthase.[76-79] Both cytosolic and plastidic DAHP synthases have been reported, and the metal ion cofactor requirements differ; the cytosolic isozyme requires Mn(II) ions and is insensitive to glyphosate while the plastidic isozyme requires Co(II) and is inhibited by the herbicide.[77] There is direct evidence for the formation of a cobalt(II)-glyphosate complex with octahedral coordination.[80] DAHP synthase has also been reported to be increased in plant tissue cultures exposed to glyphosate.[67] In the alga *Euglena gracilis*, glyphosate inhibits the conversion of shikimate to anthranilate, probably by inhibition of the EPSP synthase activity of the *arom* multifunctional enzyme, but has no effect on the DAHP synthase activity.[81] Whatever the significance of glyphosate inhibition of DAHP synthases, it is clear that intracellular flux into the shikimate pathway must continue as the accumulation of shikimic and protocatechuic acids can be provoked by glyphosate application.[78,82-86]. "Wasteful" flux of carbon into shikimic acid has been claimed to be a factor in the growth-inhibitory effects of glyphosate.[79] This may be of relevance in rapidly growing bacterial cultures, but it is unlikely that this "energy drain" contributes to phytotoxicity since shikimate can be accumulated in plants naturally.[87]

Glyphosate is phloem mobile and accumulates in accordance with source-sink relationships in meristematic areas of the treated plants.[88-90] The general physiological effects of glyphosate are consistent with its acting as a metabolic poison. Death is a slow process, requiring days or weeks; much more rapid effects , within hours, can be demonstrated at the biochemical level.[91] These effects can be explained if the turnover rates of specific proteins, and therefore sensitivity to amino acid synthesis inhibition, differ. Inhibition of phenolic biosynthesis and chlorophyll formation and effects on the levels of the phytohormone 3-indoleacetic acid have been considered as target sites but can be viewed as necessary consequences of the primary interference with aromatic amino acid biosynthesis.[86,92-102]

2.3. INHIBITORS OF BRANCHED-CHAIN AMINO ACID BIOSYNTHESIS

2.3.1. Enzymology

The branched-chain amino acids valine and leucine are synthesized from pyruvate while isoleucine is derived from one molecule of pyruvate and one molecule of α-ketobutyrate. The initial pathway enzyme for all

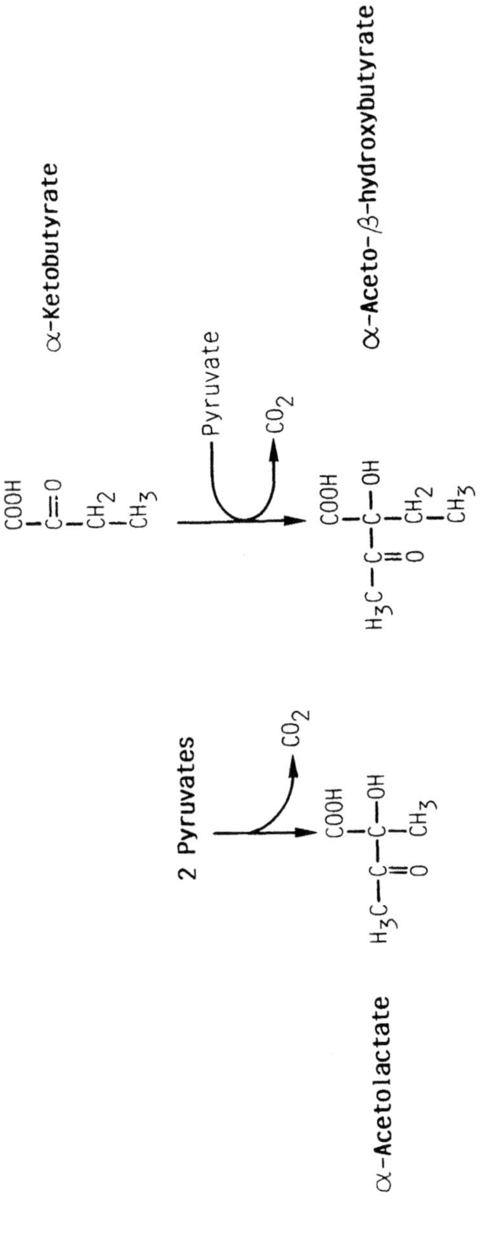

Figure 2.5. The two reactions catalyzed by acetolactate synthase.

three amino acids is acetolactate synthase (ALS; EC 4.1.3.18), which catalyzes the two reactions required (Figure 2.5).

Most of the detailed information about ALS is restricted to the microbial enzyme. In bacteria, there are isozymes important in pathway regulation.[103] The ALS II from *S. typhimurium* is a tetramer of two large (molecular weight 60,000) and two small (molecular weight 10,000) subunits.[104] Thiamine pyrophosphate (TPP) is a coenzyme, the reaction generating hydroxyethyl-TPP as an intermediate.[105,106] Flavin is also required in a nonredox capacity, possibly to prevent protonation of the hydroxyethyl-TPP.[107] Flavin may also, however, determine the equilibrium between dimeric and tetrameric forms of ALS.[108] In addition, there is a metal cofactor requirement, best met by manganese(II) or magnesium(II).[109–111] The metal ion may function in the binding of TPP to the enzyme.[110]

In green plants, branched-chain amino acids are synthesized in the chloroplast.[112] The growth regulation of ALS levels observed in leaf tissue may reflect the development of a full chloroplast complement.[113] ALS is nuclear-encoded and is transferred to the chloroplast as pre-ALS with a transit peptide.[114,115] Tobacco and *Arabidopsis* ALS genes show 70% nucleotide sequence homology and encode mature proteins which are 85% homologous.[115] Multiple isozymes may be present, although in the one successful purification of ALS from a plant source, it was found that the two apparent isoforms were different aggregation states of one polypeptide subunit.[116] As in bacteria, the end products of the pathway cooperatively inhibit the enzyme.[117,118] ALS is, however, only one of several sites of regulation for the enzymatic pathway.[119] In developing maize kernels, ALS activity in the endosperm increases rapidly during the early stages of seed formation whereas embryonic activity rises more gradually and throughout development; the ALS activity from both sources is subject to leucine/valine feedback inhibition.[120] In crude extracts, the ALS from many plant sources is unstable and difficult to purify; the recent successful expression of *Arabidopsis* ALS in *E. coli* offers the possibility of purifying relatively large amounts of a plant ALS for detailed structural investigations.[121]

2.3.2. Herbicides as ALS Inhibitors

Three distinct classes of herbicidally active chemicals inhibit ALS: sulfonylureas, imidazolinones, and triazolopyrimidine sulfonamides.[122–129] Recently, a fourth type, pyrimidinyl oxobenzoic acids, has been patented.[130]

The inhibition by the sulfonylurea herbicide sulfometuron-methyl (SMM, Oust; Figure 2.4) has been well characterized. The inhibition is complex and shows rapid and slow phases. Reaction progress curves are biphasic in the presence of the inhibitor; both stages are inhibited reversibly and competitively with respect to pyruvate.[122] With ALS II from

S. typhimurium, the K_i values for the two phases are 1.7 and 0.08 μM, respectively. With ALS from an archaebacterial *Methanococcus* species, SMM is an uncompetitive inhibitor.[131] At the molecular level, SMM appears to bind to the hydroxyethyl-TPP/ALS intermediate and inhibits the binding of pyruvate or α-ketobutyrate as the second substrate.[122]

The imidazolinone herbicide imazaquin (Scepter) shows fast and slow phases of inhibition, with K_i values of 800 and 20 μM; inhibition is uncompetitive with respect to pyruvate.[126] ALS activities from both embryonic and endosperm tissues of developing maize kernels are inhibited.[120]

Little information has been published about the 1,2,4-triazolo[1,5-*a*]-pyrimidines.[132] Their adverse effects on soybean cell growth are reversed by added valine, leucine, and isoleucine; inhibition of plant ALS is associated with noncovalent binding of the inhibitor.[133] Similarly, preliminary evidence shows that pyrimidinyl oxobenzoates are potent inhibitors of ALS.[134] Tobacco and soybean cell lines insensitive to triazolopyrimidines are also insensitive to sulfonylureas and imidazolinones, suggesting that all three classes of inhibitors have overlapping binding sites on ALS.[106] Herbicidal selectivity for triazolopyrimidines in crop varieties results from differential uptake and metabolism.[135,136]

ALS shows a high degree of primary sequence homology with pyruvate carboxylase and pyruvate oxidase; the ubiquinone cofactors of pyruvate oxidase inhibit ALS, and it has been proposed that the ubiquinone-binding site of the ancestral enzyme also is the site of both SMM and imidazolinone binding.[137] Recent evidence suggests, however, that these two types of herbicide interact differently with ALS: (a) imidazolinones cause a rapid decrease in the levels of extractable ALS activity in maize, whereas SMM does not and can protect the ALS activity from this *in vivo* effect of imidazolinones; and (b) not all imidazolinone-tolerant cell lines are insensitive to sulfonylureas.[138] Sulfonylureas and imidazolinones do not, however, show synergistic inhibition of maize ALS *in vitro*.[139]

2.3.3. ALS Inhibition and Herbicide Effects

Analysis of the primary sequences of bacterial ALS isozymes and mutant enzymes has revealed the identity of several amino acid residues essential to either enzyme activity or herbicide inhibition. In *E. coli* ALS, the substitution of alanine 26 by valine in the large subunit diminishes enzymatic activity.[140] The SMM-resistant ALS I from the same organism has a serine at this position.[140] In yeast, a mutant with SMM tolerance has an ALS with proline 192 replaced by a serine, but the corresponding proline is conserved in an SMM-insensitive *E. coli* ALS I isozyme, while being altered to a serine in an SMM-sensitive *S. typhimurium* ALS II isozyme.[140] In *Saccharomyces cerevisiae*, all nuclear and plasmid-borne SMM resistance alleles result in ALS activities highly resistant to SMM and also insensitive to valine feedback inhibition.[141] This link between herbicide insensitivity and reduced end-product feedback inhibition has also been found with enzyme from carrot cells.[142] An SMM-tolerant *Arabidopsis* mutant contains an ALS with proline 197 changed to a serine residue; this enzyme has been inserted in tobacco and other species and confers herbicide tolerance.[143] Transformation with an *Arabidopsis* ALS gene from a mutant resistant to another sulfonylurea, chlorsulfuron, renders tobacco cells resistant to the herbicide.[144]

A species of blue-green alga has been reported to be naturally resistant to SMM; the endogenous ALS activity is a hundredfold less sensitive to inhibition by SMM than is the enzyme from a species with SMM sensitivity.[145] Plant cell cultures grown in the presence of ALS inhibitors have produced many resistant cell types, none of which have been fully evaluated for changes in ALS.[146] The current evidence indicates, however, that ALS is the key target site for sulfonylurea and imidazolinone herbicides with both long- and short-term persistence in the plant tissues and the environment.[147] The phytotoxicity due to ALS blockade has been assumed to result from starvation for valine, leucine, and isoleucine, but in bacteria the buildup of α-ketobutyrate has also been claimed to be toxic *per se.*[149] This mechanism accounts for growth-inhibitory effects of sulfonylureas and imidazolinones in bacteria in which multiple ALS isozymes occur: the isozyme specialized for isoleucine biosynthesis is the principal target.[149]

2.3.4. Inhibition of Acetolactate Reductoisomerase

The succeeding enzyme to ALS in the branched-chain amino acid pathway is acetolactate reductoisomerase (EC 1.1.1.86). The experimental herbicidally active compound HOE 704, 2-methylphosphinoyl-2-hydroxyacetic acid, inhibits the reductoisomerase from bacterial and plant sources with 50% inhibition at 8 μM; the inhibition is time dependent and competitive with acetolactate.[150] The growth effects on *Lemna gibba* are alleviated by addition of the branched-chain amino acids; HOE 704 application also results in acetolactate accumulation in *L. gibba* and maize.[150] Oxalylhydroxamate inhibitors of the enzyme have been synthesized by *de novo* design as transition state analogues and have bacteriostatic and herbicidal properties.[151]

2.4. HISTIDINE BIOSYNTHESIS

The detailed biosynthetic route for histidine in plants is presently undefined; incorporation patterns and a very limited amount of enzyme studies indicate that the bacterial pathway is probably used.

The enzyme imidazole glycerol-phosphate (IGP) dehydratase (EC 4.2.1.19, Figure 2.6) is inhibited by amitrole (Figure 2.4). The enzyme from *S. typhimurium* is competitively inhibited with a K_i of 30 μM.[152] A similar inhibition has been reported for yeast and plant enzymes.[153,154] This site of action is consistent with other lines of evidence: (a) histidine reverses growth inhibition in algae[155-157]; and (b) amitrole causes the accumulation of imidazoleglycerol in algae and plant cells.[157,158]

Histidine alone, however, is not capable of reversing the growth effects of amitrole on green plants; nucleotide bases are more effective.[159-161] Purine biosynthesis has been suggested as a more important site of action than IGP dehydratase.[161] Since histidine is ultimately biosynthesized from a purine nucleotide, adenosine triphosphate, the inhibition of histidine formation may only be indirect. Amitrole causes the bleaching of newly formed plant tissue, and pigment biosynthesis has been suggested as an alternative site of herbicide action.[162] Effects of 1,2,4-triazole on cysteine biosynthesis have been noted in bacteria, but no studies have been carried out on plants to assess the relevance of this to amitrole-phytotoxicity.[163]

Tobacco cell cultures tolerant to amitrole also display tolerance to glyphosate; this dual tolerance is also observed in cultures adapted to glyphosate.[164] These curious findings may indicate that an uptake mechanism may be shared by the two herbicides or that some element of pathway regulation is common to both histidine and the aromatic amino acids. Examination of results reported with glyphosate-tolerant carrot cell cultures shows that adapted cells contain higher levels of several amino acids but, in particular, arginine, histidine, and the aromatic amino acids.[64] The phenomenon parallels that of the "general control" of amino acid biosynthesis observed in yeasts and fungi, whereby limitation of any of several individual amino acids results in increased activities of other amino acid biosynthetic pathways. Amitrole has been shown to cause such effects in a

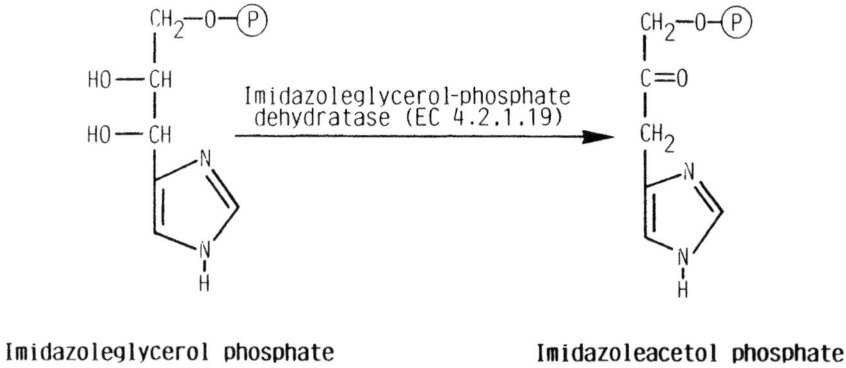

Figure 2.6. The reaction catalyzed by imidazoleglycerol-phosphate dehydratase.

Candida yeast species.[165] At the molecular level, amitrole stimulates the transcription of the *arom* gene in *Saccharomyces cerevisiae*, probably as a "cross-pathway" response to histidine starvation.[166]

2.5. GLUTAMINE SYNTHETASE

2.5.1. Enzymology

Glutamine synthetase (GS; L-glutamate: ammonia ligase, ADP forming; EC 6.3.1.2) is quite different from the enzymes considered previously in that it is always present in mammals. The basic reaction is common to all life forms (Figure 2.7). Glutamine is synthesized for incorporation into proteins but also serves as an amino or amido donor for a host of biosynthetic reactions (Figure 2.8).

The combination of GS with glutamate synthase (GOGAT, Figure 2.7) is the major mechanism for ammonia assimilation in plants; in many microorganisms, the GS/GOGAT cycle is the principal route unless the ammonia supply is sufficiently high for glutamate dehydrogenase (an enzyme with a very much higher K_m for ammonia) to function efficiently in the reductive aminating direction.[167] In plants and microbes, GS shows highly complex regulatory patterns.[168,169] Plant GS is a large octomeric enzyme with a molecular weight of 350,000–400,000 and four or more different subunits.[170] Chloroplastic and cytosolic isozymes occur in green plants, while roots appear to have only a cytosolic form.[171] Localization by immunogold techniques has shown that the chloroplastic isozyme is present in the stroma.[172] From studies of developmental changes in wheat seedlings, it has been suggested that the chloroplastic isozyme is responsible for most of the ammonia assimilation in mature leaves whereas the cytosolic GS plays a role in immature developing cells.[173] More detailed investigation, however, has shown that the chloroplastic GS and GOGAT function primarily to reassimilate ammonia liberated during the photorespiratory cycle: glutamate generated by the ferredoxin-linked GOGAT is transported to the peroxisomes and mitochondria for the transamination reactions that link glyoxylate, glycine, serine, and hydroxypyruvate; mutants lacking chloroplastic GS and GOGAT are viable under conditions that suppress photorespiration.[174]

2.5.2. Inhibitors of Glutamine Synthetase

The mechanism of the GS reaction has been extensively studied.[175–178] The initial step of the reaction involves formation of γ-glutamyl phosphate

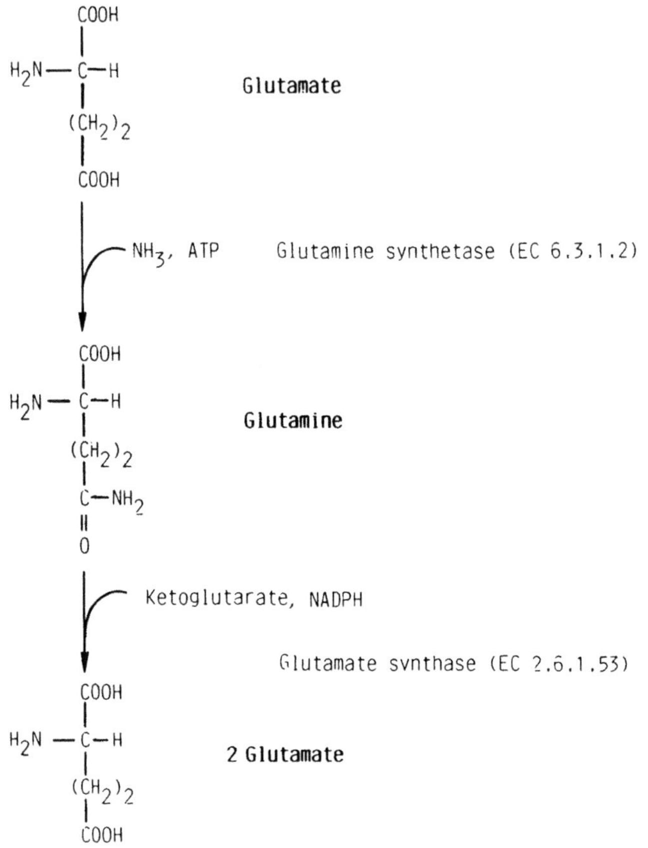

Figure 2.7. Glutamine synthetase and glutamate synthase in ammonia assimilation.

from ATP and glutamate; subsequent attack on γ-glutamyl phosphate by ammonia yields a tetrahedral adduct which breaks down to liberate glutamine.

Phosphinothricin (PPT, Figure 2.4) was derived from the natural streptomycete product bialaphos, the tripeptide L-phosphinothricinyl-L-alanyl-L-alanine.[179,180] Subsequently, the herbicidal formulation HOE 39866 containing PPT has been referred to as glufosinate. Bialaphos itself has herbicidal activity, but experiments with labeled bialaphos have shown that it breaks down *in planta* to liberate PPT.[181]

PPT has two actions: under initial rate conditions, it behaves as a reversible inhibitor of GS competitive with glutamate ($K_i = 18\ \mu M$), but in the presence of ATP a time-dependent irreversible inhibition occurs.[182] Studies with a range of substituted PPT analogues show that phosphorylation is important to the inhibition.[183] The second stage is analogous to that

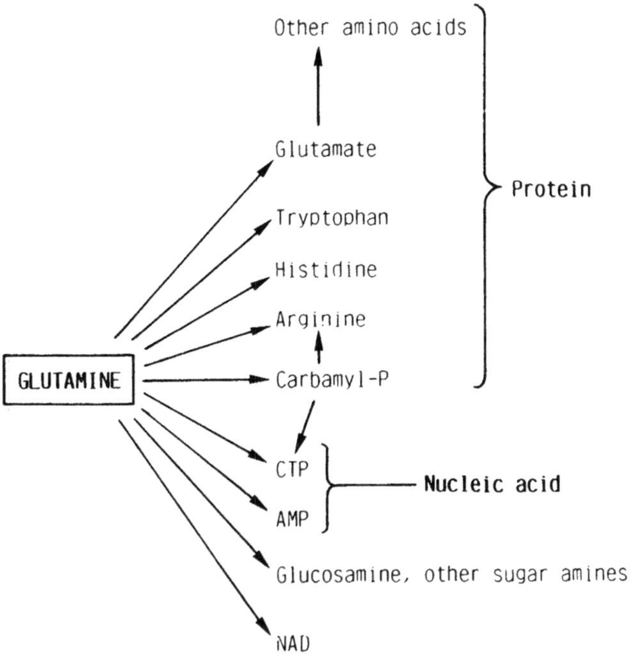

Figure 2.8. Biosynthetic role of glutamine.

which occurs with L-methionine sulfoximine, which is phosphorylated by ATP and forms a ternary complex with ADP and GS.[184-186] The inactive *E. coli* enzyme-inhibitor complex contains PPT, ADP, and inorganic phosphate; incubation at acid pH and at high ionic strength releases the ADP, phosphate, and PPT and restores enzymatic activity.[182] With higher plants, the GS-phosphorylated PPT complex may not be so labile, but it is not clear that the conditions required to reactivate the enzyme have been fully explored.[187,188]

GS activity from a wide spectrum of mono- and dicotyledonous species is inhibited by PPT, whether or not both isozymes are present; very similar kinetic parameters occur in the different plant species studied.[189,190] Neither GOGAT nor glutamate dehydrogenase is affected by PPT *in vitro.*[191]

2.5.3. Glutamine Synthetase and Phytotoxicity

Bialaphos and tabtoxinine-β-lactam, a *Pseudomonas* toxin, both cause the accumulation of high intracellular ammonia concentrations in plant tissues.[192-194] There is also a substantial body of evidence that this aspect of GS inhibition is important in the herbicidal activity of PPT.[195-198]

Increased intracellular ammonia has been implicated in the disruption of ion transport and other membrane functions following PPT application.[199,200] Short-term inhibition of photosynthesis, however, is probably the consequence of the lack of glutamate preventing amination of glyoxalate derived from the oxygenation side reaction of ribulose-bisphosphate carboxylase (RubPCase); glyoxalate itself then inhibits RubPCase.[201] In the absence of exogenous amino acids, prolonged inhibition of the synthesis of glutamine and glutamate, and therefore of the host of transamidation and transamination reactions dependent on them, is undoubtedly lethal.

Enhanced cellular levels of GS might be expected to allow the survival of treated plants, and GS gene amplification has been demonstrated in PPT-tolerant alfalfa cells; this effect is specific for the cytosolic isozyme.[202] Cells with the elevated levels of GS could not be regenerated into normal plants.[202] Protoplast fusions between the PPT-tolerant cell line and a regenerable wild-type line resulted either in normally sensitive but regenerable cells or insensitive but nonregenerable cells.[203] Success was subsequently achieved by overexpression of alfalfa GS in transgenic tobacco plants: 5-fold greater endogenous GS levels resulted in 20-fold better PPT tolerance and the regeneration of fertile plants which continued to exhibit herbicide tolerance.[204] PPT-tolerant tobacco cell lines have also been isolated which show 20-fold elevated GS and, as might be expected from the similarity of the enzyme–inhibitor interactions, are equally tolerant of methionine sulfoximine.[205]

PPT application results in a 50% decrease in GS levels in leaf tissues; glutamate dehydrogenase levels are enhanced but, as intracellular ammonia levels rise, the amination reaction catalyzed by glutamate dehydrogenase insufficiently compensates for the inhibition of GS.[198] Nitrate reductase is reduced, but proteolytic activity is unaffected.[198] A wide-ranging survey of GS levels in plants reported a 70-fold variation in extractable GS levels; there was also considerable variation in the ratios of the cytosolic and chloroplastic isozymes. The kinetics of inhibition by PPT, however, were very similar, and there was no apparent correlation between natural GS levels and herbicide susceptibility.[206] PPT uptake and transport therefore seem to be the determining factors in natural herbicide sensitivity.

2.5.4. Herbicide Selectivity

PPT, while inhibiting GS from plant, microbial, and animal sources, is nontoxic to mammalian cells because it fails to cross the blood–brain barrier and is rapidly cleared by excretion through the kidneys.[33] Inactivation of L-PPT (the D form is not metabolized) by soil bacteria occurs by three routes: *N*-acetylation, transamination, and oxidative deamination, the latter two reactions forming 2-oxo-4-[(hydroxy)(methyl)phosphinoyl]-

butyric acid.[207] The *N*-acetyltransferase from *Streptomyces hygroscopicus* has been expressed in transgenic plants and confers PPT tolerance.[208,209]

2.6. OVERVIEW: CAN INHIBITION OF AMINO ACID BIOSYNTHESIS BE PREDICTED?

The cases considered in the previous four sections show no common themes at the enzymatic or molecular levels. EPSP synthase and IGP dehydratase are not subject to feedback inhibition or other regulation whereas ALS and GS are critical points of metabolic regulation and both have a range of allosteric effectors. EPSP synthase is a simple monomeric protein with no coenzymes or cofactors; GS is a complex oligomeric enzyme and, in at least some bacteria, is subject to reversible covalent modification by adenylation.[168] ALS contains an ancestral quinone-binding domain which may be a common site for a spectrum of chemical inhibitors, including quinone analogues.[137] GS has natural toxin inhibitors; one of these, bialaphos, has given rise to a synthetic herbicide inhibitor, PPT, in a process analogous to the derivation of the pyrethroids from the natural pyrethrins.[210] More detailed investigation of allelochemicals and endogenous plant growth factors as inhibitors of amino acid biosynthesis may yield "ecologically friendly" chemical herbicides.[211]

Considering the pathways of amino acid biosynthesis in a broader context, the route to lysine (Figure 2.1) is unexplored and that to histidine has received little investigative attention. The postchorismic pathway to tryptophan is attractive for potential blockade by herbicides in that the biosynthetic enzymes in plant cells are present at very low levels, an order of magnitude less than those of the common shikimate pathway.[212–216] Chorismate mutase, the initial step in phenylalanine formation from chorismate, is another focus of current work; 1-adamantyl phosphonate is an example of an inhibitor designed on the basis of predictions of the likely transition state structure.[217] Chorismate synthase, the enzyme immediately following EPSP synthase in the shikimate pathway (Figure 2.2), is subject to an undefined regulation by tryptophan[218]; a transition state mimic has been shown to inhibit enzyme activity.[219] Phosphonate analogues of DAHP, the substrate for dehydroquinate synthase (Figure 2.2), inhibit the enzyme; unphosphorylated DAH accumulates and herbicidal effects have been reported.[220] The example of PPT has also spurred a considerable amount of synthetic organic chemistry in the search for other GS inhibitors. *S*-Phosphonomethyl homocysteines are inhibitory mimics of the tetrahedral intermediate in the enzyme-catalyzed reaction.[221] A separate line of development has recognized 4-*N*-hydroxy-L-2,4-diaminobutyric acid as an inhibitor with similar potency to PPT.[222]

An important theoretically based consideration, especially with competitive inhibitors, is that the substrates of the inhibited enzyme should not accumulate and counteract the inhibition.[223] The biological effect of glyphosate is aided by the multiple metabolic fates of PEP and the instability to phosphatase action of the second substrate, shikimate 3-phosphate.[54] Similarly, pyruvate is an important constituent of the central metabolic pathways and is unlikely to reach abnormally high concentrations if ALS is inhibited.

To a very large extent, the current generation of amino acid synthesis inhibitor herbicides reflect their derivation by empirical screening. As an adjunct to this approach, direct testing for potential inhibitors can now be based on the cloning, overexpression, and purification of the bacterial enzymes; the enzymes of the bacterial biosynthetic pathways provide good model systems for the plant target sites.[224] Of course, compounds detected as enzyme inhibitors may not always be immediately useful; for example, inhibitors of shikimate dehydrogenase were synthesized and found to have inhibitory activity *in vitro* but to be inactive at the whole-plant level.[225] A biologically more complex screen could utilize isolated intact chloroplasts, which have been shown to be able to incorporate isotopically labeled bicarbonate into alanine and aromatic and branched-chain amino acids.[112] This system might also give information about the uptake and penetration to the subcellular site of action of the test chemicals. Finally, purification of higher plant enzymes and investigation of the exact chemical structures required for potent inhibition will be of immense value for the rational construction of new herbicides and plant growth regulators.

REFERENCES

1. E. A. Bell, *FEBS Lett. 64*, 29 (1976).
2. B. J. Miflin (ed.), *The Biochemistry of Plants: Volume 5, Amino Acids and Derivatives*, Academic Press, New York (1980).
3. A. H. Fitter and R. K. M. Hay, *Environmental Physiology of Plants*, Academic Press, London (1983).
4. J. S. Titus and S. M. Kang, *Hort. Rev. 4*, 204 (1982).
5. B. J. Miflin and P. L. Lea, *Annu. Rev. Plant Physiol. 28*, 299 (1977).
6. E. Haslam, *The Shikimate Pathway*, Butterworths, London (1974).
7. U. Weiss and J. M. Edwards, *The Biosynthesis of Aromatic Compounds*, John Wiley and Sons, New York (1980).
8. T. N. Tateoka, *Bot. Mag. (Tokyo) 81*, 103 (1968).
9. P. M. Dewick and E. Haslam, *Biochem. J. 113*, 537 (1969).
10. R. Saijo, *Agric. Biol. Chem. 47*, 455 (1983).
11. V. I. Osipov and L. P. Aleksandrova, *Zh. Obshch. Biol. 47*, 79 (1986).
12. A. M. Boudet, A. Graziana, and R. Ranjeva, in: *The Biochemistry of Plant Phenolics* (C. F. Van Sumere and P. J. Lea, eds.), p. 135, Clarendon Press, Oxford (1985).
13. A. Lewendon and J. R. Coggins, *Biochem. J. 213*, 187 (1983).

14. K. D. Duncan, A. Lewendon, and J. R. Coggins, *FEBS Lett. 165*, 121 (1984).
15. H. C. Steinrucken and N. Amrhein, *Eur. J. Biochem. 143*, 341 (1984).
16. D. M. Mousdale and J. R. Coggins, *Planta 160*, 78 (1984).
17. D. M. Mousdale and J. R. Coggins, *J. Chromatogr. 367*, 217 (1986).
18. J. E. Ream, H. C. Steinrucken, C. A. Porter, and J. A. Sikorski, *Plant Physiol. 87*, 232 (1988).
19. M. B. Berlyn and N. H. Giles, *J. Bacteriol. 99*, 222 (1969).
20. J. Lumsden and J. R. Coggins, *Biochem. J. 161*, 599 (1977).
21. V. B. Patel and N. H. Giles, *Biochim. Biophys. Acta 567*, 24 (1979).
22. H. C. Steinrucken and N. Amrhein, *Eur. J. Biochem. 143*, 351 (1984).
23. J. L. Rubin, C. G. Gaines, and R. A. Jensen, *Plant Physiol. 75*, 839 (1984).
24. N. Amrhein, in: *The Shikimic Acid Pathway* (E. E. Conn, ed.), p. 83, Plenum Press, New York (1986).
25. N. Amrhein, J. Schab, and H. C. Steinrucken, *Naturwissenschaften 67*, 356 (1980).
26. H. C. Steinrucken and N. Amrhein, *Biochem. Biophys. Res. Commun. 94*, 1207 (1980).
27. M. R. Boocock and J. R. Coggins, *FEBS Lett. 154*, 127 (1983).
28. D. L. Anton, L. Hestrom, S. M. Fish, and R. H. Abeles, *Biochemistry 22*, 5903 (1983).
29. R. Bode, C. Melo, and D. Birnbaum, *Arch. Microbiol. 140*, 83 (1984).
30. E. S. Sharps, *Anal. Biochem. 140*, 183 (1984).
31. C. S. Smith, D. Pratt and G. A. Thompson, *Plant Cell Rep. 5*, 298 (1986).
32. G. Parr, S. R. Padgette, L. Brundage, and G. M. Kishore, *Fed. Proc. 46* 1977 (1987).
33. G. M. Kishore and D. M. Shah, *Annu. Rev. Biochem. 57*, 627 (1988).
34. K. S. Anderson, K. A. Johnson, A. J. Benesi, and J. A. Sikorski, *Biochemistry 28*, 1936 (1989).
35. P. J. Cassidy and F. M. Kahan, *Biochemistry 12*, 1364 (1973).
36. F. M. Kahan, J. S. Kahan, P. J. Cassidy, and H. Kropp, *Ann. N.Y. Acad. Sci. 235*, 364 (1974).
37. R. I. Zemell and R. A. Anwar, *J. Biol. Chem. 250*, 4959 (1975).
38. M. R. Boocock, Ph.D. thesis, University of Glasgow (1983).
39. K. L. Carlson and O. C. Burnside, *Weed Sci. 32*, 841 (1984).
40. S. S. Abdel-Meguid, W. W. Smith, and G. S. Bild, *J. Mol. Biol. 186*, 673 (1985).
41. Q. K. Huynh, G. M. Kishore, and G. S. Bild, *J. Biol. Chem. 263*, 735 (1988).
42. S. R. Padgette and G. M. Kishore, *Fed. Proc. 46*, 216 (1987).
43. Q. K. Huynh, *Arch. Biochem. Biophys. 258*, 233 (1987).
44. S. R. Padgette, Q. K. Huynh, S. Aykent, R. D. Sammons, J. A. Sikorksi, and G. M. Kishore, *J. Biol. Chem. 263*, 1798 (1988).
45. Q. K. Huynh, S. C. Bauer, G. S. Bild, G. M. Kishore, and J. R. Borgmeyer, *J. Biol. Chem. 263*, 11636 (1988).
46. P. Sprankle, W. F. Meggitt, and D. Penner, *Weed Sci. 23*, 229 (1975).
47. Q. K. Huynh, *J. Biol. Chem. 263*, 11631 (1988).
48. D. M. Stalker, W. R. Hiatt, and L. Comai, *J. Biol. Chem. 260*, 4724 (1985).
49. A. Schulz, D. Sost, and N. Amrhein, *Arch. Microbiol. 137*, 121 (1984).
50. D. Sost, A. Schulz, and N. Amrhein, *FEBS Lett. 173*, 238 (1984).
51. G. M. Kishore, L. Brundage, K. Kolk, S. R. Padgette, D. Rochester, Q. K. Huynh, and G. della-Cioppa, *Fed. Proc. 45*, 1506 (1986).
52. G. M. Kishore, D. M. Shah, S. R. Padgette, G. della-Cioppa, C. Gasser, D. Re, C. Hironake, M. Taylor, J. Wibbenmeyer, D. Eichholtz, N. Hoffmann, X. Delannay, R. Horsch, H. Klee, S. Rogers, D. Rochester, L. Brundage, P. Sanders, and R. T. Fraley, in: *Biotechnology for Crop Protection* (P. A. Hedin, J. J. Menn, and R. M. Hollingworth, eds.), ACS Symposium Series, No. 379, p. 37, American Chemical Society, Washington, D.C. (1985).
53. S. G. Rogers, L. A. Brand, S. B. Holder, E. S. Sharps, and M. J. Brackin, *Appl. Environ. Microbiol. 46*, 37 (1983).

54. G. Kunze, R. Bode, H. Rintala, and J. Hofmeister, *Curr. Genet. 15*, 91 (1989).
55. D. M. Shah, R. B. Horsch, H. J. Klee, G. M. Kishore, J. A. Winter, N. E. Tumer, C. M. Hironaka, P. R. Sanders, C. S. Gasser, S. Aykent, N. R. Siegel, S. G. Rogers, and R. T. Fraley, *Science 233*, 478 (1986).
56. G. S. Gasser, J. A. Winter, C. M. Hironaka, and D. M. Shah, *J. Biol. Chem. 263*, 4280 (1988).
57. G. della-Cioppa, S. C. Bauer, B. K. Klein, D. M. Shah, R. T. Fraley, and G. M. Kishore, *Proc. Natl. Acad. Sci. U.S.A. 83*, 6873 (1986).
58. L. Comai, D. Facciotti, W. R. Hiatt, G. Thompson, R. E. Rose, and D. M. Stalker, *Nature 317*, 741 (1985).
59. J. J. Fillatti, J. Kiser, R. Rose, and L. Comai, *Bio/Technology 5*, 726 (1987).
60. G. della-Cioppa, S. C. Bauer, M. L. Taylor, D. E. Rochester, and B. K. Klein, *Bio/Technology 5*, 579 (1987).
61. R. A. Jensen, *Physiol. Plant. 66*, 164 (1985).
62. H. J. Klee, Y. M. Muskopf, and C. S. Gasser, *Mol. Gen. Genet. 210*, 437 (1987).
63. R. M. Hauptmann, G. della-Cioppa, A. G. Smith, G. M. Kishore, and J. M. Widholm, *Mol. Gen. Genet. 211*, 357 (1988).
64. E. D. Nazfiger, J. M. Widholm, H. C. Steinrucken, and J. L. Killmer, *Plant Physiol. 76*, 571 (1984).
65. C. M. Smith, D. Pratt, and G. A. Thompson, *Plant Cell Rep. 5*, 298 (1986).
66. C. C. Smart, D. Johanning, G. Muller, and N. Amrhein, *J. Biol. Chem. 260*, 16338 (1985).
67. W. E. Dyer, S. C. Weller, R. A. Bressan, and K. M. Hermann, *Plant Physiol. 88*, 661 (1988).
68. A. Schulz, A. Kruper, and N. Amrhein, *FEMS Microbiol. Lett. 28*, 297 (1985).
69. A. K. Singh, M. V. Sailaja, and H. N. Singh, *FEMS Microbiol. Lett. 60*, 187 (1989).
70. O. Gottrup, P. A. O'Sullivan, R. J. Schraa, and W. H. Vanden Born, *Weed Res. 16*, 197 (1976).
71. M. A. Waldecker and D. L. Wyse, *Weed Sci. 33*, 299 (1985).
72. P. Sprankle, C. L. Sandberg, W. F. Meggitt, and D. Penner, *Weed Sci. 26*, 673 (1978).
73. K. Lund-Hoie, *Meld. Norges Landbrukshogskole 55*, 2 (1976).
74. J. K. Moore, H. D. Braymer, and A. D. Larson, *Appl. Environ. Microbiol. 46*, 316 (1983).
75. T. H. Balthazor and L. E. Hallas, *Appl. Environ. Microbiol. 51*, 432 (1986).
76. U. Roisch and F. Lingens, *Hoppe-Seyler's Z. Physiol. Chem. 361*, 1049 (1980).
77. J. L. Rubin, C. G. Gaines, and R. A. Jensen, *Plant Physiol. 70*, 833 (1982).
78. R. Bode, C. Melo Ramos, and D. Birnbaum, *FEMS Microbiol. Lett. 23*, 7 (1984).
79. R. S. Fischer, A. Berry, C. G. Gaines, and R. A. Jensen, *J. Bacteriol. 168*, 1147 (1986).
80. R. J. Ganson and R. A. Jensen, *Arch. Biochem. Biophys. 260*, 85 (1988).
81. G. S. Byng, R. J. Whitaker, and R. A. Jensen, *Can. J. Bot. 63*, 1021 (1985).
82. N. Amrhein, B. Deus, P. Gehrke, and H. C. Steinrucken, *Plant Physiol. 66*, 830 (1980).
83. J. Berlin and L. Witte, *Z. Naturforsch., C36*, 210 (1981).
84. H. Hollander-Czytko and N. Amrhein, *Plant Sci. Lett. 29*, 89 (1983).
85. N. Ishikura, S. Teramoto, Y. Takeshima, and S. Mitsui, *Plant Cell Physiol. 27*, 677 (1986).
86. J. M. Becerril, S. O. Duke, and J. Lydon, *Phytochemistry 28*, 695 (1989).
87. B. A. Bohm, *Chem. Rev. 65*, 435 (1965).
88. J. A. Gougler and D. R. Geiger, *Plant Physiol. 68*, 668 (1981).
89. P. Sprankle, W. F. Meggitt, and D. Penner, *Weed Sci. 23*, 235 (1975).
90. L. C. Haderlie, F. W. Slife, and H. S. Butler, *Weed Res. 18*, 269 (1978).
91. M. E. Foley, E. D. Nazfiger, F. W. Slife, and L. M. Wax, *Weed Sci. 31*, 76 (1983).
92. R. E. Hoagland, S. O. Duke, and C. D. Elmore, *Plant Sci. Lett. 13*, 291 (1978).
93. R. E. Hoagland, S. O. Duke, and C. D. Elmore, *Physiol. Plant. 46*, 357 (1979).
94. S. O. Duke, R. E. Hoagland, and C. D. Elmore, *Plant Physiol. 65*, 17 (1980).
95. S. Margit, C. Jolan, and K. Erzsebet, *Bot. Kozlem. 70*, 151 (1983).
96. L. M. Kitchen, W. W. Witt, and C. E. Rieck, *Weed Sci. 29*, 513 (1981).
97. L. M. Kitchen, W. W. Witt, and C. E. Rieck, *Weed Sci. 29*, 571 (1981).

98. D. J. Cole, J. C. Caseley, and A. D. Dodge, *Weed Res. 23,* 173 (1983).
99. A. Munoz-Reuda, C. Gonzalez-Murua, J. M. Becerril, and M. F. Sanchez-Diaz, *Physiol. Plant. 66,* 63 (1986).
100. T. T. Lee, *Physiol. Plant. 54,* 289 (1982).
101. T. T. Lee and T. Dumas, *Plant Physiol. 72,* 855 (1983).
102. T. T. Lee and T. Dumas, *J. Plant Growth Regul. 4,* 29 (1985).
103. H. E. Umbarger, in: *Amino Acids: Biosynthesis and Genetic Regulation* (K. M. Hermann and R. L. Somerville, eds.), p. 245, Addison-Wesley, Reading, Massachusetts (1983).
104. J. V. Schloss, D. E. Van Dyck, J. F. Vasta, and R. M. Kutny, *Biochemistry 24,* 4952 (1985).
105. L. M. Ciskanik and J. V. Schloss, *Biochemistry 24,* 3357 (1985).
106. L. M. Ciskanik and J. V. Schloss, *Fed. Proc. 45,* 1607 (1985).
107. J. V. Schloss, in: *Flavins and Flavoproteins* (R. C. Bray, P. C. Engel, and S. G. Mayhew, eds.), p. 737, de Gruyter, Berlin (1984).
108. B. K. Singh and G. K. Schmitt, *Plant Physiol. 89* (Supplement), 175 (1989).
109. L. Eoyang and P. M. Silverman, *Methods Enzymol. 166,* 435 (1988).
110. J. V. Schloss and D. E. Van Dyk, *Methods Enzymol. 166,* 445 (1988).
111. Z. Barak, J. M. Calvo, and J. V. Schloss, *Methods Enzymol. 166,* 455 (1988).
112. D. Schulze-Siebert, D. Heineke, H. Scharf, and G. Schultz, *Plant Physiol. 76,* 465 (1984).
113. B. K. Singh and G. K. Schmitt, *Plant Physiol. 89* (Supplement), 54 (1989).
114. A. V. Jones, R. M. Young, and K. Leto, *Plant Physiol. 77,* S293 (1985).
115. B. J. Mazur, C.-F. Chui, and J. K. Smith, *Plant Physiol. 85,* 1110 (1987).
116. J. Durner and P. Boger, *Z. Naturforsch., C 43,* 850 (1988).
117. B. J. Miflin, *Phytochemistry 8,* 2271 (1969).
118. B. J. Miflin, *Arch. Biochem. Biophys. 146,* 542 (1971).
119. J. K. Bryan, in: *The Biochemistry of Plants* (B. J. Miflin, ed.), Vol. 5, p. 403, Academic Press, New York (1980).
120. M. J. Muhitch, *Plant Physiol. 86,* 23 (1988).
121. J. K. Smith, J. V. Schloss, and B. J. Mazur, *Proc. Natl. Acad. Sci. U.S.A. 86,* 4179 (1989).
122. R. A. LaRossa and J. V. Schloss, *J. Biol. Chem. 259,* 8753 (1984).
123. S. C. Falco and K. S. Dumas, *Genetics 109,* 21 (1985).
124. R. S. Chaleff and C. J. Mauvais, *Science 224,* 1443 (1984).
125. T. B. Ray, *Plant Physiol. 75,* 827 (1984).
126. D. L. Shaner, P. C. Anderson, and M. A. Stidham, *Plant Physiol. 76,* 545 (1984).
127. M. J. Muhitch, D. L. Shaner, and M. A. Stidham, *Plant Physiol. 83,* 451 (1987).
128. W. A. Kleschick, M. J. Costales, J. E. Dunbar, R. W. Meikle, W. T. Monte, N. R. Pearson, S. W. Snider, and A. P. Vinogradoff, Abstracts of the 194th ACS National Meeting, New Orleans, No. AGRO162 (1987).
129. B. C. Gerwick and V. Loney, Abstracts of the 194th ACS National Meeting, New Orleans, No. AGRO163 (1987).
130. Y. Nezu, F. Ojima, S. Kusano, Y. Tokokawa, T. Miyazawa, and I. Kajiwara, European Patent Application 223406 (1987).
131. R. Y. Xing and W. B. Whitman, *J. Bacteriol. 169,* 4486 (1987).
132. W. A. Kleschick and B. C. Gerwick, in: *Prospects for Amino Acid Biosynthesis Inhibitors in Crop Protection and Pharmaceutical Chemistry* (L. G. Copping, J. Dalziel, and A. D. Dodge, eds.), British Crop Protection Council Monograph No. 42, p. 139 (1989).
133. M. V. Subramanian, V. Loney, and L. Pao, in: *Prospects for Amino Acid Biosynthesis Inhibitors in Crop Protection and Pharmaceutical Chemistry* (L. G. Copping, J. Dalziel, and A. D. Dodge, eds.), British Crop Protection Council Monograph No. 42, p. 97 (1989).
134. T. R. Hawkes, in: *Prospects for Amino Acid Biosynthesis Inhibitors in Crop Protection and Pharmaceutical Chemistry* (L. G. Copping, J. Dalziel, and A. D. Dodge, eds.), British Crop Protection Council Monograph No. 42, p. 131 (1989).

135. C. C. Hodges, J. Avalos, P. J. McCall, L. E. Stafford, and P. R. Yackovitch, Abstracts of the 194th ACS National Meeting, New Orleans, No. AGRO164 (1987).
136. M. V. Subramanian and M. Kozuch, *Plant Physiol. 89* (Supplement), 176 (1989).
137. J. V. Schloss, L. M. Ciskanik, and D. E. Van Dyk, *Nature 331*, 360 (1988).
138. D. Shaner, B. Singh, K. Newhouse, M. Stidham, and G. Schmitt, *Biochemistry 28*, 1935 (1989).
139. G. K. Schmitt and B. K. Singh, *Plant Physiol. 89* (Supplement), 53 (1989).
140. N. Yadav, R. E. McDevitt, S. Benard, and S. C. Falco, *Proc. Natl. Acad. Sci. U.S.A. 83*, 4418 (1986).
141. S. N. Maiti, M. W. Zink, and G. H. Rank, *Can. J. Microbiol. 34*, 680 (1988).
142. B. Rathinasabapathi, D. Williams, and J. King, *Plant Physiol. 89* (Supplement), 190 (1989).
143. J. K. Smith, C. J. Mauvais, S. Knowlton, and B. J. Mazur, in: *Biotechnology for Crop Protection* (P. A. Hedin, J. J. Menn, and R. M. Hollingworth, eds.), ACS Symposium Series, No. 379, p. 25, American Chemical Society, Washington, D.C. (1985).
144. G. W. Haughn, J. Smith, B. Mazur, and C. Somerville, *Mol. Gen. Genet. 211*, 266 (1988).
145. D. Friedberg and J. Seijffers, *Arch. Microbiol. 150*, 278 (1988).
146. P. C. Anderson and M. Georgeson, in: *The Biochemical Basis of Herbicide Action*, 27th Harden Conference Abstracts, Wye College, Ashford, U.K. (1986).
147. J. M. Gabard, P. J. Charest, V. N. Iyer, and B. L. Miki, *Plant Physiol. 91*, 574 (1989).
148. R. A. LaRossa, T. K. Van Dyk, and D. R. Smulski, *J. Bacteriol. 169*, 1372 (1987).
149. D. Scheel and J. E. Casida, *Pestic. Biochem. Physiol. 23*, 398 (1985).
150. A. Schulz, P. Sponemann, H. Kocher, and F. Wengenmayer, *FEBS Lett. 238*, 375 (1988).
151. J. V. Schloss and A. Aulabaugh, *Biochemistry 28*, 1936 (1989).
152. J. L. Hilton, P. C. Kearney, and B. N. Ames, *Arch. Biochem. Biophys. 112*, 544 (1965).
153. T. Klopotowski and D. Hulanicka, *Acta Biochim. Polon. 10*, 209 (1963).
154. A. Wiater, T. Klopotowski, and G. Bagdasarian, *Acta Biochim. Polon. 18*, 309 (1971).
155. P. J. Casselton, *Nature 204*, 93 (1964).
156. P. J. Casselton, *Physiol. Plant. 19*, 411 (1966).
157. J. N. Siegel and A. C. Gentile, *Plant Physiol. 41*, 670 (1966).
158. M. E. Davies, *Phytochemistry 10*, 783 (1971).
159. K. A. Sund, E. C. Putala, and H. N. Little, *J. Agric. Food Chem. 8*, 210 (1960).
160. F. T. Wolf, *Nature 193*, 901 (1962).
161. J. L. Hilton, *J. Agric. Food Chem. 17*, 182 (1969).
162. D. R. Heim and I. M. Larrinua, *Plant Physiol. 89* (Supplement), 175 (1989).
163. D. Hulanicka, T. Klopotowski, and D. A. Smith, *J. Gen. Microbiol. 72*, 291 (1972).
164. S. R. Singer and C. N. McDaniel, *Plant Physiol. 78*, 411 (1985).
165. R. Bode, P. Casper, and G. Kunze, *Biochem. Physiol. Pflanzen 178*, 457 (1983).
166. K. Duncan, R. M. Edwards, and J. R. Coggins, *FEBS Lett. 241*, 83 (1988).
167. B. J. Miflin and P. J. Lea, in: *The Biochemistry of Plants, Vol. 5, Amino Acids and Derivatives* (B. J. Miflin, ed.), p. 169, Academic Press, New York (1980).
168. E. R. Stadtman, A. Ginsburg, J. E. Ciardi, J. Yeh, S. B. Hennig, and B. M. Shapiro, *Adv. Enzyme Regul. 8*, 99 (1970).
169. G. R. Stewart, A. F. Mann, and P. A. Fentem, in: *The Biochemistry of Plants, Vol. 5, Amino Acids and Derivatives* (B. J. Miflin, ed.), p. 271, Academic Press, New York (1980).
170. S. F. McNally, B. Hirel, P. Gadal, A. F. Mann, and G. R. Stewart, *Plant Physiol. 72*, 23 (1983).
171. M. Lara, H. Porta, J. Padilla, J. Folch, and F. Sanchez, *Plant Physiol. 76*, 1019 (1984).
172. J. R. Botella, J. P. Verbelen, and V. Valpuesta, *Plant Physiol. 88*, 943 (1988).
173. A. K. Tobin, S. M. Ridley, and G. R. Stewart, *Planta 163*, 544 (1985).
174. A. C. Kendall, S. W. J. Bright, N. P. Hall, A. J. Keys, P. J. Lea, J. C. Turner, and R. M. Wallsgrove, in: *Progress in Photosynthesis Research* (J. Biggens, ed.), Vol. 3, p. 629, Martinus Nijhoff, Dordrecht (1987).

175. R. H. McParland, J. G. Guevara, R. R. Becker, and H. J. Evans, *Biochem. J. 153,* 597 (1976).
176. A. Meister, in: *The Enzymes* (P. D. Boyer, ed.), Vol. 10, p. 699, Academic Press, New York (1974).
177. T. D. Meek, K. A. Johnson, and J. J. Villafranca, *Biochemistry 21,* 2158 (1982).
178. T. D. Meek and J. J. Villafranca, *Biochemistry 19,* 5513 (1980).
179. E. Bayer, K. H. Gugel, H. Hagele, M. Hagenmaier, S. Jessipow, W. A. Konig, and H. Zahner, *Helv. Chim. Acta 55,* 224 (1972).
180. K. Tachibana, T. Watanabe, Y. Sekizawa, and T. Takematsu, *J. Pestic. Sci. 11,* 33 (1986).
181. A. Wild and C. Ziegler, *Z. Naturforsch., C 44,* 97 (1989).
182. J. A. Colanduoni and J. J. Villafranca, *Bioorg. Chem. 14,* 163 (1986).
183. E. W. Logusch, D. M. Walker, J. F. McDonald, and J. E. Franz, *Biochemistry 28,* 3043 (1989).
184. J. D. Gass and A. Meister, *Biochemistry 9,* 1380 (1970).
185. F. C. Wedler, Y. Sugiyama, and K. E. Fisher, *Biochemistry 21,* 2168 (1982).
186. J. J. Villafranca, D. E. Ash, and F. C. Wedler, *Biochemistry 15,* 544 (1976).
187. R. Manderscheid and A. Wild, *J. Plant Physiol. 123,* 135 (1986).
188. M. Hopfner, G. Reifferscheid, and A. Wild, *Z. Naturforsch., C 43,* 194 (1988).
189. M. Leason, D. Cunliffe, D. Parkin, P. J. Lea, and B. J. Miflin, *Phytochemistry 21,* 855 (1982).
190. M. A. Acaster and P. D. J. Weitzman, *FEBS Lett. 189,* 241 (1985).
191. A Wild and R. Manderscheid, *Z. Naturforsch., C 39,* 500 (1984).
192. S. Imai, H. Seto, T. Sasaki, T. Tsuruoka, H. Ogawa, A. Satoh, S. Inouye, T. Niida, and N. Otake, *J. Antibiot. 38,* 687 (1985).
193. T. J. Knight, R. D. Durbin, and P. J. Langston-Unkefer, *J. Bacteriol. 169,* 1954 (1987).
194. P. J. Langston-Unkefer, A. C. Robinson, T. J. Knight, and R. D. Durbin, *J. Biol. Chem. 262,* 1608 (1987).
195. P. J. Lea, K. W. Joy, J. L. Ramos, and M. G. Guerrero, *Phytochemistry 23,* 1 (1984).
196. A. Wild, H. Sauer, and W. Ruhle, *Z. Naturforsch., C 42,* 263 (1987).
197. H. Sauer, A. Wild, and W. Ruhle, *Z. Naturforsch., C 42,* 270 (1987).
198. M. Lacuesta, B. Gonzalez-Moro, C. Gonzalez-Murua, P. Aparicio-Tejo, and A. Muñoz-Rueda, *J. Plant Physiol. 134,* 304 (1989).
199. H. Kocher, *Aspects Appl. Biol. 4,* 227 (1983).
200. G. D. Trogisch, H. Kocher, and W. R. Ullrich, *Z. Naturforsch., C 44,* 33 (1989).
201. C. Ziegler and A. Wild, *Z. Naturforsch., C 44,* 103 (1989).
202. G. Donn, E. Tischer, J. A. Smith, and H. M. Goodman, *J. Mol. Appl. Genet. 2,* 621 (1984).
203. M. Deak, G. Donn, A. Feher, and D. Dudits, *Plant Cell Rep. 7,* 158 (1988).
204. P. Eckes, P. Schmitt, W. Daub, and F. Wengenmayer, *Mol. Gen. Genet. 217,* 263 (1989).
205. Y. Ishida, T. Hiyoshi, M. Sano, and T. Kumashiro, *Plant Sci. 63,* 227 (1989).
206. S. M. Ridley and S. F. McNally, *Plant Sci. 39,* 31 (1985).
207. K. Bartsch and C. C. Tebbe, *Appl. Environ. Microbiol. 55,* 711 (1989).
208. M. De Block, J. Botterman, M. Vandewiele, J. Dockx, C. Thoen, V. Gossele, N. R. Movva, L. Tompson, M. Van Montagu, and J. Leemans, *EMBO J. 6,* 2513 (1987).
209. C. J. Thompson, N. R. Movva, R. Tizard, R. Crameri, J. E. Davies, M. Lauwereys, and J. Botterman, *EMBO J. 6,* 2519 (1987).
210. G. S. Hartley and T. F. West, *Chemicals for Pest Control,* Pergamon Press, Oxford (1969).
211. M. F. Balandrin, J. A. Klocke, E. S. Wurtele, and W. M. Bollinger, *Science 228,* 1154 (1985).
212. J. M. Widholm, *Physiol. Plant. 25,* 75 (1971).
213. J. M. Widholm, *Biochim. Biophys. Acta 320,* 217 (1973).
214. C. N. Hankins, M. T. Largen, and S. E. Mills, *Plant Physiol. 57,* 101 (1976).
215. W. Grosse, *Z. Pflanzenphysiol. 80,* 463 (1976).
216. D. M. Mousdale and J. R. Coggins, *FEBS Lett. 205,* 328 (1986).
217. P. A. Bartlett, in: *The Shikimic Acid Pathway* (E. E. Conn, ed.), p. 119, Plenum Press, New York (1986).

218. H. Bickel and G. Schultz, *Phytochemistry 18*, 498 (1979).
219. P. A. Bartlett, U. Maitra, and P. M. Chouinard, *J. Am. Chem. Soc. 108*, 8068 (1986).
220. D. L. Pompliano, L. M. Reimer, S. Myrvold, and J. W. Frost, *J. Am. Chem. Soc. 111*, 1866 (1989).
221. E. W. Logusch, *Tetrahedron Lett. 29*, 6055 (1988).
222. S. Fushiya, K. Maeda, T. Funayama, and S. Nozoe, *J. Med. Chem. 31*, 480 (1988).
223. A. Cornish-Bowden, *FEBS Lett. 203*, 3 (1986).
224. J. R. Coggins, in: *Biotechnology and Crop Improvement and Protection* (P. R. Day, ed.), British Crop Protection Council Monograph No. 34, p. 101 (1986).
225. A. C. Baillie, J. R. Corbett, J. R. Dowsett, and P. McCloskey, *Pestic. Sci. 3*, 113 (1972).

3

Lipid Synthesis

JOHN L. HARWOOD

3.1. THE NATURE AND DISTRIBUTION OF PLANT LIPIDS

Plant lipids can be divided into major groups along structural lines. The main membrane components are phosphoglycerides and glycosylglycerides while sterols, carotenoids, and sphingolipids are present in certain membranes in much smaller amounts. The surface layers of different plant tissues consist of complex mixtures of molecules which include many lipids which contribute to the wax, cutin, or suberin structures. Finally, the major storage lipids of plants are the triacylglycerols, with wax esters being present in some species, notably jojoba (*Simmondsia chinensis* L.). Herbicides are known which interfere with sterol, carotenoid, acyl lipid, and wax/cutin/suberin synthesis. The first two groups of compounds are dealt with in Chapters 5 and 4, respectively. In this chapter, consideration will be given to herbicides which inhibit fatty acid production or the further metabolism of these molecules in acyl lipids or in the generation of surface layers.

First, a short account needs to be given of the main membrane acyl lipids and the composition of the different surface layers. As a generalization, plant (and algal) membranes can be divided into two types with regard to their lipid composition. Most membranes contain the phosphoglyceride/protein matrix so familiar to biochemists dealing with animal or bacterial systems. In startling contrast, the photosynthetic thylakoid

JOHN L. HARWOOD • Department of Biochemistry, University of Wales College of Cardiff, Cardiff CF1 1ST, United Kingdom.

membranes contain little phosphoglyceride but, instead, have glycosylglycerides as their main lipids.[1] These glycosylglycerides comprise monogalactosyldiacylglycerol (MGDG), digalactosyldiacylglycerol (DGDG), and sulfoquinovosyldiacylglycerol (SQDG) or plant sulfolipid. The only phosphoglyceride found in significant quantities in thylakoids generally is phosphatidylglycerol (PG). The structures of the major chloroplast thylakoid lipids are shown in Figure 3.1.

Thylakoids show only limited variation in composition, as exemplified in Table 3.1. The two galactosylglycerides are the main lipids and together represent about 70% of the total, with MGDG predominating over DGDG. This is interesting from the viewpoint of membrane structure since MGDG does not form bilayers when purified and added to aqueous solutions; instead, it gives rise to hexagonal II phases.[5] Physical experiments on membrane structure, however, have shown that mixtures of MGDG with other bilayer-forming lipids (such as DGDG) can generate membranes *in vitro* which are fully capable of maintaining ion gradients,[6] albeit with some differences with respect to artificial phosphoglyceride membranes.[7] The reason for the ubiquitous and predominant presence of MGDG in thylakoids is not clear although it has been speculated that it may be useful to "package" bulky membrane protein complexes.[5] A function in stabilizing the highly curved thylakoid regions on the ends of granal stacks has not been confirmed by experiments on transmembrane asymmetry.[8]

The third glycosylglyceride is the plant sulfolipid (SQDG), which is negatively charged at physiological pH. The function of this lipid in photosynthetic membranes is unknown,[5] although its presence in all thylakoids examined would imply that it has a vital function or functions. Interestingly, the other major thylakoid lipid, phosphatidylglycerol, is also negatively charged. Although some analyses of certain isolated thylakoid fractions have sometimes reported significant amounts of other phospholipids (notably, phosphatidylcholine),[1] phosphatidylglycerol can be regarded as the typical phosphoglyceride of photosynthetic membranes. Indeed, in cyanobacteria, phosphatidylglycerol is the only phospholipid.[4]

While discussing the lipid composition of thylakoids, a few points should be made about their acyl content. Some generalizations can be made, and these will be discussed in relation to two "typical" examples of 16:3 and 18:3 plants, respectively (Table 3.2).* First, the thylakoids of higher plant chloroplasts contain notably high amounts of α-linolenate. Even in the negatively charged SQDG and PG, about half of the total acyl groups

* 16:3 Plants are defined as those which have hexadecatrienoic acid esterified in their MGDG fraction while 18:3 plants have α-linolenic acid. These differences arise from differences in the fatty acid metabolism of such plants.[9] Because 16:3 plants resemble cyanobacteria in their metabolism of fatty acids, such species are said to have "procaryotic metabolism" while 18:3 plants have "eucaryotic metabolism."[10]

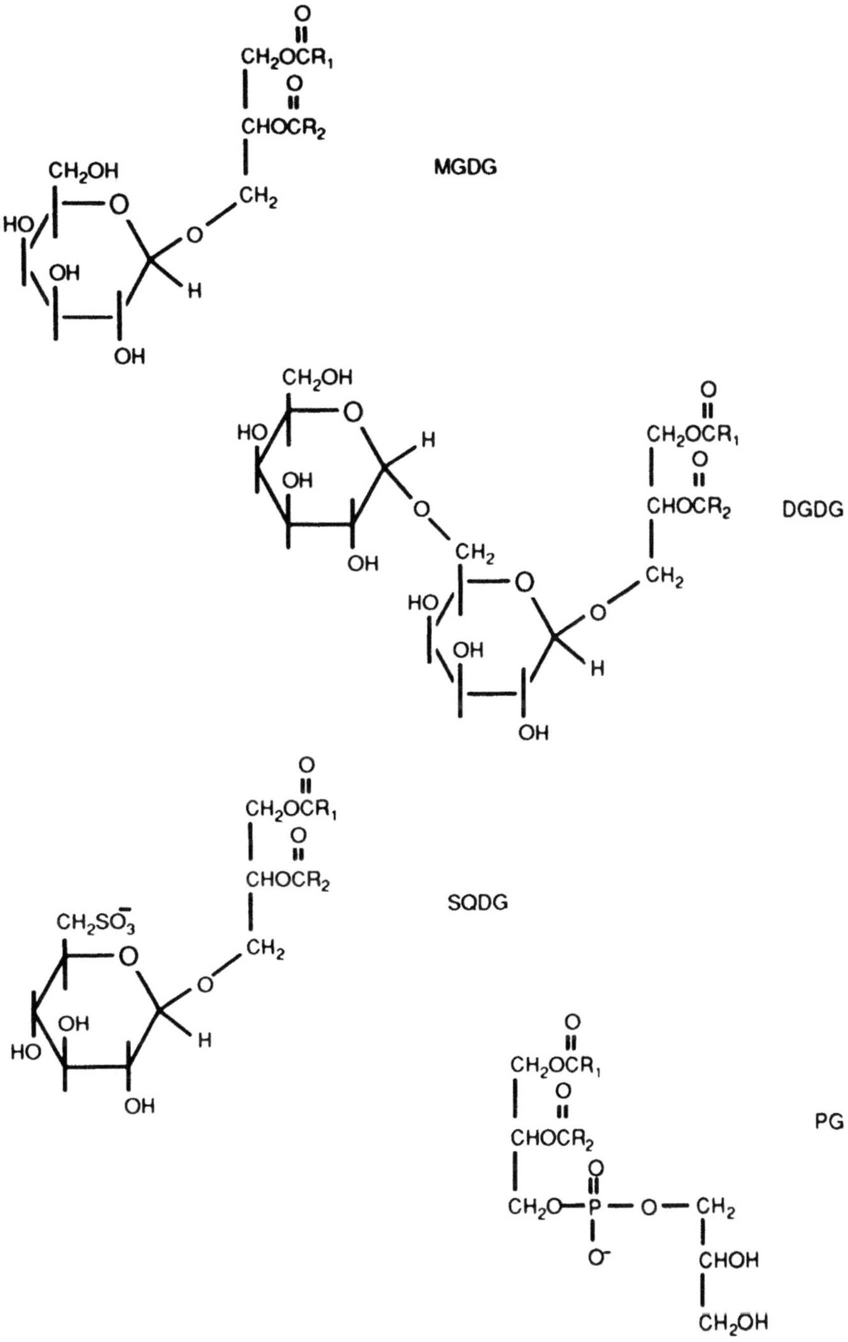

Figure 3.1. Structures of the major chloroplast thylakoid lipids.

Table 3.1. Acyl Lipid Compositions of Thylakoids from Higher Plants, Green Algae, and Cyanobacteria

Species	Acyl lipid composition (mol %)			
	MGDG	DGDG	SQDG	PG
Spinach	45	25	7	13
Pea	41	28	9	12
Barley	44	27	12	10
Maize	53	27	10	7
Wheat	45	36	8	10
Chlamydomonas reinhardtii[a]	55	20	13	6
Anabaena variabilis[b]	54	17	11	17
Anacystis nidulans[b]	55	12	11	22

[a] Ref. 3.
[b] Ref. 4.

are made up of linolenate. In MGDG, the percentage can rise to about 95%. Of the two galactosylglycerides, MGDG always has a higher proportion of trienoates than DGDG, in which palmitate is a minor though significant component. SQDG contains palmitate as a major component, such that in some plants this lipid class may have a significant amount of disaturated molecular species,[11] thus rendering these plants potentially susceptible to frost damage. PG has, perhaps, the most striking fatty acid composition, with large amounts of the unusual *trans*-Δ 3-hexadecenoic acid.[1] Molecular species of PG containing palmitate and *trans*-Δ 3-hexadecenoate are referred to as "disaturated" because *trans* unsaturated fatty acids have physical properties rather similar to those of saturated

Table 3.2. The Fatty Acid Compositions of the Main Chloroplast Thylakoid Lipids from Spinach, 16:3 Plant, and Barley, 18:3 Plant

Plant		Fatty acid composition (% of total)					
		16:0	16:1[a]	16:3 (n − 3)	18:1 (n − 9)	18:2 (n − 6)	18:3 (n − 3)
Spinach	MGDG	1	—	25	1	2	72
	DGDG	4	—	5	2	2	87
	SQDG	39	Trace	Trace	1	7	53
	PG	11	32	—	2	4	47
Barley	MGDG	4	—	—	1	3	90
	DGDG	9	2	—	3	7	78
	SQDG	32	3	—	2	5	55
	PG	18	27	—	2	11	38

[a] 16:1 is *cis*-Δ9-16:1 (palmitoleic acid) for all lipids except PG, where it is *trans*-Δ3-16:1.

acids. Plants with large amounts of such species of PG, together with dipalmitoyl species, are believed to show phase separation of lipid on cooling to around freezing temperatures; consequently, they are frost sensitive, and this phenomenon has been rationalized in terms of the membrane properties of lipids.[12] The sensitivity of plants to frost damage has implications for some herbicides, such as the substituted pyridazinones, which inhibit desaturation reactions.

The chloroplast is surrounded by two envelope membranes. Techniques for separating these membranes from each other have allowed their analysis. These studies revealed that, while the inner envelope has a lipid composition similar to that of thylakoids, the outer envelope contains little glycosylglyceride and is similar to endoplasmic reticulum.[13]

Extrachloroplast membranes contain different amounts of four main phosphoglycerides—phosphatidylcholine (PC), phosphatidylinositol (PI), phosphatidylethanolamine (PE), and PG. Because individual plant tissues vary appreciably in their levels of these lipids, it is difficult to describe the composition of, say, plasma membrane in precise terms. Moreover, there is the added problem that not all the subcellular isolations reported in the literature have been characterized adequately! Nonetheless, a few general remarks can be made. First, the plasma membrane appears to contain appreciable quantities of plant steroids (as well as phospholipids) stigmasterol, and β-sitosterol, being particularly important. Second, although glycolipids and neutral lipids have been reported, their presence in isolated fractions may be due to contamination.

Mitochondria contain an outer membrane which has PC and PE as its main components while the inner membrane is the unique site of diphosphatidylglycerol (DPG, cardiolipin) in plant cells.[14] Microbodies, such as those of potato or the glyoxysomes of castor bean, have a rather simple lipid pattern with PC, PE, and PG usually predominating. On the other hand, microsomal fractions, which are assumed to contain mainly endoplasmic reticulum, have a mixture of PC, PE, PG, and PI, with PC predominating. Some examples of the phospholipid compositions of different plant membranes are given in Table 3.3.

Extrachloroplastic membranes are more saturated than thylakoids. Linoleate is usually the major polyunsaturated component rather than linolenate. In addition, palmitate and oleate are much more important constituents than in photosynthetic membranes. Almost all extrachloroplastic membranes have palmitate, oleate, and linoleate as their main fatty acids. Some examples of the patterns for the main phosphoglyceride, PC, are given in Table 3.4.

Some herbicides, such as ethofumesate or the substituted thiocarbamates, seem to affect the structure of surface coverings such as leaf wax. The composition of these coverings—wax, cutin, suberin—is very complex

Table 3.3. Phosphoglyceride Patterns of Different Membrane Fractions from Plants

Fraction	Phosphoglyceride pattern (% of total)				
	PC	PE	PI	PG	DPG
Plasma membrane (potato)	21	39	12	n.d.[a]	2
Mitochondria (cauliflower)	44	34	7	11	4
Mitochondria (sycamore)	43	35	6	13	3
Peroxisome (potato)	64	20	4	15	n.d.
Peroxisome (castor bean)	48	36	8	4	n.d.
Microsomal (potato)	45	33	16	2	Trace
Microsomal (pea)	67	14	19	Trace	n.d.

[a] n.d.: Not detected.

and ill defined. Kolattukudy and his co-workers[15] have made some sophisticated analyses in an effort to clarify the component parts and the overall structures of cutin and suberin. The reader is referred to the above reference for more detail. One feature which should be stressed here, though, is that a major portion of cuticular wax, cutin, and suberin has very long chain ($>C_{18}$) constituents. These are either very long chain fatty acids or are derived from these compounds. The essential hydrophobicity of such long hydrocarbon chains is, of course, one of the main properties of surface layers and can cause problems for the entry of many herbicides when they are applied in the field. This aspect of the efficacy of herbicides and their commercial utilization is discussed in Chapters 8 and 9.

3.2. BASIC PATHWAYS OF FATTY ACID AND ACYL LIPID METABOLISM

De novo formation of fatty acids in plants requires the concerted action of acetyl-CoA carboxylase and fatty acid synthetase. In most plants, these

Table 3.4. Fatty Acid Compositions of Phosphatidylcholine in Different Membrane Fractions from Higher Plants

Fraction	Fatty acid composition (% of total)					
	16:0	16:1	18:0	18:1	18:2	18:13
Spinach chloroplast	12	—	—	9	16	58
Barley chloroplast	31	5	5	9	16	34
Pea microsomes	4	1	4	46	43	2
Sycamore mitochondria	25	—	2	4	46	23
Potato mitochondria	12	—	3	Trace	64	21

enzyme complexes result in palmitate and stearate as the major products although shorter chain acids are produced by certain seed tissues. The source of carbon for fat synthesis arises ultimately from photosynthesis. In nonphotosynthetic tissues, sucrose is usually used, but in leaves acetyl-CoA is formed mainly from pyruvate via the pyruvate dehydrogenase/decarboxylase complex.[16] Spinach, and possibly other plants, may generate acetyl-CoA in plastids by an indirect route. This pathway appears to use the mitochondrial pyruvate dehydrogenase to form acetyl-CoA, which is hydrolyzed to acetate within mitochondria. Free acetate is rather permeable and is able to cross organelle membranes easily. It can, therefore, pass into chloroplasts, where it is activated to acetyl-CoA by a thiokinase. The physiological significance of this pathway in most plants is not established (chloroplast pyruvate dehydrogenase being sufficiently active),[16] but it does allow [^{14}C]acetate to be used generally as a fatty acid precursor in biochemical experiments.

Acetyl-CoA carboxylase can be regarded as catalyzing the first committed step for *de novo* fatty acid formation. The reaction involves the ATP-dependent formation of malonyl-CoA from acetyl-CoA and bicarbonate. Recent studies using a variety of biotin-dependent enzymes have shed light on the detailed mechanism of such reactions. The first partial reaction is the carboxylation of the biotin group which is attached to the ε-amino group of a lysine residue of the biotin carboxyl carrier protein (BCCP). This carboxylation itself occurs in two steps using a carboxyphosphate intermediate.[17] Transfer of the carboxyl residue from biotin to acetyl-CoA substrate is catalyzed by a carboxyltransferase enzyme. Type I biotin carboxylases share these general features in common, and it is the specificity of the carboxyltransferase reaction which confers specificity on the whole complex.

Two general types of acetyl-CoA carboxylase have been found in nature. In *E. coli*, the enzyme exists as a multienzyme complex from which the biotin carboxylase, BCCP, and the carboxyltransferase can be isolated. By contrast, acetyl-CoA carboxylase from animals is a high-molecular-mass multifunctional protein containing distinct domains which correspond to BCCP and biotin carboxylase or carboxyltransferase activity. For many years, it was thought that the plant acetyl-CoA carboxylase was similar to the multienzyme complex of *E. coli* or, alternatively, represented an intermediate form. However, these suggestions were based on the partial purification of proteins which had been degraded by proteinases. It seems now that the plant acetyl-CoA carboxylase is a high-molecular-mass (200–240 kDa) multifunctional protein.[9] The evidence is firm for seed tissues but less so for leaves.

Purification procedures for acetyl-CoA carboxylase usually use an affinity step with avidin attached to the column matrix. This allows a more rapid purification to take place and is important since, even with proteinase inhibitors, the plant enzyme is very unstable. Perhaps because of this instability, there has been much less work on the enzymatic characteristics of the acetyl-CoA carboxylase from plants compared with that from animals or *E. coli.* In particular, the regulation of the enzyme is unclear. This control is important since there is quite a lot of evidence that the overall carbon flux to fatty acids (and lipids) may be regulated by acetyl-CoA carboxylase.[9] Net increases in carboxylase protein, presumably by increased translation due to elevated mRNA, have been shown to accelerate fatty acid production in developing oilseeds.[18] However, the increase in leaf acetyl-CoA carboxylase activity in the light may be due to a number of alterations in chloroplast constituents.[19] The increase in stromal pH and ATP and Mg^{2+} concentrations and the decrease in ADP concentration which occur on photosynthesis *in vivo* have been shown to increase acetyl-CoA carboxylase activity *in vitro* 24-fold.[19] This is about the same amount that the carboxylase activity is raised in leaves on illumination. Therefore, the plant acetyl-CoA carboxylase may be subject to long-term and short-term regulation. There is, however, absolutely no evidence of a role for tricarboxylic acids in such control.[9]

Fatty acid synthetase in plants is a type II enzyme with separate proteins catalyzing the partial reactions and a dissociable acyl carrier protein (ACP).[9] Most of the component enzymes have been purified to homogeneity from at least one plant tissue.[9,20] The reactions catalyzed are indicated in the upper half of Figure 3.2. A few remarks on the general characteristics of the enzymes will be given here, but more information will be found in appropriate reviews.[9,20,21]

Acetyl-CoA:ACP transacylase is a thiolactomycin-sensitive enzyme which catalyzes what has often been regarded as the slowest of all the partial reactions. However, the recent observation that in spinach leaves[22] and all other plant tissues examined[23] there is an acetoacetyl-ACP synthetase which bypasses this reaction (and which has much higher activity) sheds doubt on the physiological importance of acetyl-CoA:ACP transacylase. The acetoacetyl-ACP synthetase, sometimes called the short-chain condensing enzyme, is cerulenin insensitive[22] and can, therefore, be differentiated easily from the β-ketoacyl-ACP synthetase I, which is the main condensing enzyme. The latter has been purified recently to homogeneity for the first time.[24]

It is well known that the two-carbon units for condensation in *de novo* fatty acid synthesis are provided as malonate. A specific transacylase transfers the malonate from coenzyme A to ACP. Malonyl-CoA:ACP transacylases have been purified from a number of sources including leaf and seed tissues as well as the cyanobacterium *Anabaena variabilis.*[9,21] The enzymes usually have masses of about 40 kDa and appear to use a random sequence mechanism. Tissue-specific isoforms have been reported in some instances though their significance is not known.

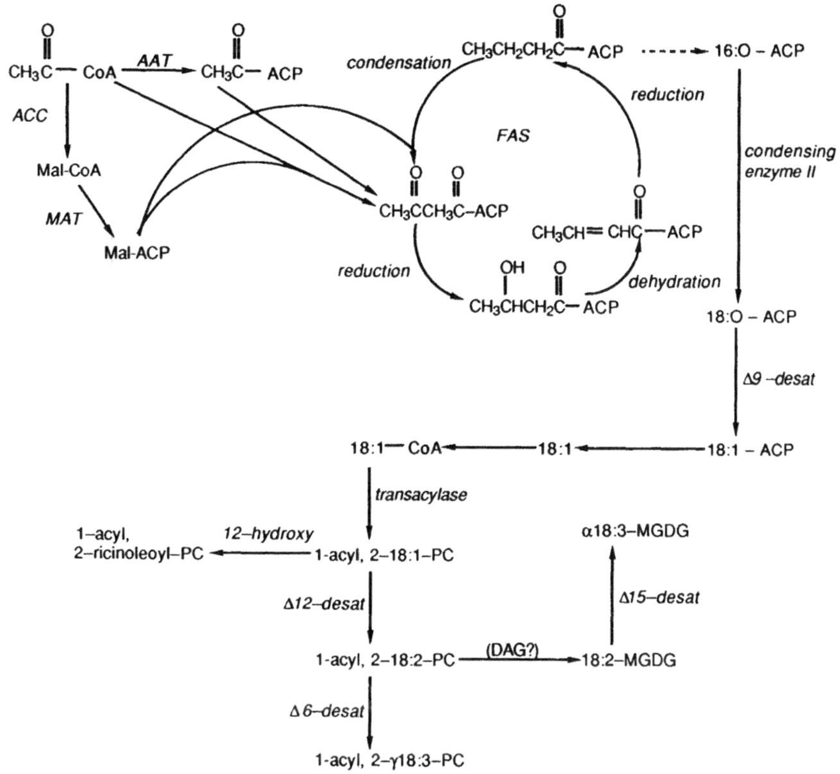

Figure 3.2. The main reactions involved in plant fatty acid synthesis.

Once condensation has taken place, the β-ketoacyl-ACP undergoes a cycle of reduction, dehydration, and reduction to yield a new fatty acid. The enzymes concerned are β-ketoacyl-ACP reductase, β-hydroxyacyl-ACP dehydrase, and enoyl reductase, and detailed information is provided in recent reviews.[9,20,21] Both of the reductases have been reported in isoforms from several tissues. Thus, in avocado, an NADH form of β-ketoacyl-ACP reductase has been resolved from an NADPH form.[25] In seed tissues, two forms of enoyl reductase appear to be present: they differ in the chain-length activity and also in whether they use NADH only or use NADPH in preference to NADH; only the NADH form has been purified from leaves.[9,20]

β-Ketoacyl-ACP synthetase I can condense substrates up to 14 carbons in length, the end product of its action being 16 carbons in length. It is well known, however, that stearate is the major product of most fatty acid synthetases. The production of stearate is dependent on the activity of another condensing enzyme, termed β-ketoacyl-ACP synthetase II. Apart from its chain-length specificity, the above enzyme differs from β-ketoacyl-ACP synthetase I in being inhibited by arsenite but not by cerulenin. The relative activities of condensing enzymes I and II will clearly control the ratio of palmitoyl- to stearoyl-ACP produced *de novo.*

Once fatty acid synthetase has produced long-chain saturated fatty acids, these products can be desaturated or elongated. For desaturation, three desaturase complexes are principally involved. These insert double bonds in the Δ9, Δ12, and Δ15 positions, respectively. The Δ9-desaturase operates with acyl-ACP substrates and, like fatty acid synthetase, is present in plastids. The only Δ9-desaturase which has been purified to any degree is that from developing safflower seeds.[26] The enzyme is active with 16- and 18-carbon substrates, forming palmitoleate and oleate, respectively. The source of electrons for this mixed-function oxidase appears to be reduced ferredoxin from the photosynthetic electron transport chain.[26]

Once oleate has been formed, in the majority of plants it appears to be exported to the extraplastidic compartment. This movement requires hydrolysis of oleoyl-ACP by a stromal thioesterase followed by formation of oleoyl-CoA on the chloroplast envelope.[9,27] Oleoyl-CoA is then used to acylate lipids on the endoplasmic reticulum. The esterification to the *sn*-2 position of PC is particularly important since 1-acyl-2-oleoyl-PC is the major substrate for the Δ12-desaturase.[27,28] The linoleate generated by this desaturation can then be further desaturated on PC in oilseeds or returned to the chloroplast.[9] The latter case will naturally occur in leaf tissue, where linoleate is esterified to MGDG in order to provide the main substrate for the Δ15-desaturase.[29,30] Little is known about the enzymatic characteristics of the Δ12- and Δ15-desaturases,[9,27] not least because no purification has yet been achieved and, in addition, the Δ15-desaturase is particularly labile.

In addition to the general pathway outlined above (see Figure 3.3), some desaturation of palmitoleate and oleate seems to be carried out completely within the chloroplast. Thus, plants operating the so-called "procaryotic" pathway of lipid metabolism[10] retain the bulk of their newly formed acyl chains within the plastid. This is because the glycerol-3-phosphate acyltransferase activity is much greater than the acyl-ACP thioesterase activity (Figure 3.4). Such channeling of acyl chains results in rapid synthesis of phosphatidate, which can then be catabolized by phosphatidate phosphohydrolase[31] to yield the diacylglycerol needed for galactosylglyceride and PG synthesis. The unique features of "procaryotic" metabolism include a localization of 16-carbon acids at the *sn*-2 position of chloroplast lipids together with the formation of hexadecatrienoate, which is esterified to MGDG. By contrast, the "eucaryotic" pathway yields chloroplast lipids with palmitate concentrated at the *sn*-1 position while hexadecatrienoate is absent from such plants[19] (Figure 3.4).

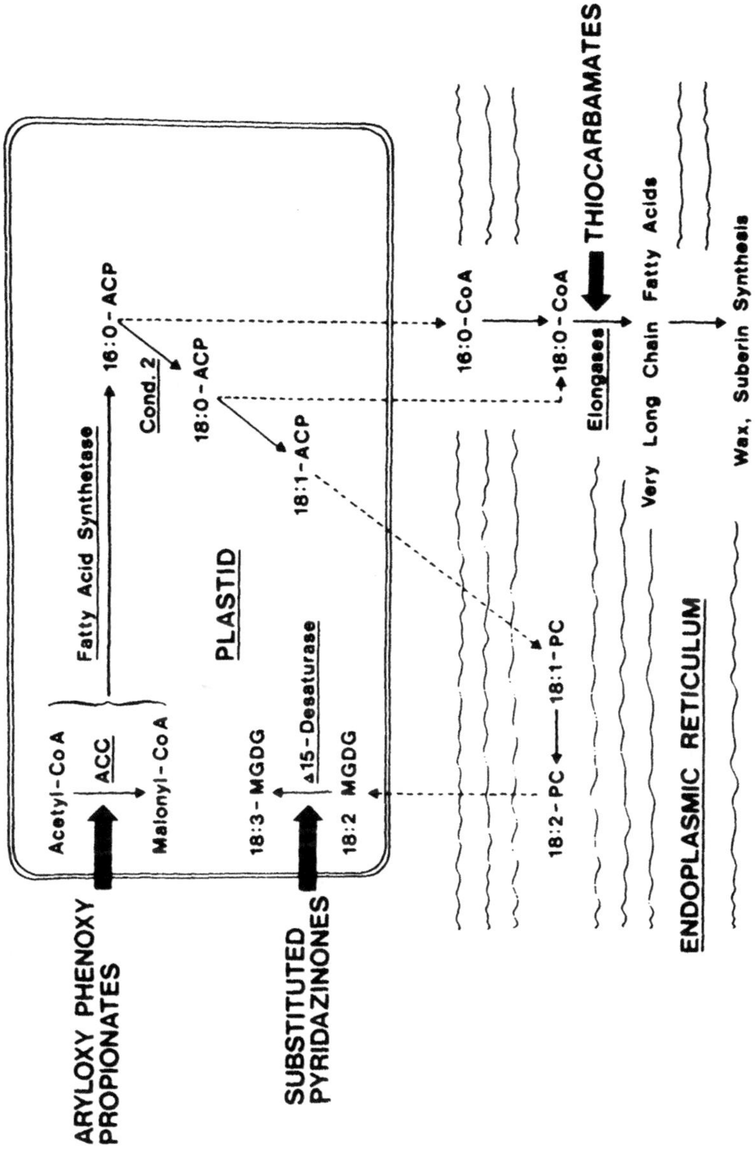

Figure 3.3. Compartmentation of fatty acid synthesis in leaves showing the sites of action of some herbicide classes.

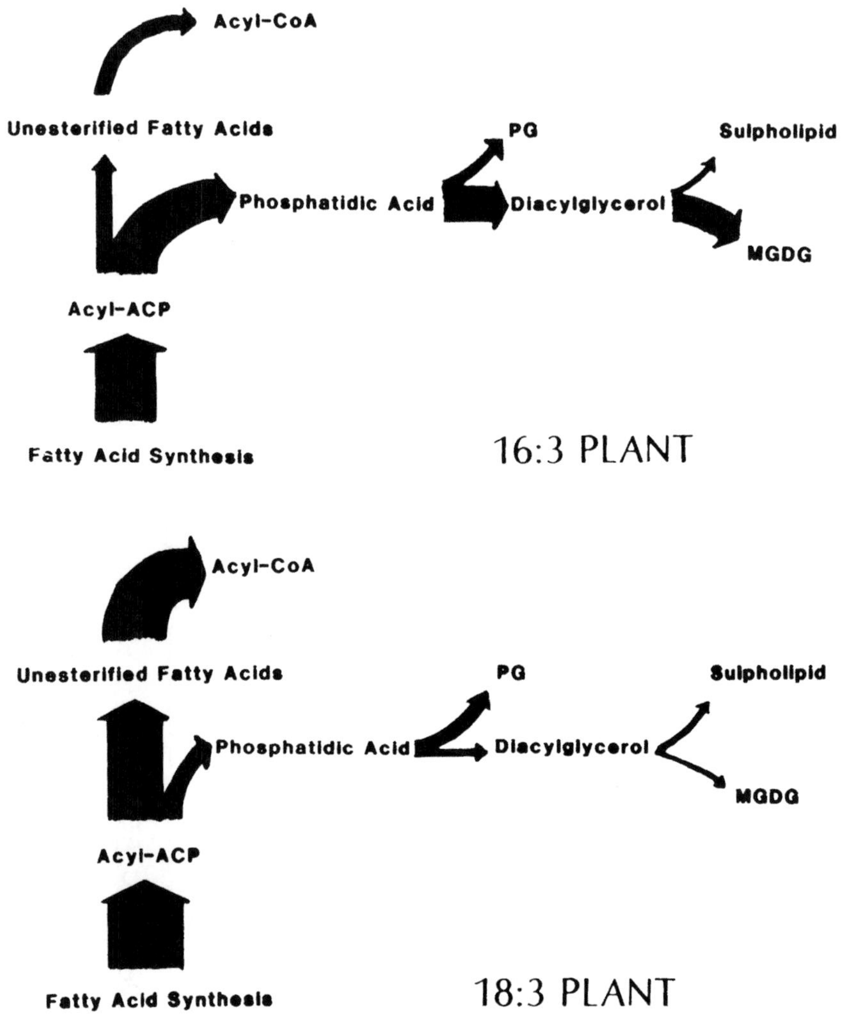

Figure 3.4. Generalized differences in metabolism by the "procaryotic" (16:3 plant) and "eucaryotic" (18:3 plant) pathways.

In addition to the Δ9-, Δ12- and Δ15-desaturases discussed above, plants contain other desaturation systems. Probably of most note insofar as herbicides are concerned is the formation of *trans*-Δ3-hexadecenoate. The substrate for this desaturation appears to be 1-acyl-2-palmitoyl-PG, with the *trans*-Δ3-hexadecenoate formed found exclusively at the *sn*-2 position of PG.[9]

Instead of desaturation, fatty acids can undergo elongation (Figures 3.3 and 3.4). The enzymes responsible are located on the endoplasmic reticulum and use malonyl-CoA as the two-carbon addition unit.[9] As in the case of the mammalian fatty acid elongases, there seem to be several chain-length-specific systems,[32] and acyl-CoA substrates are functional in, at least, some tissues.[9,33] Recently, there has been limited success in solubilizing the elongase activities from leek,[34,35] and a ketoacyl intermediate has been demonstrated.[36] As mentioned in Section 3.1, the fatty acid elongases are particularly important for the synthesis of precursors for cutin, suberin, and surface wax formation.

No herbicide has yet been demonstrated to have a specific action against one of the enzymes responsible for acyl lipid formation. Nevertheless, because many of the reactions involved (particularly in glycosylglyceride synthesis) have unique features, or are not found in significant amounts

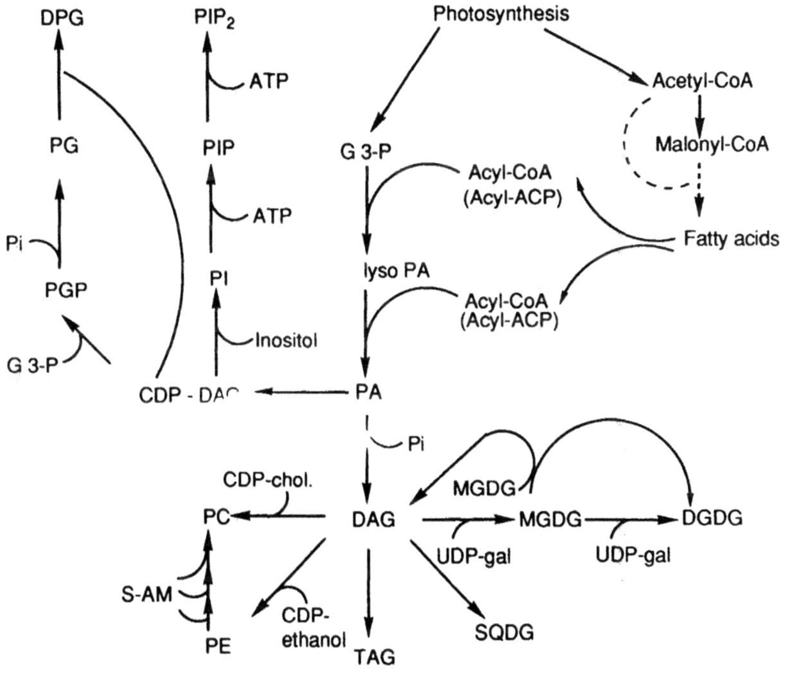

Figure 3.5. Major pathways for acyl lipid synthesis in plants. Abbreviations: G 3-P, Glycerol 3-phosphate; lyso PA, 1-monoacyl phosphatidic acid; PA, phosphatidic acid; DAG, diacylglycerol; TAG, triacylglycerol; UDP-gal, UDP-galactose; CDP-chol., CDP-choline; CDP-ethanol., CDP-ethanolamine; S-AM, *S*-adenosylmethionine; CDP-DAG, CDP-diacylglycerol; PIP, phosphatidylinositol 4-phosphate; PIP_2, phosphatidylinositol 4,5-bisphosphate; PGP, phosphatidylglycerol phosphate; DPG, diphosphatidylglycerol.

outside plants, they offer attractive targets for herbicides in the future. A summary of the main pathways for phosphoglyceride, acylglyceride, and glycosylglyceride synthesis is given in Figure 3.5. For further information on acyl lipid metabolism, see Ref. 37 and references therein.

3.3. SUBSTITUTED PYRIDAZINONES

Substituted pyridazinones were developed as potential photosynthetic electron transport inhibitors. For this action, they have I_{50} activities in the range 10^{-5}–10^{-10} *M*.[38] It was soon discovered, however, that many of these compounds (Figure 3.6) also reduced the formation of thylakoid pigments, chlorophylls, and carotenoids. In each case, the mechanism of phytotoxicity was found to vary. In some cases, there was evidence for metabolism of the pyridazinone to a more herbicidal metabolite while in others the destruction of photosynthetic function was due to an inhibition of carotenoid (and chlorophyll) synthesis.[38]

The inhibition of photosynthesis by substituted pyridazinones led to their use in studies on the development of chloroplast structure and function. While carrying out such experiments in barley and mustard, Hilton *et al.*[39] noted that Sandoz 6706 also caused a reduction in linolenic acid levels. As previously noted (Table 3.2), α-linolenate is particularly prevalent in chloroplast lipids. These authors postulated that impaired chloroplast development could be due to a lack of normal amounts of galactosylglycerides and extended their studies by the use of Sandoz 9785. The latter is an interesting pyridazinone because it has minimal action on pigment formation.[38] Sandoz 9785, however, was found to have a significant effect on the distribution of acyl groups in different lipid classes; treatment resulted in a decrease in α-linolenate content accompanied by a commensurate rise in linoleate proportions (Table 3.5). Although this action on linolenate levels was the only observed effect on acyl lipids, not all plant tissues were susceptible.[40] Nevertheless, the absence of inhibition of chloroplast pigments showed that any reduction in α-linolenate occurred independently of oxidation-induced damage. In addition, the herbicide effects were examined in relation to low-temperature-induced increases in membrane unsaturation.[40,41] At lower environmental temperatures, the proportion of unsaturated fatty acids increases, either because of increased oxygen solubility or through an adaptive mechanism involving membrane fluidity.[42] This increase in α-linolenate proportions on transfer to lower growth temperatures is prevented by Sandoz 9785 treatment. In fact, linolenate levels were well below those of untreated tissue grown at higher temperatures. Furthermore, Sandoz 9785 treatment led to a prevention of the ability of winter wheat cv. Kharkov to acquire frost tolerance.[41]

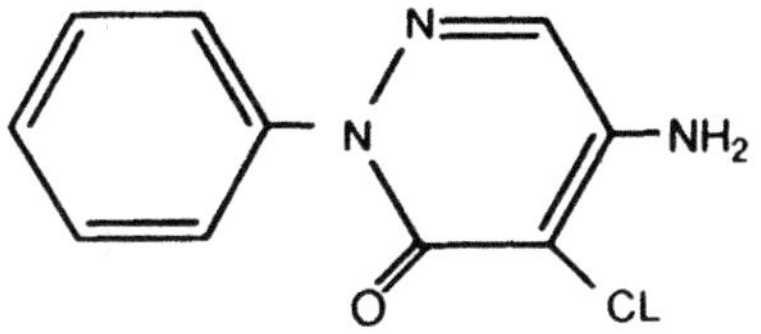

PYRAZON

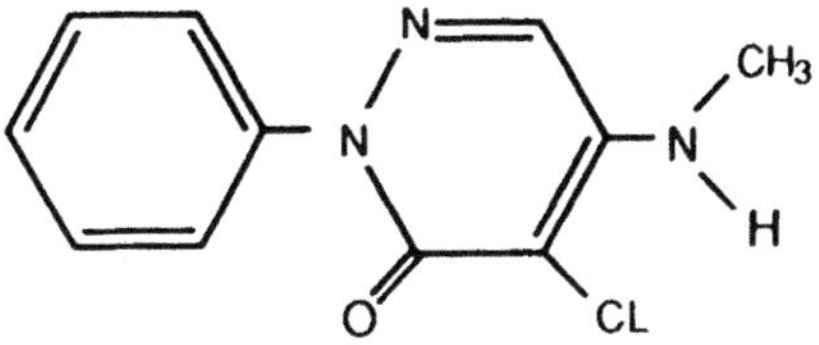

133-410 H

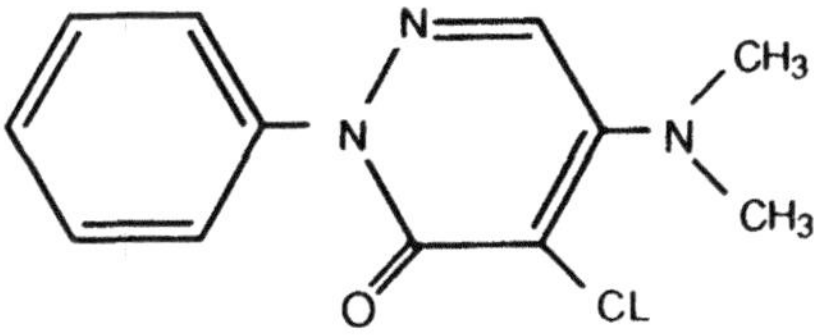

9785

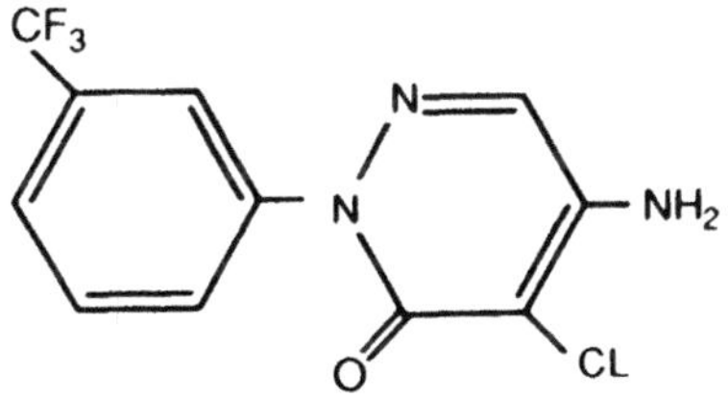

9774

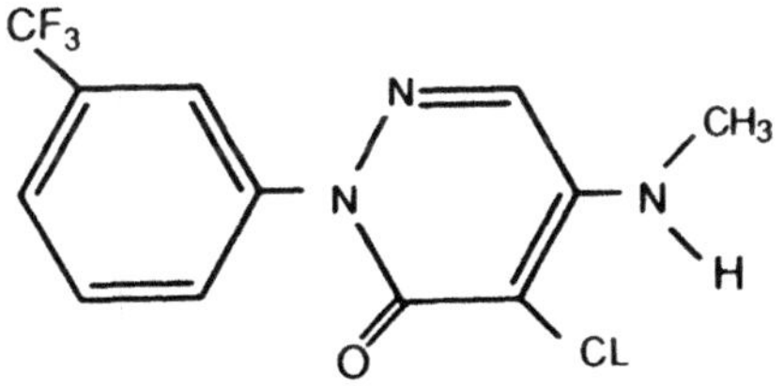

NORFLURAZON

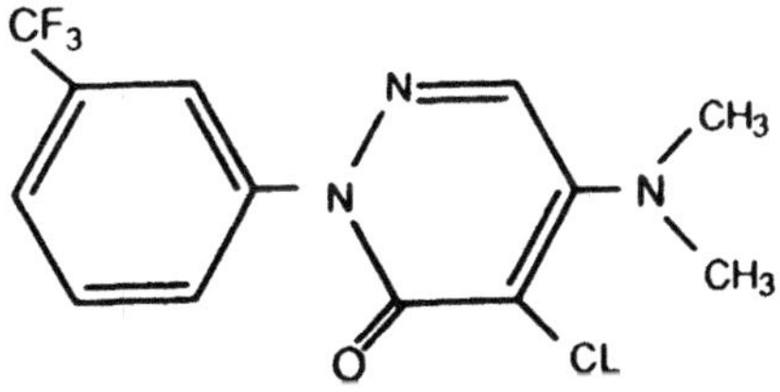

6706

Figure 3.6. Structures of some pyridazinones.

Table 3.5. Action of the Substituted Pyridazinone Sandoz 9785 on the Relative Properties of Linoleate and α-Linolenate in Plant Tissues[a]

Tissue	Sandoz 9785	Fatty acid composition (% of total)				
		16:0	18:0	18:1	18:2	18:3
Wheat roots	−	18	1	5	33	44
	+	19	1	3	55	22
Wheat shoots	−	18	1	4	32	45
	+	19	1	5	49	26
Barley shoots	−	10	1	2	10	75
	+	10	1	2	40	47

[a] Refs. 40 and 41.

α-Linolenate is particularly prevalent in galactosylglycerides (Table 3.2), and effects of substituted pyridazinones which reduce linolenate levels are most easily seen within this lipid class. Interestingly, individual structures show different potency toward the inhibition of pigment synthesis as opposed to linolenate formation (Table 3.6). Pyrazon itself caused little reduction of either pigment or linolenate formation, in spite of being a potent electron transport inhibitor. Sandoz 133-410H caused a moderate inhibition of both pigment and linolenate synthesis, while Sandoz 9789 and Sandoz 6706 were potent inhibitors. Sandoz 9774 was reported to be much

Table 3.6. Relative Effectiveness of Different Substituted Pyridazinones on Pigment Synthesis, Photosynthetic Electron Transport, and Galactolipid Fatty Acid Patterns[a]

Herbicide	Pigment synthesis (% inhib.)		Electron transport (conc. for I_{50})	MGDG 18:2/18:3	DGDG 18:2/18:3
	Chlorophylls	Carotenoids			
Pyrazon	7	12	$7 \times 10^{-6}\,M$	0.42	0.25
Sandoz 133-410H	40	40	—	0.93	0.50
Sandoz 9785	0	2	$1 \times 10^{-5}\,M$	3.82	3.32
Sandoz 9774	55	43	$1 \times 10^{-5}\,M$	2.30	0.33
Sandoz 9789	96	99	—	1.67	1.80
Sandoz 6706	99	99	$4 \times 10^{-6}\,M$	1.70	1.40
None (control)	—	—	—	0.32	0.17

[a] Data taken from St. John.[38] Photosynthetic electron transport was measured by ferricyanide reduction via the Hill reaction in isolated barley chloroplasts. Pigment synthesis was evaluated at a herbicide concentration of $10^{-4}\,M$ while fatty acid composition was at $10^{-5}\,M$, both in wheat shoots.

more effective toward MGDG than DGDG, but the reason for this is not at all obvious. The results for Sandoz 9785 were particularly clear cut, with no detectable action on pigment synthesis but a marked effect on the linoleate/linolenate ratio[38] (Table 3.6). It has been concluded that substituted pyridazinones have at least three sites of action—electron transport, pigment accumulation, and α-linolenate synthesis. Although individual actions on one process could be as a result of inhibition of another, individual cases are seen where the susceptibilities of the three sites of action are different. Consequently, it has been concluded that the multiple actions displayed by pyridazinone herbicides are usually independent.[38]

The significant effects of Sandoz 9785 on fatty acid, but not pigment, synthesis made this pyridazinone a useful compound to use in further experiments. Its action on frost hardiness[41] was noted also in other cereal varieties. Application of Sandoz 9785 either as a preemergence herbicide in the field or through watering solutions in the greenhouse led to a reduced α-linolenate percentage in all seedlings tested. Although germination and early seedling growth were unaffected, herbicide application reduced overwinter survival.[43] Variations in susceptibility were probably due to differences in herbicide uptake, movement, and metabolism (see later).

When Sandoz 9785 was used against a dicotyledon, *Vicia faba* L., it was found to be less effective.[44] Thus, at 10^{-4} *M*, no lipid changes were noted whereas a similar concentration caused significant changes in the linoleate/linolenate ratios in chloroplast and nonchloroplast lipids of barley.[44] The same workers also used Sandoz 6706 and Sandoz 9774. Sandoz 6706 showed its most potent effects toward the *trans*-Δ3-hexadecenoate content of PG (see Table 3.2). A simultaneous increase in palmitate content suggested that the Δ3-desaturase which acts at the *sn*-2 position of PG was being inhibited.[9] The linoleate/linolenate ratio of barley, but not broad bean, was also changed by Sandoz 6706.[44] Electron microscopic examination of treated plants showed that Sandoz 9785 caused significant changes in thylakoid structure. Few intergranal membranes were noted, and, although grana were present, in the most severely affected tissues their number was reduced; large vacuolar inclusions were also observed. With Sandoz 6706, the major effect appeared to be an almost complete absence of grana; at the highest concentration (10^{-3} *M*), the single thylakoid membranes which were present were highly disorganized.[44,45]

Other laboratories have also carried out simultaneous measurements of the lipid or lipid metabolism changes associated with pyridazinone-induced in plastid morphology. Davies and Harwood[46] examined the effects of Sandoz 6706 and Sandoz 9785 on lipid metabolism and membrane structure in greening barley. Changes seen in chloroplast structures included a slowing of normal differentiation on greening and the appearance of large short stacks following treatment with Sandoz 9785; Sandoz 6706 caused similar and even more obvious effects. Little normal stacking was observed in leaves treated with Sandoz 6706 during 8 h of illumination compared with untreated leaves. Due to the multiple effects of Sandoz 6706 on susceptible plants, however, it was difficult to correlate the alterations in thylakoid morphology with specific lipid changes. The more specific action of Sandoz 9785 may mean that the lack of normal stacking in treated plants was caused by α-linolenate deficiency.[46]

Rather similar experiments were carried out in another laboratory, and, again, Sandoz 9785 was found to cause changes in thylakoid structure in addition to significantly reducing α-linolenate levels. Specifically, the ratio of appressed to nonappressed (exposed) membranes was increased from 2.5 in controls to 3.4 in treated barley. It was concluded, however, that there was generally little evidence in plants for a correlation of α-linolenate content and membrane appression.[47] In a further study with Sandoz 9785 using pea and wheat, some changes in chloroplast morphology were seen, and it was thought possible that these were connected to reduced α-linolenate as well as alterations in photosynthetic parameters.[48–50]

The purported site of action of Sandoz 9785 on linoleate desaturation was tested more directly using radiolabeled precursors. In addition, the range of photosynthetic tissues examined was increased by the inclusion of *Azolla* and *Chlorella* spp. and *Anabaena cylindricum.*[51,52] In higher plants, Sandoz 9785 caused no detectable change in the total uptake or incorporation of radioactivity from [^{14}C]acetate into lipids; in lower plants, however, the compound caused large reductions in the incorporation into total fatty acids. This inhibition was accompanied by changes in the proportions of radiolabeled fatty acids. Such alterations could not, however, be shown to include the Δ12-desaturation of linoleate because neither this acid nor α-linolenate were labeled in the duration of the experiment. Six higher plant species were examined, and Sandoz 9785 caused a large inhibition of linolenate labeling from [^{14}C]acetate in barley and ryegrass leaves and cucumber cotyledons but had much less effect in maize, pea, or spinach leaves.[52] Further experiments with [1-^{14}C]oleate or [1-^{14}C]linoleate confirmed that linoleate desaturation was a site of action in susceptible species (Table 3.7).

For the six plant species, PC and MGDG were the major acyl lipid classes labeled. When the labeling patterns of PC and MGDG were examined, however, it was found that Sandoz 9785 had no effect on the proportion of radiolabeled acids in PC. In contrast, the herbicide caused reciprocal changes in the linoleate and α-linolenate proportions in MGDG. In fact, the high labeling of α-linoleate in MGDG and inhibition of its formation were cited as evidence for the use of MGDG as a substrate in

Table 3.7. Action of Sandoz 9785 on the Desaturation of Linoleate to α-Linolenate in Radiolabeling Experiments Using Barley

Precursor	Concentration of Sandoz 9785 (m*M*)	Fatty acid labeling (% of total ^{14}C-acids)				
		16:0	18:0	18:1	18:2	18:3
[1-^{14}C]Acetate	0	17	4	15	41	21
	0.1	18	4	14	54	11
[1-^{14}C]Oleate	0	—	—	77	17	3
	0.1	—	—	79	19	Trace
[1-^{14}C]Linoleate	0	—	—	—	64	36
	0.1	—	—	—	82	18

Δ15-desaturation.[51,52] Experiments were carried out with [1-^{14}C]linoleate and [1-^{14}C]linolenate in order to eliminate the possibility that linoleate was desaturated as part of a different substrate and that the labeled linoleate and α-linolenate were merely transferred preferentially to MGDG. The results showed clearly that radiolabeled oleate, linoleate, and linolenate were incorporated in a similar manner, with PC being the most labeled lipid class; there was no evidence for a preferential transfer of α-linolenate to MGDG.[51,52] All the evidence pointed to MGDG being involved in linoleate desaturation and to this process being inhibited by Sandoz 9785.

Later experiments by Willemot and co-workers[53] confirmed these conclusions. In addition, these authors noticed that spinach leaves were relatively unaffected by Sandoz 9785. By contrast, disks from spinach leaves were sensitive, perhaps because in leaves there was a detoxification process.[53] Isolated spinach chloroplasts were insensitive to Sandoz 9785, and it was suggested that this might be due to a need for Sandoz 9785 to be activated before it could inhibit Δ15-desaturation.[53]

The selectivity of Sandoz 9785 may be explained by different rates of uptake and/or metabolism, as were observed for ^{3}H-labeled herbicide.[52] Thus, in pea, which was relatively insensitive to Sandoz 9785, there was a very high rate of herbicide metabolism, presumably to less active compounds. In contrast, good rates of uptake and only moderate rates of metabolism were observed in cucumber and ryegrass, which were sensitive.

Arabidopsis thaliana is a much studied plant these days because of its small genome. The species belongs to the "16:3 type" of plants and shows "procaryotic" as well as "eucaryotic" fatty acid metabolism (see Section 3.2). Sandoz 9785 reduced the amounts of 18:3/16:3, di-18:3, and 18:3/16:0 MGDG molecular species,[54] reductions in the 18:3/16:3 and 18:3/16:0 MGDG species being much the greater; both arise by the "procaryotic" pathway. Radiolabeling studies showed that the molecular species of PC were unaffected, agreeing with the general conclusion that Sandoz 9785 has no effect on this phospholipid.[52]

In conclusion, one of the possible sites of action of the substituted pyridazinones occurs at the level of fatty acid desaturation. In particular, desaturations involving linoleate[52] or hexadecadienoate[54] substrates are inhibited when MGDG is the substrate. Some pyridazinones also inhibited the formation of *trans*-Δ3-hexadecenoate, the unusual acyl group of chloroplast PG. There is evidence that pyridazinone treatment of sensitive plants gives rise to structural changes in chloroplast membranes, but the link of these alterations to inhibition of fatty acyl desaturation is unproven.

3.4. THIOCARBAMATES AND THEIR SAFENERS

Thiocarbamate herbicides, which are normally applied preemergence, exert their herbicidal action at early stages of seedling growth. Grass and some broadleaf weeds are selectively controlled in corn, cotton, soybean, and other crops,[55] shoots being more sensitive than roots.[56] In addition, a primary effect of these herbicides appears to be due to inhibition of cell elongation rather than cell division.[57]

Several thiocarbamates including EPTC, diallate, CDEC, and triallate (Figure 3.7) have been reported to inhibit lipid metabolism.[55,58–62] In cabbage, diallate and EPTC inhibited the formation and deposition of surface wax[63]; diallate altered the proportions of very long chain fatty acids and reduced the primary alcohol content.[63] In peas, CDEC altered wax formation but not the total amounts of wax formed whereas diallate and EPTC significantly reduced wax formation.[64] EPTC reduced the cuticle thickness of the petiole and altered the fatty acid content in sicklepod leaflets.[65,66]

Very long chain fatty acids are the precursors of many wax constituents, and Harwood and Stumpf[67] examined the effect of thiocarbamates on fatty acid elongation. They found that diallate, CDEC, and EPTC inhibited the incorporation of radioactivity from [1-^{14}C]acetate into very long chain fatty acids of germinating peas. The labeling of other fatty acids was unaffected by 10^{-5} *M* concentrations, indicating that the effects were rather specific (Table 3.8). Further work by others[68] confirmed that diallate, CDEC, and EPTC affected only the formation of surface lipids and not the membrane lipids in pea leaves. These results all suggested that fatty acid elongation was a target site for thiocarbamate action. In order to test this hypothesis, a second system for making very long chain fatty acids was utilized. Aged potato slices can be induced to label C_{20}–C_{26} acids at high rates because these compounds are precursors of suberin. Diallate, triallate, and EPTC specifically reduced the formation of very long chain fatty acids in such

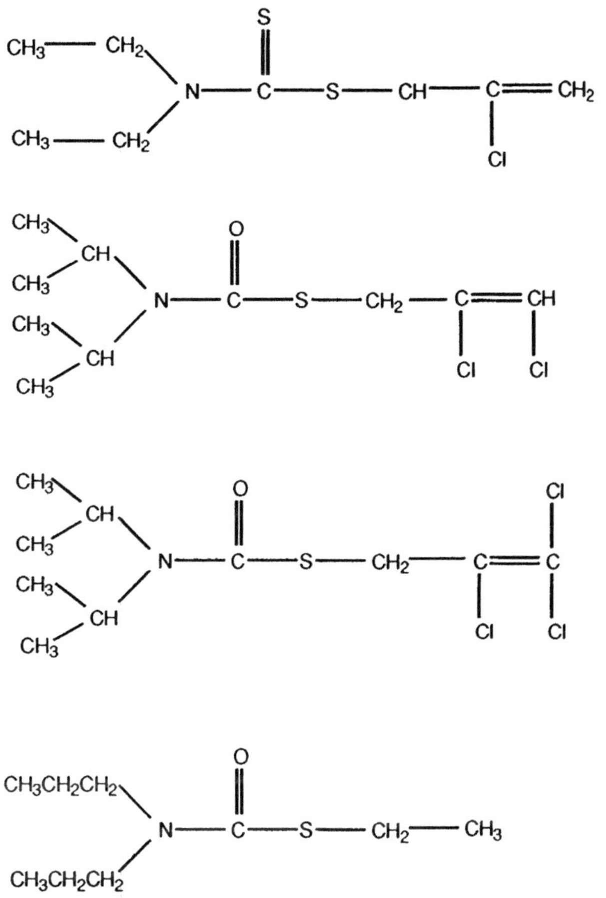

CDEC

Diallate

Triallate

EPTC

Figure 3.7. Structures of some thiocarbamates that affect lipid metabolism.

Table 3.8. Effect of Thiocarbamates on the Labeling of Fatty Acids from [1-^{14}C]Acetate in Germinating Pea Seeds[a]

	Total labeling (% of control)	Fatty acid labeling (% of total)	
		C_{16}–C_{18} acids	C_{20}–C_{24} acids
Untreated	100	89	11
+10^{-5} *M* EPTC	118	99	1
+10^{-5} *M* CDEC	118	95	5
+10^{-5} *M* Diallate	134	96	4

[a] Ref. 67.

Table 3.9. Action of Three Thiocarbamates on the Labeling of Very Long Chain Fatty Acids from [1-^{14}C]Acetate by Aged Potato Slices[a]

	Fatty acid labeling (% total)					
	16:0	18:0	18:1	20:0	22:0	24:0
Control	31	7	36	9	7	3
$+10^{-5}$ *M* EPTC	33	7	42	11	1	Trace
$+10^{-5}$ *M* Diallate	43	6	43	2	1	n.d.[b]
$+10^{-5}$ *M* Triallate	40	5	39	9	1	n.d.

[a] Ref. 69.
[b] n.d.: Not detected.

tissues[69] (Table 3.9), giving further support to the hypothesis that elongases are an important target for thiocarbamate action.

Various actions of thiocarbamates on lipid contents, lipid labeling, membrane constituents, and growth regulators have been reported (see Ref. 59). In addition, several antidotes or safeners which increase the tolerance of crop plants to thiocarbamates have been shown to reverse the effects of these herbicides on lipid metabolism.[70,71]

In order to examine the apparently selective inhibition of very long chain fatty acid synthesis *in vivo*, experiments have been carried out in our laboratories using microsomal fractions; these contain endoplasmic reticulum, which is the site of fatty acid elongation (Section 3.2).

Using [^{14}C]malonyl-CoA as precursor, microsomal fractions from pea seeds, germinated for 24 h, are able to label arachidic and, sometimes, behenic acids. At a concentration of 10^{-5} *M*, both diallate and triallate inhibited labeling of these fatty acids; it was only at 10^{-4} *M* that a different pattern of labeled fatty acids was seen (Table 3.10). Thus, the rather selective

Table 3.10. Effect of the Direct Addition of Diallate on the Incorporation of Radioactivity from [2-^{14}C]Malonyl-CoA into Fatty Acids by a Microsomal Fraction from Germinating Seeds[a]

Concentration of diallate (*M*)	Total fatty acid labeling (% of control)	Fatty acid labeling (% of total)			
		<16	16:0	18:0	VLCFAs[b]
0	100	11	28	40	13
10^{-5}	55	12	27	40	14
0	100	16	30	37	10
10^{-4}	45	15	31	39	5

[a] Data from Refs. 61 and 72.
[b] VLCFAs: Very long chain fatty acids.

inhibition of the synthesis of these acids by thiocarbamates found *in vivo*[61] was lost.

This difference between *in vivo* and *in vitro* results could be explained by assuming that in whole tissues the herbicides are converted to active derivatives (such as sulfoxides) which are themselves more active and selective.[73] In our laboratory, pea seeds were treated with herbicide during the germination period and microsomal fractions prepared for incubation without further addition of herbicides. Pretreatment of pea seeds had little effect on the total incorporation of radioactivity from [2-^{14}C]malonyl-CoA into fatty acids (Table 3.11), but there was little, or no, labeling of very long chain fatty acids. It is possible that thiocarbamates are converted to an active metabolite which remains bound to the endoplasmic reticulum during fractionation or that the herbicide (or its metabolite) prevents synthesis of elongase proteins during germination.

Further experiments were carried out in which exogenous fatty acids were added to increase elongation. Addition of both stearate and nonadecanoate enhanced elongation, but addition of thiocarbamates *in vitro* still produced nonselective inhibition. However, when herbicide-pretreated seeds were used to isolate microsomal fractions, addition of exogenous fatty acids allowed labeling of the elongation product, albeit at reduced rates.[71] This result suggested that the elongase proteins were still synthesized in treated seeds. Since direct addition of thiocarbamates to microsomes inhibits the synthesis of all fatty acids, it seems most likely that a metabolite is responsible for the selective effects of thiocarbamates *in vivo*.

A number of chemicals have been developed which act as antagonists to various thiocarbamate herbicides. One such compound, 5-aminobenzotriazole (ABT), has been reported to neutralize the effect of diallate in plants. Preliminary studies were carried out to determine if ABT caused any effects on fatty acid labeling in test systems. ABT produced some inhibition of total fatty acid labeling in leaf blades from several different

Table 3.11. Effect of Pretreatment of Pea Seeds with Diallate on the Subsequent Ability of Isolated Microsomal Fractions to Synthesize Fatty Acids[a]

	Total fatty acid synthesis (dpm/mg protein)	Fatty acid labeling (% total)			
		<16	16:0	18:0	VLCFAs[b]
None	8940	13	28	33	12
Diallate (10^{-4} *M*)	7825	17	32	36	n.d.[c]

[a] Ref. 72.
[b] VLCFAs: Very long chain fatty acids.
[c] n.d.: None detected.

Table 3.12. Action of the Safener 5-Aminobenzotriazole in Reversing the Effects of Diallate on Fatty Acid Synthesis in Leaf Blades of Monocotyledons[a]

Plant	Addition	Total synthesis (% of control)	Fatty acid labeling (% of total)		
			C_{16}	C_{18}	VLCFAs[b]
Barley	None	100	25	33	42
	+Diallate	64	39	48	13
	+Diallate + ABT	72	30	34	36
Oat	None	100	33	47	20
	+Diallate	82	39	52	9
	+Diallate + ABT	77	32	48	20

[a] Diallate was used at 10^{-5} *M* and 5-aminobenzotriazole at 50 mg/liter; labeling was from [1-^{14}C]acetate,[72] using leaf blades from 9-day-old etiolated plants.
[b] VLCFAs: Very long chain fatty acids.

monocotyledons but had no effect on the pattern of labeling. At a concentration of 50 mg/liter, ABT was able to completely reverse the selective inhibition of very long chain fatty acid (VLCFA) labeling caused by 10^{-5} *M* diallate (Table 3.12) or triallate.[71] These results suggest that the effects on fatty acid metabolism could be related to a primary safening action of ABT. Studies using maize cell cultures in which thiocarbamate herbicides have been shown to inhibit lipid synthesis revealed that this inhibition could be counteracted by safeners before such compounds had a measurable effect on other parameters such as glutathione levels.[74,75] This suggested that the safener may act at a primary site of thiocarbamate action. Although not yet proven, it seems likely that ABT is an effective antidote toward diallate and triallate in monocotyledons because it may enhance the total metabolism of these herbicides to less active products.

In summary, thiocarbamates have been shown to specifically affect very long chain fatty acid (and wax/cutin/suberin) synthesis in a variety of plant tissues *in vivo.* This effect is most likely due to the formation of a more active metabolite (possibly a sulfoxide) and can be reversed by safening agents. It seems probable that effects on fatty acid elongation are important in the mechanism of action of thiocarbamates.

3.5. ARYLOXYPHENOXYPROPIONATES AND CYCLOHEXANEDIONES

The aryloxyphenoxypropionates and cylcohexanediones are two important classes of grass-selective herbicides which have been developed

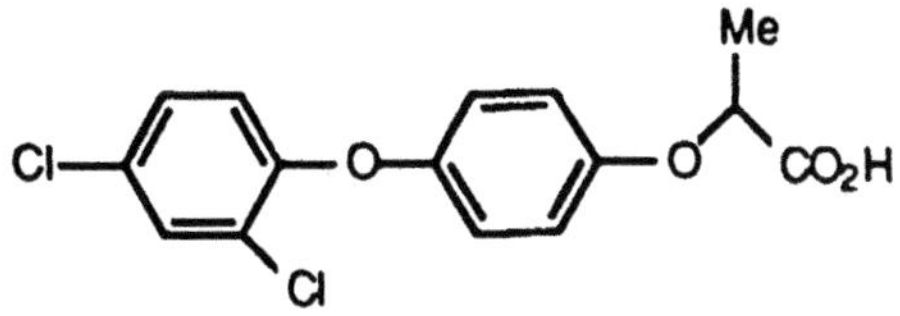

DICLOFOP

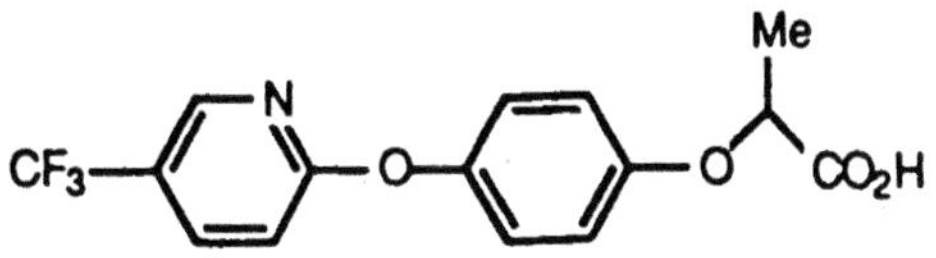

FLUAZIFOP

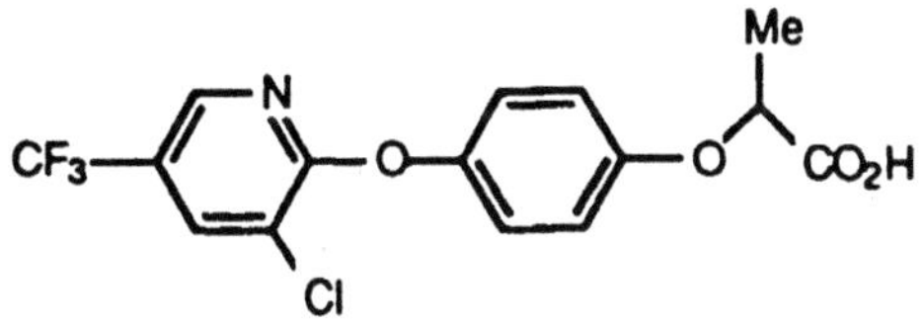

HALOXYFOP

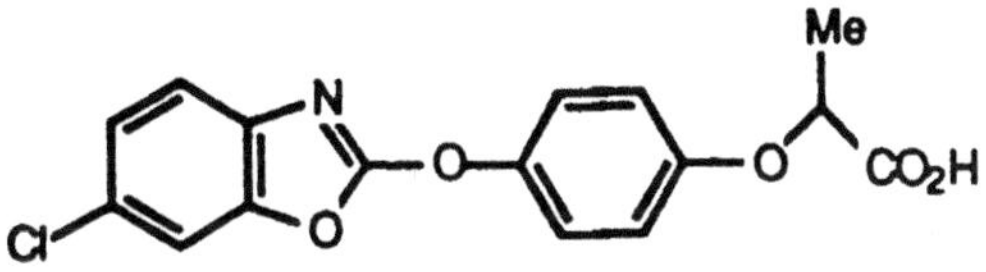

FENOXAPROP

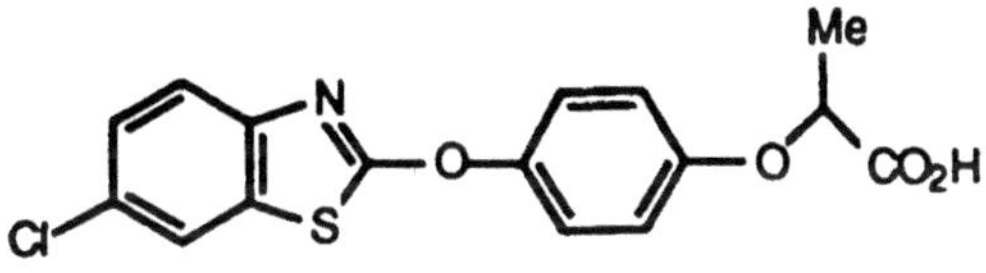

FENTHIAPROP

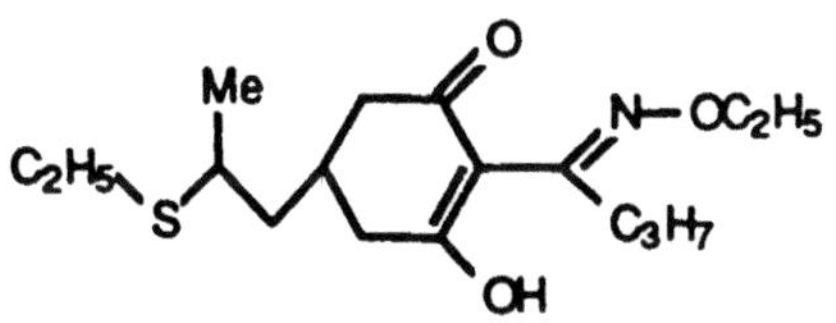

SETHOXYDIM

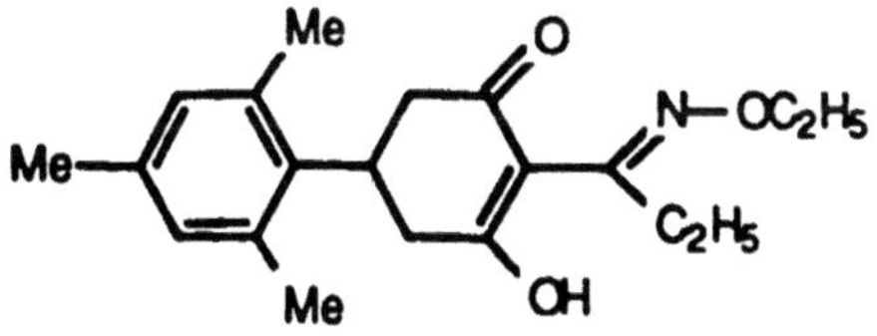

TRALKOXYDIM

Figure 3.8. Structures of some aryloxyphenoxypropionates and cyclohexanediones that have been shown to inhibit acetyl-CoA carboxylase in sensitive species.

Table 3.13. Effect of Fluazifop (10^{-4} M) on the Distribution of ^{14}C Following Treatment with ^{14}C-Precursors[a]

Precursor	Treatment	Total incorporation (% of control)	Distribution of label (%) 16:0	18:0	Unsat. C_{18}	VLCFAs[b]
[1-^{14}C]Acetate	Control	100	13	2	71	3
	+Fluazifop	11 ± 5	18	12	32	33
[2-^{14}C]Malonate	Control	100	4	8	13	75
	+Fluazifop	41 ± 5	n.d.[c]	16	n.d.	84

[a] Ref. 78.
[b] VLCFAs: Very long chain fatty acids.
[c] n.d.: Not detected.

fairly recently; examples are seen in Figure 3.8. The action of these compounds on lipid metabolism has been thoroughly reviewed recently,[8,76] and only a summary of known facts will be made here together with some very recent observations. For full references to previous work, the reader is referred to various reviews.[8,60–62,73,76,77]

Early work on aryloxyphenoxypropionates failed to show any metabolic effects except those on acyl lipid synthesis. The general inhibition of labeling of all acyl lipid classes but not that of sterols or terpenoids led to the idea that *de novo* fatty acid synthesis was being reduced. In fact, there is little effect on fatty acid elongation (e.g., Table 3.13). Of the two enzyme systems involved in *de novo* synthesis, it has now been established that the aryloxyphenoxypropionates and cyclohexanediones only affect acetyl-CoA carboxylase. Moreover, this carboxylase from dicotyledons (resistant) appears to be different from that in monocotyledons (sensitive) because the selectivity of herbicides (including stereospecificity) is retained during *in vitro* measurements (Table 3.14).

Comparisons of relative activities of aryloxyphenoxypropionates on acetyl-CoA carboxylase measured *in vitro* with results from whole-plant

Table 3.14. Stereospecificity of Herbicide Action on Acetyl-CoA Carboxylase Activity[a]

Source of enzyme	Herbicide (isomer)	I_{50} value
Maize leaves	Haloxyfop (*R*)	1 μM
	Haloxyfop (*RS* mixture)	2 μM
Barley leaves	Fluazifop (*R*)	4 μM
	Fluazifop (*S*)	>1 mM[b]

[a] Data for haloxyfop from Ref. 79 and for fluazifop from Ref. 80.
[b] Corrected for contamination of stereoisomer by the *R*-isomer.

studies of toxicity show rather good agreement.[81] For example, the I_{50} values against the maize enzyme range from 24 n*M* for quizalofop to 2.8 μM for fluazifop; this correlates well with whole-plant activities. The only exception in a group of five herbicides was diclofop, which was about three times more active at the enzyme level; *in vivo*, the compound is hydroxylated to a less active molecule.[81] The selectivity of individual species toward cyclohexanediones and aryloxyphenoxypropionates has been studied further.[82] Within monocotyledons, only the Poaceae seem to be sensitive toward these herbicides. About 20 different species were studied at the whole-plant or the isolated chloroplast level, and even in plants closely related to Poales (Restionales, Commelinales, Cyperales, Juncales) there was little inhibition except at very high herbicide levels. There are some exceptions to the broad generalization that monocotyledons (or, at least, Poaceae) are sensitive to cylohexanediones and aryloxyphenoxypropionates, and dicotyledons resistant. These particular cases may be due to specific metabolic properties or to some altered enzyme forms. Thus, it is well established that the physiological resistance of wheat to diclofop is due to inactivation of the herbicide rather than to any property of its acetyl-CoA carboxylase.[76,77] *Poa annua* has been reported to be rather resistant to cycloxydim and, indeed, the I_{50} value for inhibition of fatty acid synthesis in isolated chloroplasts was also much higher than for a sensitive species, such as *P. pratensis.*[82] In contrast, chloroplasts from both species were highly sensitive to the aryloxyphenoxypropionate diclofop. Thus, *P. annua* is one species showing a difference in sensitivity to the two classes of herbicides at the subcellular level. It is uncertain whether this finding reflects the relative ability of the chloroplasts to metabolize or transport the herbicides or indicates that the acetyl-CoA carboxylase from *P. annua* is modified.

Just as wheat appeared to be tolerant toward diclofop at the whole-plant but not the enzyme level, because of selective inactivation, so various fescue grasses have been shown to be tolerant. At the chloroplast level, the sensitivity of *Festuca ovina* and *F. rubra* toward diclofop was comparable to that of the sensitive *F. arundinacea.*[82] The resistance to diclofop *in vivo* appears to be due to metabolism of the herbicide to less active compounds, as in wheat.[76,77] By contrast, *F. ovina* and *F. rubra* are also resistant to cycloxydim, but, in that case, fatty acid synthesis by isolated chloroplasts is also relatively unaffected (I_{50} about 100 μM). There may be some differences in the acetyl-CoA carboxylase protein structure in these species that renders cycloxydim binding less effective than usual.

In view of the large difference in structure between cyclohexanediones and aryloxyphenoxypropionates, it would be rather surprising if all acetyl-CoA carboxylases showed equal sensitivity or resistance against these herbicide classes. It has also been found that *Lolium multiflorum* L. is resistant toward diclofop but sensitive toward sethoxydim and that this difference is also found for the acetyl-CoA carboxylase.[83] It was concluded, therefore, that *L. multiflorum* contains an acetyl-CoA carboxylase that retains good binding toward sethoxydim but not diclofop.

Acetyl-CoA carboxylase is a very important enzyme in metabolism, and it is not unexpected that its inhibition should be so catastrophic for sensitive plants. The enzyme catalyzes the first committed step in fatty acid, and, hence, acyl lipid, synthesis; in addition, its product, malonyl-CoA, is used for a significant number of other syntheses.[84] Like all biotin-containing carboxylases,[17] acetyl-CoA carboxylase carries out its reaction in two main parts. The first (common to type I biotin enzymes) involves the ATP-dependent carboxylation of biotin which is attached to a particular protein or domain, known as biotin carboxyl carrier protein (BCCP) (Figure 3.9). This part of the overall reaction occurs stepwise via the generation of a carboxyphosphate intermediate[17] and is catalyzed by biotin carboxylase. The second part of the overall reaction involves movement of the carboxylated biotin moiety to the active site of a carboxyltransferase enzyme. This enzyme then catalyzes the stepwise transfer of the carboxyl moiety to acetyl-CoA, forming malonyl-CoA (Figure 3.9). There is much conservation in the amino acid sequences of biotin enzymes,[85] and the mechanism of biotin carboxylation is essentially identical for the various enzymes even though some of them catalyze rather different reactions from those catalyzed by the type I enzymes. It is the carboxyltransferase step which appears to be distinct, and it would seem logical to seek a site of action for the aryloxyphenoxypropionates and cyclohexanediones on this enzyme step.

Plant acetyl-CoA carboxylases appear most likely to be high-molecular-mass multifunctional proteins.[9] Therefore, the biotin carboxylase, biotin carboxyl carrier protein, and carboxyltransferase will represent specific domains on the same protein, probably arranged as in the pyruvate carboxylase.[17] The question of where grass-selective herbicides bind to this large (200–240 kDa) protein has been addressed recently. Enzyme kinetics have been used to determine the nature of the binding characteristics.[86] Diclofop and clethodim were selected as representative inhibitors and were used with a partially purified acetyl-CoA carboxylase from wheat. Two types of experiments were performed. First, graphs of the effect of the herbicides on velocity versus substrate concentration curves were made. Second, the action of clethodim or diclofop was examined in assays for the two partial reactions.

Inhibition by clethodim or diclofop was found to be much more sensitive to acetyl-CoA concentration than to either ATP or bicarbonate. Furthermore, the kinetic characteristics were also different in that inhibition of the overall reaction by clethodim or diclofop was noncompetitive against ATP but nearly competitive against acetyl-CoA and nearly noncompetitive

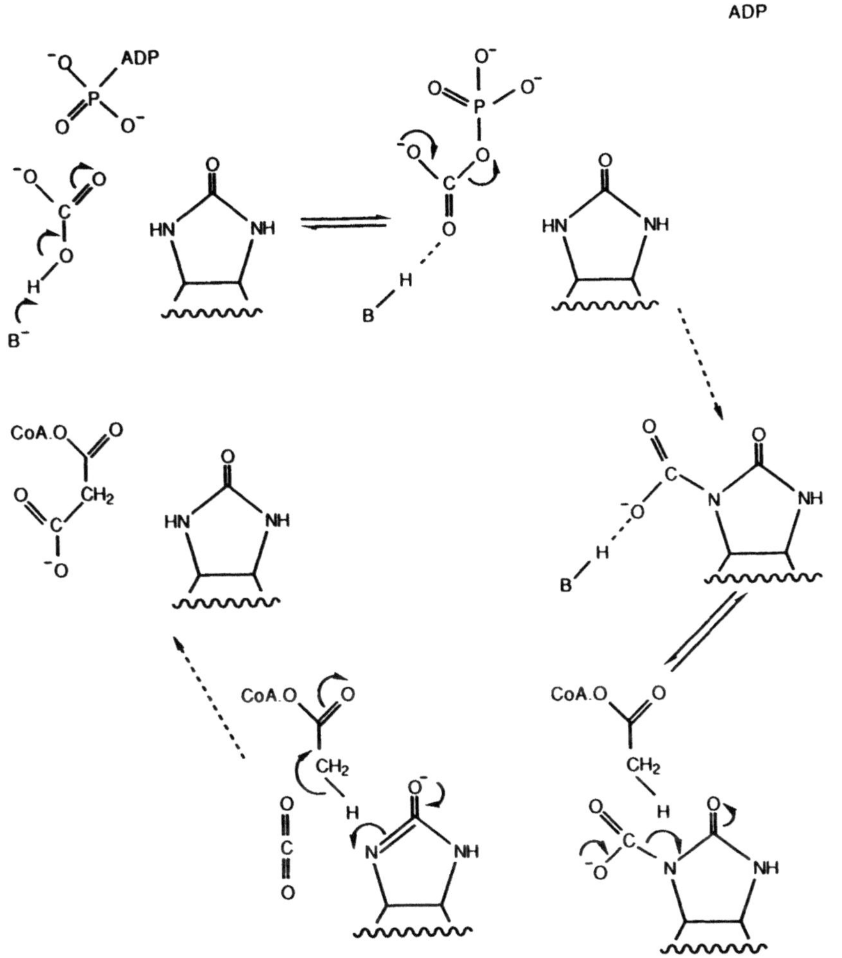

Figure 3.9. Simplified mechanism for acetyl-CoA carboxylase. Based on information in Ref. 17.

against bicarbonate (Figure 3.10). The product of the reaction, malonyl-CoA, was (like the herbicides) also nearly competitive against acetyl-CoA.[86] By analogy with the malonyl-CoA results, the herbicides appear to interfere with the carboxyltransferase partial reaction. Similar experiments with fluazifop have produced similar results and conclusions (K. A. Walker, S. M. Ridley, T. Lewis, and J. L. Harwood, unpublished data).

Acetyl-CoA carboxylase is an enzyme having a ping-pong mechanism, and it is possible to measure isotope exchange between substrates and products for each half-reaction in the absence of the substrates for the other

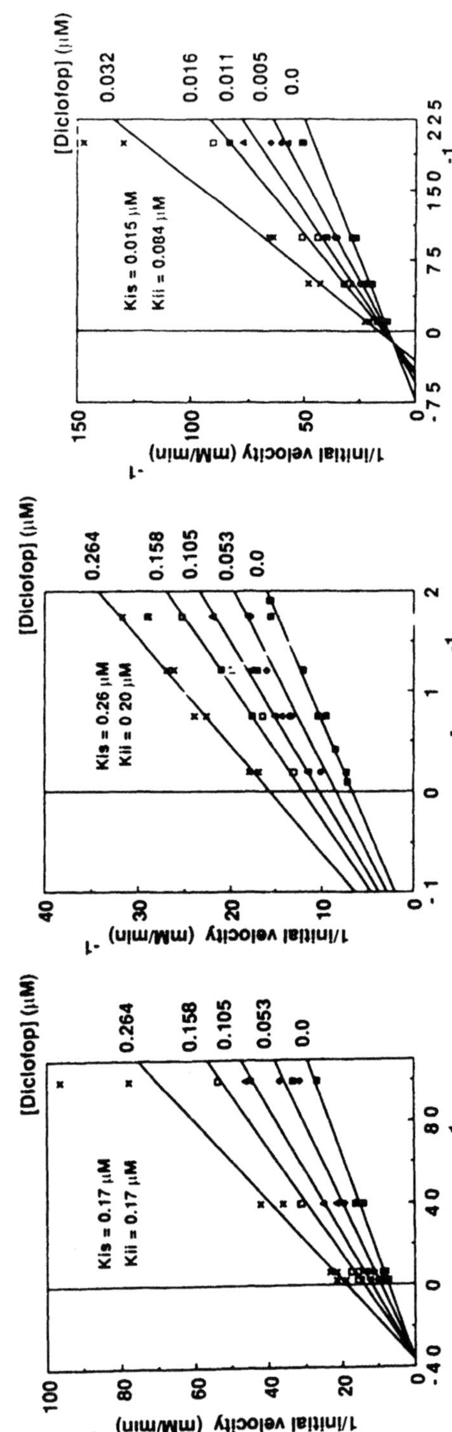

Figure 3.10. Double-reciprocal plots of the inhibition of wheat acetyl-CoA carboxylase by diclofop versus different substrates.[86]

half-reaction.[87] Thus, exchange of label between ATP and [^{32}P]orthophosphate can be measured in the presence of Mg^{2+} and bicarbonate for the first partial reaction. Since avidin is a specific inhibitor of biotin enzymes, this compound can be used to distinguish exchange catalyzed by biotin carboxylase from that due to kinases or ATPases.[86] The carboxyltransferase reaction was studied by examining the exchange of label between [^{14}C]acetyl-CoA and malonyl-CoA and also by [^{14}C]malonyl-CoA decarboxylation (Table 3.15). These individual activities are based on well-established work for the animal acetyl-CoA carboxylase and other biotin-containing enzymes.[88] For the first partial reaction, the I_{50} values for the two herbicides were at least 200 times higher than for the whole reaction. In contrast, the I_{50} values for the second half-reaction were only 2–20 times those for the whole reaction. Thus, the measurements of half-reactions confirmed the enzyme kinetic data in showing that clethodim and diclofop inhibited the carboxyltransferase partial reaction.

Clethodim and diclofop displayed very similar kinetic behavior, and it seemed likely that these two types of herbicide could share a common binding site. In order to confirm this, the concentration of the two inhibitors was varied at fixed, subsaturating levels of the substrate. If the inhibitors exclude each other from a common binding site, then this so-called double-inhibition method[89] should yield a series of parallel lines. Such a result was obtained for clethodim and diclofop (Figure 3.11) and suggests that cyclohexanediones interfere with the binding of aryloxyphenoxypropionates and vice versa. Similar experiments also showed that clethodim, diclofop, and malonyl-CoA can exclude each other from a common binding site.[86]

A number of biotin-containing enzymes have now been demonstrated in plants.[90] If cyclohexanediones and aryloxyphenoxypropionates interfere with the carboxyltransferase reaction of acetyl-CoA carboxylase, one would

Table 3.15. Inhibition of Wheat Acetyl-CoA Carboxylase Partial Reactions by Clethodim or Diclofop[a]

Reaction	I_{50} values (μM) Clethodim	Diclofop
Complete reaction	0.26 ± 0.02	0.07 ± 0.01
Biotin carboxylase		
ATP–P_i exchange	420 ± 170	460 ± 80
Carboxylation of free biotin	13,500 ± 1400	7400 ± 1300
Carboxyltransferase		
Acetyl-CoA–Malonyl-CoA exchange	4.10 ± 0.40	1.30 ± 0.05
Malonyl-CoA decarboxylation	0.38 ± 0.04	0.14 ± 0.02

[a] Ref. 87.

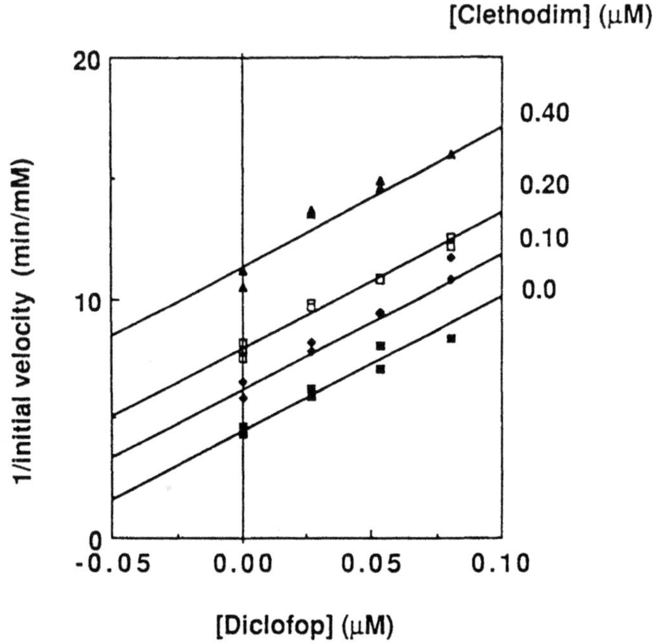

Figure 3.11. Double-inhibition plot of clethodim versus diclofop for wheat acetyl-CoA carboxylase.[86]

not expect them to inhibit other biotin carboxylases. In agreement with this, it has been shown that methylcrotonyl-CoA carboxylase from maize is insensitive to haloxyfop and tralkoxydim.[81]

A third class of chemical has recently been shown to produce similar effects to those of the aryloxyphenoxypropionates and cyclohexanediones.[91] This herbicide, PP600 {3-isopropyl-6-[*N*-(2,2-dimethylpropyl)acetamido]-1,3,5-triazine-2,4(1*H*,3*H*)-dione}, was shown to inhibit fatty acid synthesis in monocotyledons, but not dicotyledons, having an I_{50} value of about 1 μM for barley. Although acetyl-CoA carboxylase was inhibited by PP600, the enzyme was much less sensitive than whole-tissue preparations, raising the possibility that PP600 is metabolized to a more active metabolite *in vivo.*[91]

In summary, the cyclohexanediones and aryloxyphenoxypropionates are herbicidal because they inhibit the carboxyltransferase partial reaction of acetyl-CoA carboxylase in sensitive plants. Their selectivity appears to be based on structural differences in the target enzyme, and this raises fundamental questions about protein sequence and conformation as well as offering some attractive possibilities for gene manipulation.[92]

3.6. OTHER HERBICIDES WHICH AFFECT LIPIDS

3.6.1. Ethofumesate

Ethofumesate is a commonly used preemergence herbicide for the control of grasses and other weeds in sugar beet and onion crops.[93] It has also proven useful in the control of annual bluegrass, which sometimes infests Bermuda (turf) grass.[94]

Ethofumesate has been found to affect both the quantity and quality of the epicuticular wax in leaves of cabbage,[95] sugar beet,[95] and onion.[96] The main components of the wax, namely, the C_{29} ketone and C_{20} alkane, were found to be depleted by 90% and 95%, respectively.

In view of the previously observed connection between thiocarbamate-induced wax changes and fatty acid synthesis (Section 3.4), experiments have been carried out to see if ethofumesate has similar effects. In several test systems using intact tissues, ethofumesate was found to inhibit production of very long chain fatty acids (Table 3.16). Although the effect was concentrated within the C_{20}–C_{24} acid products, there was also a significant reduction of *de novo* synthesis.

The action of ethofumesate was also examined with microsomal fractions from germinating peas. In such preparations, the herbicide was found to inhibit the labeling of all fatty acids but with a strong preference toward very long chain products (Table 3.17). As a further check of the effect of ethofumesate, pretreated pea seeds were germinated, and incubated microsomes were isolated from such tissues in the absence of further herbicide. With such preparations, the synthesis of very long chain fatty acids was again observed to be particularly affected.

In summary, ethofumesate appears to inhibit wax synthesis through preventing the formation of the precursor very long chain fatty acids. While the compound has parallels with thiocarbamates, its selective

Table 3.16. Effect of Ethofumesate (10^{-4} *M*) on Very Long Chain Fatty Acid Labeling in Pea Seeds and Barley Leaves[a]

Tissue	Treatment	Total synthesis (% of control)	Fatty acid labeling (% of total)				
			16:0	C_{18}	20:0	22:0	24:0
Pea seeds	Control	100	29 ± 1	48 ± 1	11 ± 1	8 ± 1	4 ± 1
	+ Ethofumesate	46	34 ± 1	55 ± 1	8 ± 1	4 ± trace	1 ± trace
Barley leaves	Control	100	32 ± 1	41 ± 1	7 ± trace	12 ± 1	9 ± 1
	+ Ethofumesate	49	40 ± 1	54 ± 1	4 ± 1	3 ± 1	n.d.[b]

[a] Ref. 72.
[b] n.d.: Not detected.

Table 3.17. Action of Ethofumesate (10^{-4} *M*) on Fatty Acid Synthesis by Microsomal Fractions from Germinating Pea Seeds[a]

Preparation	Total synthesis (% of control)	Fatty acid labeling (% of total)				
		12:0	14:0	16:0	18:0	20:0
Direct addition						
Control	100	9	17	36	29	11
+ Ethofumesate	57	8	16	39	35	4
Pretreatment						
Control	100	5	10	45	34	7
+ Ethofumesate	79	6	10	45	25	4

[a] Ref. 72.

action on elongation does not seem to involve metabolism of the herbicide (cf. Section 3.4).

3.6.2. Chloroacetamides

Chloroacetamide herbicides are applied for the control of preemergence grass and some broadleaf weeds. Their mode of action, however, is not understood. It has been reported that lipid synthesis is one of the target sites[97] although such an effect is not seen with all tissues;[98] inhibition of cuticular wax formation has also been found.[99]

Recently, the action of two chloroacetamides, alachlor and metazachlor, were investigated in short-term experiments using the green alga *Scenedesmus acutus.*[100] The most pronounced and rapid effect of the herbicides on [^{14}C]acetate incorporation was on that into acyl lipids. The authors concluded that inhibition of acetate metabolism was the first phytotoxic disturbance and led, in turn, to impairment of membrane and thylakoid development. Using the same experimental system, the pattern of endogenous fatty acids was examined, and palmitate and oleate were found to accumulate while stearate and α-linolenate were greatly decreased.[101] The site of action of the chloroacetamides was not identified, but the data suggest that chain lengthening of palmitate to stearate and the desaturation of oleate are impaired.

3.6.3. Other Herbicides

Several other herbicides (for example, triazines) have been found to inhibit acyl lipid synthesis, but their effects may be due to secondary effects following a more acute action elsewhere in metabolism. For further details of such compounds, the reader is referred to previous reviews.[8,60,62,73,77]

3.7. FUTURE PROSPECTS

The demonstration that graminaceous herbicides, such as the cyclohexanediones, show toxicity as well as selectivity by inhibiting an important step in acyl lipid synthesis initiates an important chapter in herbicide biochemistry. We can now anticipate the prospect that other fundamental steps in lipid metabolism may be important future herbicide targets. Given the fundamental differences which exist between the synthesis of acyl lipids in plants and that in bacteria or animals, the development of new herbicides with minimal environmental implications should be forthcoming. In addition, as we discover more about the modes of action of known herbicides as well as the molecular details of their target enzymes, systematic design of improved compounds should be possible. Rather than having a range of different photosynthesis inhibitors as our main herbicides, an increasing importance for agrochemicals which act on metabolism can be predicted.

REFERENCES

1. J. L. Harwood, in: *The Biochemistry of Plants* (P. K. Stumpf and E. E. Conn, eds.), Vol. 4, p. 1, Academic Press, New York (1980).
2. C. Hitchcock and B. W. Nichols, *Plant Lipid Biochemistry*, Academic Press, London (1971).
3. L. Mendiola-Morgenthaler and W. Eichenberger, in: *Structure, Function and Metabolism of Plant Lipids* (P.-A. Siegenthaler and W. Eichenberger, eds.), p. 405, Elsevier, Amsterdam (1984).
4. N. Murata and N. Nishida, in: *The Biochemistry of Plants* (P. K. Stumpf and E. E. Conn, eds.), Vol. 9, p. 315, Academic Press, New York (1987).
5. K. Gounaris, J. Barber, and J. L. Harwood, *Biochem. J. 237*, 313 (1986).
6. A. A. Foley, A. P. R. Brain, P. J. Quinn, and J. L. Harwood, *Biochim. Biophys. Acta 939*, 430 (1989).
7. M. S. Webb and B. R. Green, *Biochim. Biophys. Acta 984*, 41 (1989).
8. J. L. Harwood, in: *Topics in Photosynthesis* (N. R. Baker and M. P. Percival, eds.), Vol. 10, in press, Elsevier, Amsterdam (1990).
9. J. L. Harwood, *Annu. Rev. Plant Physiol. 39*, 101 (1988).
10. P. G. Roughan, in: *The Metabolism, Structure and Function of Plant Lipids* (P. K. Stumpf, J. B. Mudd, and W. D. Nes, eds.), p. 247, Plenum Press, New York (1987).
11. D. G. Bishop, J. R. Kenrick, T. Kondo, and N. Murata, *Plant Cell Physiol. 27*, 1593 (1986).
12. N. Murata, O. Ishizaki, and I. Nishida, in: *Plant Membranes: Structure, Assembly and Function* (J. L. Harwood and T. J. Walton, eds.), p. 223, The Biochemical Society, London (1988).
13. A.-J. Dorne, J. Joyard, M. A. Block, and R. Douce, *J. Cell Biol. 100*, 1690 (1985).
14. R. Bligny and R. Douce, *Biochim. Biophys. Acta 617*, 254 (1980).
15. P. E. Kolattukudy, in: *The Biochemistry of Plants* (P. K. Stumpf and E. E. Conn, eds.), Vol. 4, p. 571, Academic Press, New York (1980).
16. B. Leidvogel, in: *The Metabolism, Structure and Function of Plant Lipids* (P. K. Stumpf, J. B. Mudd, and W. D. Nes, eds.), p. 509, Plenum Press, New York (1987).
17. J. R. Knowles, *Annu. Rev. Biochem. 58*, 195 (1989).
18. E. Turnham and D. N. Northcote, *Biochem. J. 212*, 223 (1983).

19. B. J. Nikolau and J. C. Hawke, *Arch. Biochem. Biophys. 228*, 86 (1984).
20. P. K. Stumpf, in: *The Biochemistry of Plants* (P. K. Stumpf and E. E. Conn, eds.), Vol. 9, p. 121, Academic Press, New York (1987).
21. J. L. Harwood, in: *The Metabolism, Structure and Function of Plant Lipids* (P. K. Stumpf, J. B. Mudd, and W. D. Nes, eds.), p. 465, Plenum Press, New York (1987).
22. J. G. Jaworski, R. C. Clough, and S. Barnum, *Plant Physiol. 90*, 41 (1989).
23. M. Walsh, W. Klopfenstein, and J. L. Harwood, unpublished results.
24. R. W. Mackintosh, D. G. Hardie, and A. R. Slabas, *Biochim. Biophys. Acta 1002*, 114 (1989).
25. I. Caughey and R. G. O. Kekwick, *Eur. J. Biochem. 123*, 553 (1982).
26. T. M. McKeon and P. K. Stumpf, *J. Biol. Chem. 257*, 12141 (1982).
27. J. G. Jaworski, in: *The Biochemistry of Plants* (P. K. Stumpf and E. E. Conn, eds.), Vol. 9, p. 159, Academic Press, New York (1987).
28. S. Stymne and A. K. Stobart, in: *The Biochemistry of Plants* (P. K. Stumpf and E. E. Conn, eds.), Vol. 9, p. 175, Academic Press, New York (1987).
29. P. G. Roughan, J. B. Mudd, T. T. McManus, and P. K. Stumpf, *Biochem. J. 8*, 571 (1979).
30. A. V. M. Jones and J. L. Harwood, *Biochem. J. 190*, 851 (1980).
31. J. L. Harwood and M. J. Price-Jones, in: *Phosphatidate Phosphohydrolase* (D. N. Brindley, ed.), Vol. 2, p. 1, CRC Press, Boca Raton, Florida (1987).
32. K. A. Walker and J. L. Harwood, *Biochem. J. 237*, 41 (1986).
33. C. Cassagne, R. Lessire, J. Bessoule, and J. Moreau, in: *The Metabolism, Structure and Function of Plant Lipids* (P. K. Stumpf, J. B. Mudd, and W. D. Nes, eds.), p. 481, Plenum Press, New York (1987).
34. G. P. Agrawal and P. K. Stumpf, *Arch. Biochem. Biophys. 240*, 154 (1985).
35. R. Lessire, J.-J. Bessoule, and C. Cassagne, *FEBS Lett. 187*, 314 (1985).
36. R. Lessire, J.-J. Bessoule, and C. Cassagne, *Biochim. Biophys. Acta 1006*, 35 (1989).
37. J. L. Harwood, *Crit. Rev. Plant Sci. 8*, 1 (1989).
38. J. B. St. John, in: *Biochemical Responses Induced by Herbicides* (D. E. Moreland, J. B. St. John, and F. D. Hess, eds.), p. 97, American Chemical Society, Washington, D.C. (1982).
39. J. L. Hilton, J. B. St. John, M. N. Christiansen, and K. H. Norris, *Plant Physiol. 48*, 171 (1971).
40. J. B. St. John and M. N. Christiansen, *Plant Physiol. 57*, 257 (1976).
41. C. Willemot, *Plant Physiol. 60*, 1 (1977).
42. J. L. Harwood, in: *Structure, Function and Metabolism of Plant Lipids* (P.-A. Siegenthaler and W. Eichenberger, eds.), p. 543, Elsevier, Amsterdam (1984).
43. J. B. St. John, M. N. Christiansen, E. N. Ashworth, and W. S. Gentner, *Crop Sci. 19*, 65 (1979).
44. M.-U. Khan, N. W. Lem, W. R. Chandorkar, and W. Williams, *Plant Physiol. 64*, 300 (1979).
45. M.-U. Khan, D. J. Chapman, N. W. Lem, W. R. Chandorkar, and W. P. Williams, in: *Advances in the Biochemistry and Physiology of Plant Lipids* (L.-A. Appelqvist and C. Liljenberg, eds.), p. 415, Elsevier, Amsterdam (1979).
46. A. O. Davies and J. L. Harwood, *J. Exp. Bot. 34*, 1089 (1983).
47. R. M. Leech, C. A. Walton, and N. R. Baker, *Planta 165*, 277 (1985).
48. R. M. Mannan and S. Bose, *Photochem. Photobiol. 41*, 63 (1985).
49. R. M. Mannan and S. Bose, *Plant Physiol. 80*, 264 (1986).
50. R. M. Mannan and S. Bose, *Plant Physiol. 85*, 17 (1987).
51. D. J. Murphy, J. L. Harwood, J. B. St. John, and P. K. Stumpf, *Biochem. Soc. Trans. 8*, 119 (1980).
52. D. J. Murphy, J. L. Harwood, K. A. Lee, F. Roberto, P. K. Stumpf, and J. B. St. John, *Phytochemistry 24*, 1923 (1985).
53. C. Willemot, C. R. Slack, J. Browse, and P. G. Roughan, *Plant Physiol. 70*, 78 (1982).

54. H. A. Norman and J. B. St. John, in: *The Metabolism, Structure and Function of Plant Lipids* (P. K. Stumpf, J. B. Mudd, and W. D. Nes, eds.), p. 429, Plenum Press, New York (1987).
55. E. P. Fuerst, *Weed Technol. 1*, 270 (1987).
56. H. Shibayama and J. F. Worley, *Weed Sci. 24*, 276 (1976).
57. E. P. Fuerst and J. W. Gronwald, *Weed Sci. 34*, 354 (1986).
58. C. M. Rivera and D. Penner, *Res. Rev. 70*, 45 (1979).
59. C. Fedtke, *Weed Res. 27*, 22 (1987).
60. S. O. Duke, in: *Weed Physiology* (S. O. Duke, ed.), Vol. 11, p. 91, CRC Press, Boca Raton, Florida (1985).
61. J. L. Harwood, K. A. Walker, K. Abulnaja, and S. M. Ridley, *Proc. Br. Crop Prot. Conf.—Weeds*, p. 159 (1987).
62. J. L. Harwood, S. M. Ridley, and K. A. Walker, in: *Herbicides and Plant Metabolism* (A. D. Dodge, ed.), p. 73, Cambridge University Press, Cambridge (1989).
63. W. A. Gentner, *Weed Sci. 14*, 27 (1966).
64. G. G. Still, D. G. Davis, and G. L. Zander, *Plant Physiol. 46*, 307 (1970).
65. R. E. Wilkinson and W. S. Hardcastle, *Weed Sci. 17*, 335 (1969).
66. R. E. Wilkinson, *Plant Physiol. 53*, 269 (1974).
67. J. L. Harwood and P. K. Stumpf, *Arch. Biochem. Biophys. 142*, 281 (1971).
68. P. E. Kolattukudy and L. Brown, *Plant Physiol. 53*, 903 (1974).
69. P. Bolton and J. L. Harwood, *Phytochemistry 15*, 1507 (1976).
70. R. E. Wilkinson, *Pestic. Biochem. Physiol. 8*, 208 (1978).
71. R. E. Wilkinson and A. E. Smith, *Weed Sci. 23*, 100 (1975).
72. K. O. Abulnaja and J. L. Harwood, unpublished data.
73. C. Fedtke, *Biochemistry and Physiology of Herbicide Action*, Springer-Verlag, Heidelberg (1982).
74. G. Ezra and J. Gressel, *Pestic. Biochem. Physiol. 17*, 48 (1982).
75. G. Ezra, E. Krochaml, and J. Gressel, *Pestic. Biochem. Physiol. 18*, 107 (1982).
76. K. A. Walker, S. M. Ridley, T. Lewis, and J. L. Harwood, *Rev. Weed. Sci. 4*, 71 (1989).
77. H. H. Hoppe, in: *Target Sites for Herbicide Action* (P. Boger and G. Sandmann, eds.), CRC Press, Boca Raton, Florida (1989).
78. K. A. Walker, S. M. Ridley, and J. L. Harwood, *Biochem. J. 254*, 811 (1988).
79. J. Secor and C. Cseke, *Plant Physiol. 86*, 10 (1988).
80. K. A. Walker, S. M. Ridley, T. Lewis, and J. L. Harwood, *Biochem. J. 254*, 307 (1988).
81. J. Secor, C. Cseke, and W. J. Owen, *Proc. Brighton Crop Prot. Conf.—Weeds*, p. 145 (1989).
82. H. K. Lichtenthaler, K. Kobek, and M. Focke, *Proc. Brighton Crop Prot. Conf.—Weeds*, p. 173 (1989).
83. J. W. Gronwald, C. V. Eberlein, K. J. Betts, K. M. Rosow, N. J. Ehlke, and D. L. Wyse, *Plant Physiol. 89* (Supplement) Abstr. 685, 115 (1989).
84. J. L. Harwood, *Proc. Brighton Crop Prot. Conf.—Weeds*, p. 155 (1989).
85. D. Samols, C. G. Thornton, V. L. Murfit, G. K. Kumar, F. C. Haase, and H. G. Wood, *J. Biol. Chem. 263*, 6461 (1988).
86. A. R. Rendina, J. D. Beaudoin, A. C. Craig-Kennard, and M. K. Breen, *Proc. Brighton Crop Prot. Conf.—Weeds*, p. 163 (1989).
87. W. Cleland, *J. Biol. Chem. 248*, 8353 (1973).
88. J. Moss and M. Lane, *Adv. Enzymol. 35*, 321 (1971).
89. D. Northrup and W. Cleland, *J. Biol. Chem. 249*, 5820 (1974).
90. B. J. Nikolau, J. Croxdale, T. H. Ulrich, and E. S. Wurtele, in: *The Metabolism, Structure and Function of Plant Lipids* (P. K. Stumpf, J. B. Mudd, and W. D. Nes, eds.), p. 517, Plenum Press, New York (1987).
91. K. A. Walker, J. L. Harwood, T. Lewis, and S. M. Ridley, *Proc. Brighton Crop Prot. Conf.—Weeds*, p. 463 (1989).

92. J. L. Harwood, *Trends Biochem. Sci. 13*, 330 (1988).
93. J. R. C. Leavitt, D. N. Duncan, D. Penner, and W. F. Meggitt, *Plant Physiol. 61*, 1034 (1978).
94. S. W. Bingham, *Proc. South. Weed Sci. Soc. 36*, 123 (1983).
95. D. N. Duncan, W. F. Meggitt, and D. Penner, *Weed Sci. 30*, 195 (1982).
96. B. Rubin, U. Adler, R. Varsano, and H. D. Rabinowitich, *Ann. Appl. Biol. 108*, 365 (1986).
97. J. D. Mann and M. Pu, *Pestic. Biochem. Physiol. 7*, 122 (1977).
98. J. M. Mellis, P. Pillai, D. E. Davies, and B. Truelove, *Weed Sci. 30*, 299 (1982).
99. E. Ebert, *Weed Res. 22*, 305 (1982).
100. H. Weisshaar and P. Boger, *Pestic. Biochem. Physiol. 28*, 286 (1987).
101. H. Weisshaar, G. Retzlaff, and P. Boger, *Pestic. Biochem. Physiol. 32*, 212 (1988).

4

Carotenoid Biosynthesis

PETER M. BRAMLEY

4.1. INTRODUCTION

The carotenoids represent one of the most important and widespread groups of pigments in nature. The structures of some 500 of them are known, and they are responsible for many of the yellow and red colors of flowers, fruits, birds, insects, and other animals.[1-3] Carotenoids are synthesized *de novo* in all organisms except for animals, where the pigments are of dietary origin.[4] Their universal presence in photosynthetic tissues is only noticeable at the onset of leaf senescence, when the chlorophylls disappear. The dramatic changes in color of ripening fruits also reflect the disappearance of chlorophyll and the concomitant, massive increase in carotenoids.

The carotenoids fulfill vital functions in green tissues, and so inhibition of their synthesis, or their destruction by peroxidation (see Chapter 5), leads to bleaching of the tissue and subsequent cell death. The purpose of this chapter is to explain the mode of action of bleaching herbicides which inhibit carotenoid biosynthesis in plants. Present and future experimental approaches which should enhance our current understanding of carotenoid biosynthesis and its regulation are also discussed in relation to the development of a new generation of bleaching herbicides.

4.2. STRUCTURE AND NOMENCLATURE

Carotenoids are isoprenoids which generally consist of eight isoprene units joined together so that the linking of the units is reversed at the center

PETER M. BRAMLEY • Department of Biochemistry, Royal Holloway and Bedford New College, University of London, Egham, Surrey TW20 0EX, United Kingdom.

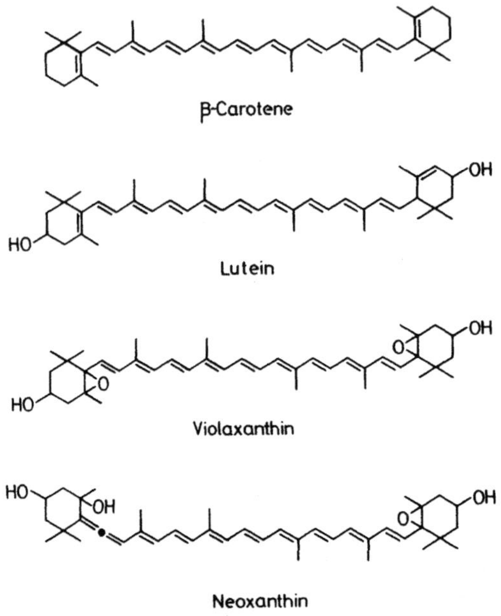

Figure 4.1. Structures of major chloroplast carotenoids.

of the molecule. The main feature of the symmetrical tetraterpenoid is a polyene chain which may extend from 3 to 15 conjugated double bonds. The length of the chromophore determines the absorption spectrum, and hence color, of the molecule, for example, the yellows and reds of unsaturated carotenoids such as *β*-carotene (*β*,*β*-carotene; Figure 4.1) and lycopene

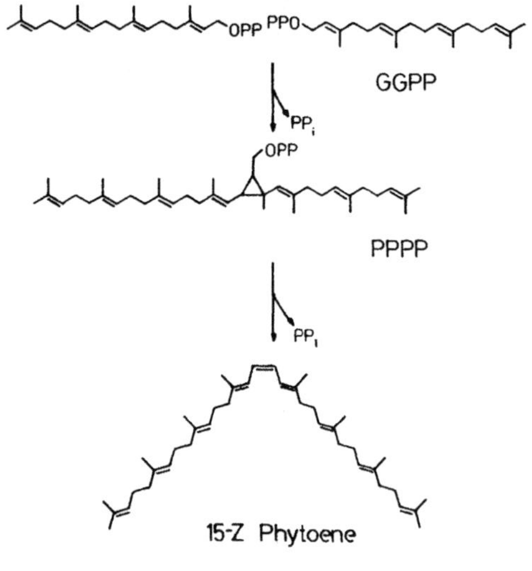

Figure 4.2. The formation of 15-*Z* phytoene from GGPP.

(ψ,ψ-carotene) and the lack of color of phytoene (7,8,11,12,7′,8′,11′,12′-octahydro-ψ,ψ-carotene; Figure 4.2). The absorption characteristics are often used in the identification of carotenoids (see Section 4.5.1). Cyclization of the carbon skeleton, at one or both ends of the chromophore, occurs (e.g., β-carotene; Figure 4.1), while xanthophylls are formed from the hydrocarbon carotenes by the introduction of oxygen functions, especially hydroxy, oxo, and epoxy groups [e.g., lutein (3,3′-dihydroxy-β,ε-carotene) and violaxanthin (5,6,5′,6′-diepoxy-5,6,5′,6′-tetrahydro-β,β-carotene-3,3′-diol) (Figure 4.1)]. Common carotenoids have well-established trivial names and also IUPAC-recommended systematic nomenclature.[5,6] Trivial names are used throughout this chapter; the corresponding semisystematic names are quoted at the first mention of each carotenoid.

The structures of carotenoids can be found in Ref. 7, and regular reviews on advances in carotenoid chemistry are published by The Chemical Society of Great Britain (see Ref. 8). The proceedings of the triennial symposium on carotenoids are also published (e.g., Ref. 9).

4.3. OCCURRENCE AND FUNCTIONS IN PLANTS

4.3.1. Distribution and Subcellular Location

The same four major carotenoids, β-carotene, lutein, violaxanthin, and neoxanthin (5′,6′-epoxy-6,7-didehydro-5,6,5′,6′-tetrahydro-β,β-carotene-3,5,3′-triol), accumulate in the chloroplasts of all green leaves (Figure 4.1), with minor amounts of α-carotene (β,ε-carotene), zeaxanthin (β,β-carotene-3,3′-diol), antheraxanthin (5,6-epoxy-5,6-dihydro-β,β-carotene-3,3′-diol) and lutein 5,6-epoxide (3,3′-dihydroxy-5,6-epoxy-α-carotene). Although quantitative differences between species are found, the carotenes account for about 25% of total leaf carotenoids, while lutein, the main xanthophyll, is about 45% of the total. The xanthophylls are unesterified except in senescing leaves. The distribution of carotenoids in flowers, fruits, and roots of plants can be found in several comprehensive reviews[1-3] and will not be discussed here.

In photosynthetically active chloroplasts, the carotenoids are found as part of photosynthetic pigment–protein complexes (PPCs) in the thylakoid membranes. The core complex, CCI, contains one β-carotene molecule per 40 chlorophyll *a*, whereas the light-harvesting complex LHCI is associated with lutein, violaxanthin, and neoxanthin. β-Carotene is also located in CCII,[10] while LHCII contains xanthophylls.[11,12] The separation, identification, and nomenclature of the PPCs has been discussed in detail in several recent reviews.[13-16]

There are small amounts of carotenoids in the chloroplast envelope,[17,18] as well as in the envelope of amyloplasts.[19-21] The carotenoid composition of these two envelopes is virtually the same, but they contain a higher xanthophyll : carotene ratio than thylakoids. Carotenes are found in plastoglobuli of photosynthetic tissues,[22] and there is a report of their presence in microsomes and mitochondria of nonplastid plant tissues.[23] Nongreen plant tissues such as flowers and fruits contain large quantities of carotenoids, typically located in chromoplasts, which are themselves of various types and derived from chloroplasts.[24] Dark-grown plants have etioplasts instead of chloroplasts, and these contain carotenoids, predominantly xanthophylls,[25] predominantly bound to membranes.[26] Algae also contain an array of carotenoids in their chloroplasts,[27] and those of the Chlorophyta, for example, *Chlorella*, are similar to the major carotenoids of higher plants. Under conditions of stress, however, the carotenoid content can alter, with the accumulation of extraplastidic pigments, typically ketocarotenoids in oil droplets or the cell wall.[28]

4.3.2. Functions of Carotenoids in Higher Plants

There is no doubt that carotenoids play vital roles in photosynthetic tissues. Probably their major function, and certainly the best documented, is the photoprotection of the photosynthetic apparatus. Carotenoid molecules are able to quench both triplet chlorophyll (^{3}chl) and singlet oxygen ($^1O_2$). The former is produced under conditions of high light intensity, while the latter is created by the transfer of energy from ^{3}chl to oxygen. Singlet oxygen is deexcited to triplet oxygen ($^3O_2$) by the carotenoid molecule to produce triplet carotenoid, which is in turn deactivated by heat transfer. Thus, the accumulation of highly reactive species of oxygen is prevented. Full details of these mechanisms can be found in other reviews.[29-31] All carotenoid molecules witth nine or more conjugated double bonds are almost equally protective, but below nine quenching becomes increasingly ineffective.[32]

Energy transfer between chlorophyll and carotenoids occurs by two pathways: a static route involving no intramolecular motion, and a dynamic mechanism which involves movement of the two chromophores.[33] Sufficient carotenoid molecules are normally present to deexcite the ^{3}chl and $^1O_2$ present, but if this concentration is reduced, for example, by inhibition of carotenoid biosynthesis, then $^1O_2$ accumulates, and rapid destruction of membranes occurs by peroxidation.[34]

A second function of carotenoids is to act as accessory light-harvesting pigments. They absorb light and pass on the excitation energy by singlet–singlet energy transfer, via antenna chlorophyll, to the reaction centers (reviewed in Refs. 35–37). Therefore, alteration of the carotenoid content of thylakoids would decrease the efficiency of light harvesting.

Additional functions ascribed to carotenoids include a structural role in enabling thylakoids to form partitions[38,39] and a biochemical role as a precursor of the plant hormone abscisic acid.[40]

4.4. BIOSYNTHESIS

The pathways of carotenoid biosynthesis have been the subject of numerous investigations over the past 35 years, using both photosynthetic and nonphotosynthetic tissues and organisms. In this section, only the steps from geranylgeranyl pyrophosphate (GGPP) to the carotenoids will be discussed, together with aspects of the regulation and enzymology of the pathways in plants. Comprehensive reviews on carotenoid biosynthesis can be found elsewhere,[2,41–43] while the early steps from acetate to GGPP are described in Chapter 5. The role of the plastids in the early steps has been reviewed recently.[44]

4.4.1. Pathways

4.4.1.1. The Formation of Carotenes

The first reaction of isoprenoid biosynthesis that is specific for the carotenoids is the head-to-head condensation of two molecules of GGPP into phytoene, via the intermediate prephytoene pyrophosphate (PPPP; Figure 4.2). The conversion of PPPP into phytoene involves the loss of a hydrogen atom to yield, in plants, the 15-*Z* isomer. This reaction has been demonstrated in several cell extracts, such as tomato plastids,[42] chloroplasts of *Triticum sativum*, and plastids of *Capsicum annuum*.[45,46] The alternative, all-*E* isomer of phytoene is not found in higher plants.

The desaturation of phytoene occurs in a stepwise sequence to form phytofluene (7,8,11,12,7′,8′-hexahydro-ψ,ψ-carotene), ζ-carotene (7,8,7′,8′-tetrahydro-ψ,ψ-carotene), neurosporene (7,8-dihydro-ψ,ψ-carotene), and lycopene (Figure 4.3). At each stage, two hydrogen atoms are lost by *trans* elimination from adjacent positions to extend the polyene chromophore by two double bonds.[47] Desaturation can be carried further to 3,4,3′4′-tetrahydrolycopene, as demonstrated in spinach chloroplasts.[48]

Since phytoene is the 15-*Z* isomer, whereas the fully desaturated carotenoids in plants are all-*E*, an isomerization of the 15–15′ bond must occur in the sequence. It is thought to take place at the phytofluene stage in tomato plastids[49] and *Capsicum* fruits,[50] although it occurs at the ζ-carotene level in a strain of *Scenedesmus*.[51] The direct, *in vitro* conversion of 15-*Z* phytoene into unsaturated carotenes has been demonstrated with tomato fruit,[49] *Capsicum*,[52] and daffodil chromoplasts.[53]

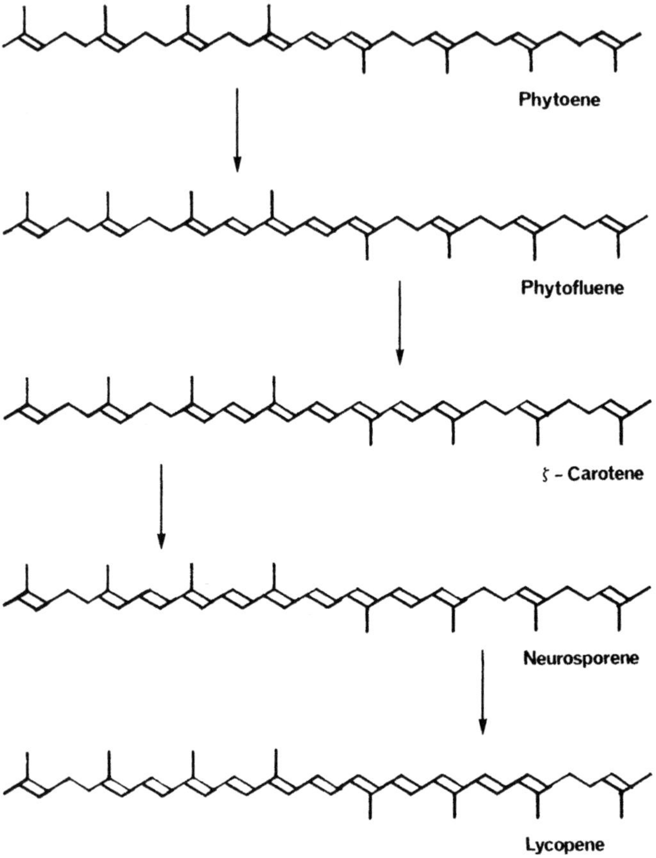

Figure 4.3. The desaturation sequence from phytoene to lycopene in higher plants. For convenience, all the structures are shown in the all-*E* configuration (see Section 4.4.1.1).

Many carotenoids contain alicyclic rings, and in higher plants two types are found, the β and the ε, as typified by β-carotene and α-carotene, respectively (see Figure 4.4). These rings are formed by separate pathways, probably by proton loss from alternative positions in the same carbonium ion precursor,[54,55] which has been created by proton attack at C-2 of the acyclic precursor. These reactions have been explored in a series of elegant experiments with D_2O, which have also elucidated the stereochemistry of H^+ attack at C-2 and the behavior of the methyl substituents at C-1 (see Ref. 43). The pathways from acyclic to cyclic carotenes in plants are shown in Figure 4.4. Some of these conversions have been demonstrated with cell extracts of higher plant tissues, especially the conversion of lycopene to β-carotene.[56–58]

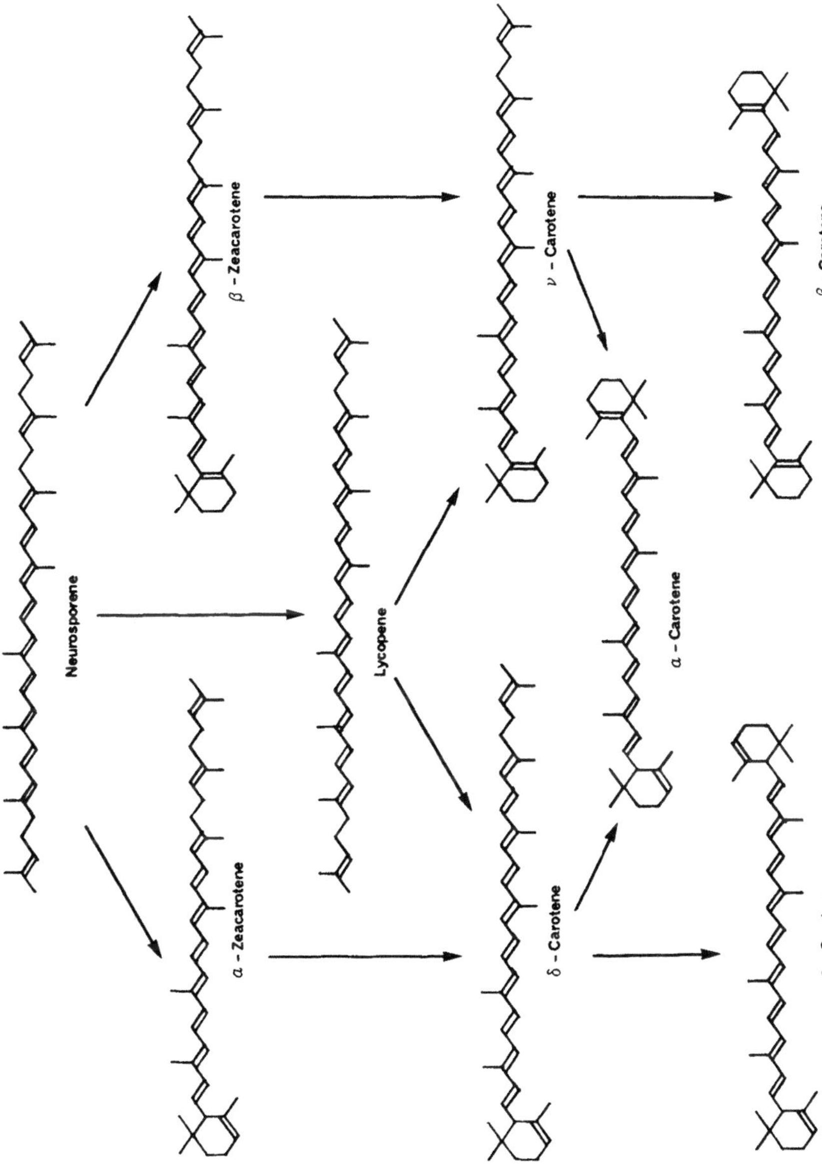

Figure 4.4. Cyclization reactions in higher plants.

4.4.1.2. Xanthophyll Biosynthesis

Although oxygen functions are inserted into carotenes at early stages of the pathway (e.g., 1,2-epoxyphytoene),[59] most xanthophylls are formed by oxidations at the end of the pathway.

Alicyclic carotenoids with oxygen substituents on one or both rings are common in plants [e.g., lutein (Figure 4.1) and zeaxanthin]. Hydroxylation at C-3 of the β ring proceeds by the replacement of the 3-*pro-R*-hydrogen by —OH.[60] The stereochemistry of these reactions is discussed by Britton.[43]

The 5,6-epoxycarotenoids such as violaxanthin, neoxanthin, and antheraxanthin are also widely distributed. Zeaxanthin is epoxidized to antheraxanthin, which in turn is epoxidized to form violaxanthin in the thylakoid membrane. These three xanthophylls constitute the "violaxanthin cycle," found in the chloroplast.[61] Although aspects of its enzymology have been elucidated (see Section 4.4.2.4), the physiological role of the cycle is a matter of debate.[62–64] The formation of violaxanthin in the chloroplast envelope is thought to be separate from the violaxanthin cycle.[65] The formation of some xanthophylls *in vitro* has been demonstrated with a few higher plant systems (see Ref. 66). Details of the biosynthesis of the other xanthophylls found in plants can be found in several reviews.[41–43]

4.4.2. Enzymology

Compared to many other biosynthetic pathways, little is known about the properties of the enzymes responsible for carotenogenesis or about the regulation of their synthesis and activity. The slow progress in this area, despite the concerted efforts of several groups, stems largely from the inherent instability of the membrane-bound proteins *in vitro*, their very low abundance in plant cells, and the technical difficulties associated with assays of enzyme activities (see Ref. 66). The most extensively studied systems from plant tissues are from the plastids of tomato and *Capsicum* fruits and the flowers of *Narcissus*.

4.4.2.1. Phytoene Synthase

Studies in the 1960s and 1970s on the cofactor requirements for phytoene synthesis by cell extracts of higher plants produced conflicting results (see Ref. 66). More definitive experiments over the past few years have clarified some of these anomalies. In particular, the enzyme has been purified to homogeneity from the chromoplast stroma of *Capsicum annuum*, to yield a single protein, M_r 47,500, with an absolute requirement for Mn^{2+}, but no other cofactors.[67] This enzyme catalyzes a two-step, kinetically coupled reaction (Figure 4.2), with K_m values of 0.3 and 0.27 μM for GGPP and PPPP, respectively. The reaction is inhibited by inorganic phosphate.

A partially purified phytoene synthase has also been isolated from tomato fruit plastids.[68,69] This complex also required Mn^{2+} for activity. Attempts at further purification, however, resulted in the loss of phytoene-synthesizing activity, but a PPPP synthase was isolated,[70] albeit contaminated with other proteins.[71]

Phytoene synthase is located in the stroma of *Capsicum* chromoplasts, chloroplasts, and amyloplasts, with the highest activity in the nonphotosynthetic plastids.[72] It has also been found in the stroma of pea fruits[73] and *Triticum* leaves.[44] In contrast, it is reported to be located on the chloroplast envelope of spinach[48] and to be a peripheral membrane protein of the inner membrane of *Narcissus* chromoplasts.[74]

4.4.2.2. Phytoene Desaturation

To date, there has been no report of the purification of a protein capable of metabolizing phytoene to unsaturated carotenes, and so the only information available on the enzymology of phytoene desaturation in plants has been derived from studies of crude preparations, typically from chromoplasts.

Although it appears from *in vivo* studies that the conversion of 15-*Z* phytoene to all-*E* lycopene involves four desaturations and one isomerization (see Section 4.4.1.1), the number of enzymes catalyzing these conversions is open to question. The use of ^{14}C-labeled 15-*Z* and all-*E* phytoenes and phytofluenes as substrates in cell extracts of tomato fruit plastids[49] and *Capsicum* chromoplasts[57] has shown that isomerization occurs at the stage of phytofluene in these tissues, but no isomerase has been isolated. It may be that a strained carotene molecule, present in an enzyme complex, is the true intermediate, as suggested by Goodwin.[75] It is pertinent to note that experiments with geometric isomers of carotenes *in vitro* must be carried out very carefully, since chemical isomerization readily occurs, and this may result in anomalies between *in vitro* incorporation experiments and *in vivo* structural analyses, such as in the case of studies on *Tangella* tomatoes.[76–78] A recent report on a CHAPS-solubilized membrane preparation of *Narcissus* chromoplasts has demonstrated a series of *Z* to *E* isomerizations *in vitro*, although there is no evidence of such reactions occurring in the intact flower.[79]

Several cell extracts convert [^{14}C] phytoene into lycopene (see Ref. 66), and there are suggestions, but little direct proof, that O_2, NADP, FAD, and/or cytochromes are necessary for these conversions (see Ref. 66). There is experimental evidence to suggest that, at least in higher plants, green algae, and cyanobacteria, two enzymes are required—the so-called phytoene desaturase, which converts 15-*Z* phytoene into all-*E* ζ-carotene, and then ζ-carotene desaturase, which forms lycopene from ζ-carotene via neurosporene (Figure 4.3).

The first line of evidence comes from studies of carotene mutants. Mutations which involve a deletion in the synthesis of chloroplast carotenoids are normally lethal unless plants can be grown under dim light or heterotrophically in darkness. Two types of mutants have been characterized: one from tomato fruits and maize that accumulates phytoene,[80,81] and a second group from maize, barley, *Chlorella*, and tomato fruits that is blocked at ζ-carotene (reviewed in Ref. 82). If these mutations are in structural genes, then it is likely that two enzymes are responsible for the desaturation sequence. Secondly, *in vitro* carotenogenic activities of *Narcissus* and *Aphanocapsa* point to separate enzymes being involved. The conversions of phytoene to ζ-carotene and ζ-carotene to lycopene in solubilized extracts of *Narcissus* chromoplasts have different cofactor requirements and involve *trans* hydrogen elimination in the first sequence, but *cis* elimination in the second. The two reactions are also affected differently by norflurazon.[79] Inhibition studies on cell extracts of *Aphanocapsa* are also indicative of two desaturase enzymes in this organism (see Section 4.6.1). In the fungus *Phycomyces*, however, genetic evidence has shown that the desaturation sequence is catalyzed by multiple copies of the *carB* gene product.[83]

Whatever number of proteins are responsible for carotene desaturation, there is general agreement that the structural genes are nuclear rather than chloroplastidic[84] and that the proteins are synthesized on 80S ribosomes prior to transport into the plastid.

An alternative approach to the partial characterization of the desaturase enzymes has been reported recently. The *crtI* (desaturase) gene from *Rhodobacter capsulatus* has been sequenced[85] and the amino acid sequence of the enzyme predicted.[86] A fusion protein, M_r 150,000, was prepared and polyclonal antibodies were raised. Western blotting against several photoautotrophic organisms has revealed proteins, which are thought to be the corresponding phytoene desaturases, of 64 kDa in all but *Bumilleriopsis*, which exhibited a protein band at 55 kDa.[87] This study is the first to pinpoint the size of the plant phytoene desaturase. Unfortunately, but predictably, the protein is enzymatically inactive in *Escherichia coli*, but it may be possible to reconstitute activity using appropriate techniques.[88]

4.4.2.3. Cyclization Reactions

No carotene cyclase enzymes has been purified, but cell extracts of tomato fruits and *Capsicum* chromoplasts have shown them to be membrane bound (see Ref. 66). Several different cyclizations have been demonstrated with a variety of tomato fruits, and the formation of α-, β-, γ-, and δ-carotenes requires FAD.[69] The conversion of lycopene to β-carotene in *Capsicum* chromoplasts is optimal at pH 6.8 and has no nucleotide cofactor requirement. Under optimal conditions, the formation of γ-carotene was not observed.[45]

4.4.2.4. Xanthophyll Formation

Very few enzymatic studies on xanthophyll biosynthesis in plants have been reported. The conversion of zeaxanthin into antheraxanthin and violaxanthin by lettuce chloroplasts requires molecular oxygen and NADPH, with pH optima at 7.8 and 7.4 in the light and dark, respectively.[89] The hydroxylation of β-carotene to β-cryptoxanthin (3-hydroxy-β-carotene) also requires NADPH and is probably catalyzed by a mixed-function oxygenase, as the hydroxy group originates from O_2.[90] The 5,6- and 5′,6′-epoxy groups are probably formed by a similar mechanism.

4.4.3. Regulation of Carotenoid Biosynthesis

The formation of carotenoids in plants is directly associated with the development of the chloroplast. Since carotenoids are an integral component of the PPCs (see Section 4.3.1), their synthesis must be coordinated with other processes in chloroplast development, such as the formation of chlorophylls, membrane lipids, and proteins.

Light is the major regulatory factor in chloroplast development. At least three photoreceptors are known to operate in plants: phytochrome, the blue/UV receptor (often called "cryptochrome"), and protochlorophyll(ide). The chemical nature and properties of these receptors are discussed elsewhere.[91,92] During the formation of chloroplasts from proplastids, it is believed that the reaction center complexes are made first, followed by the light-harvesting complexes.[93,94] The level of light received by the leaves affects the relative amounts of PPCs and their carotenoid compositions. The so-called "sun" and "shade" chloroplasts are present in plants grown under high and low light intensities, respectively.[95,96]

There are conflicting reports about whether the chloroplast is completely autonomous with respect to carotenoid biosynthesis or whether early biosynthetic intermediates are imported into the chloroplast. For example, Buggy *et al.*[97] reported that isolated chloroplasts can form carotenoids from mevalonic acid (MVA), with MVA itself being formed intrachloroplastidically from acetyl-CoA. More recently, the formation of β-carotene from ^{14}C-labeled bicarbonate by isolated spinach chloroplasts, under photosynthetic conditions, has been reported.[98] In contrast, however, Kreuz and Kleinig have concluded that isopentenyl pyrophosphate must be imported into the chloroplast.[99] Further details of these studies are contained in two recent reviews.[44,100] There is also a dispute as to the intracellular location of the carotenogenic enzymes, with one group claiming them to be associated with the plastid envelope,[48] while another has found the conversion of phytoene to β-carotene to occur on the thylakoid membranes.[101]

Importation of the proteins post-translationally, from the cytoplasm, and subsequent segregation into functional compartments presumably involves an energy-dependent process and the processing of amino-terminal transit peptides, as shown for other chloroplast proteins.[102–104] It has been proposed that nuclei and chloroplasts communicate through a bidirectional regulatory mechanism, the basis of evidence being the presence of chloroplasts in plant tissue and the expression of some nuclear genes which code for chloroplast proteins.[105] Secondly, results with carotenoid-deficient plants such as maize suggest that active chloroplasts are needed to activate gene expression of some nuclear genes.[106] Intraorganelle targeting may be different for the enzymes of carotene and xanthophyll synthesis,[107] and lutein formation could be independent from that of β-carotene and other xanthophylls.[108]

Clearly, it is not yet possible to understand completely the regulation of carotenogenesis in chloroplasts, although it must be complex (see Refs. 43 and 107). However, three phases of carotenoid biosynthesis within the chloroplast can be distinguished[43]: (i) bulk synthesis during initial development; (ii) formation in the mature chloroplast during turnover; and (iii) synthesis in response to environmental stimuli, for example, light. These phases may not be controlled in the same ways, so that a general scheme for regulation cannot be presented.

4.5. EXPERIMENTAL APPROACHES AND ANALYTICAL TECHNIQUES IN STUDIES ON BLEACHING HERBICIDES

A variety of techniques has been used to identify the mode of action of bleaching herbicides on carotenoid biosynthesis and to discover the functional groups on such compounds which are prerequisites for inhibitory activity. The trivial names of herbicides are used in the following sections, and their systematic names are listed in the Appendix at the end of the book.

4.5.1. Carotenoid Analyses of Higher Plants, Algae, and Cyanobacteria

In order to establish that a compound is inhibiting carotenoid biosynthesis at a particular step, it is necessary to extract, identify, and quantify the carotenoids from control and treated tissue. Traditionally, this has been achieved by extracting the lipids, separating the carotenoids chromatographically, and identifying them by their absorption spectra, polarities, and, perhaps, mass spectra. Such protocols are described in detail in several reviews.[109–111] A recent analytical development has been the use of high-pressure liquid chromatography (HPLC), preferably with a diode array detector, for carotenoid analyses.[112,113] This technique has now been successfully adapted for the routine screening of carotenoid inhibitors[114–116] and can be automated by the use of an autosampler. The carotenoids from plant and algal tissues can be separated in one chromatographic run, as shown in Figure 4.5. The resolution given by HPLC has also led to the discovery of novel carotenes in herbicide-treated cells, for example, hydroxyphytoene[114] and hydroxylated cyclic carotenes.[116]

Several tissues have been used in such investigations. Higher plant seedlings, grown in a light/dark cycle or in total darkness, have often been

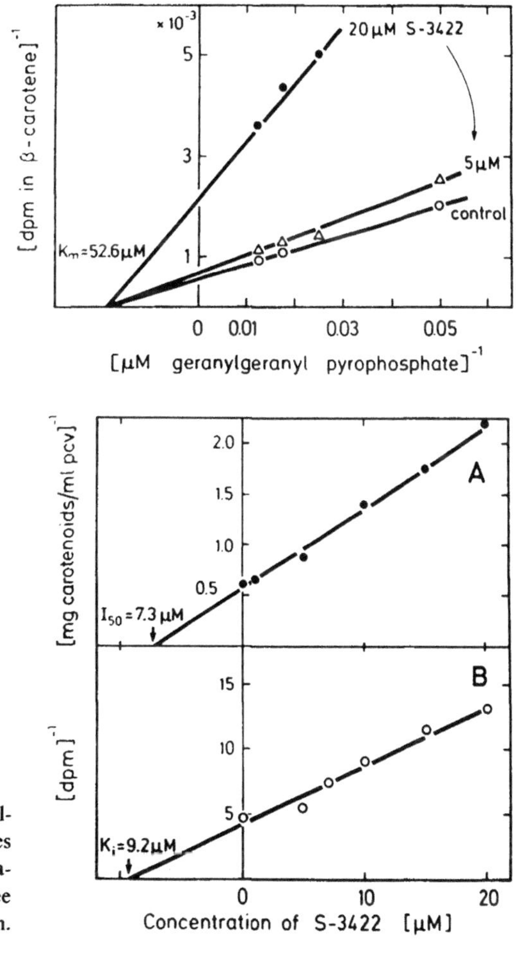

Figure 4.5. Determination of cellular I_{50} (A), K_i (B), and K_m (C) values for the inhibition of carotene formation by S-3422 in *Aphanocapsa*. See Ref. 134. Reproduced with permission.

used. Alternatively, green algae such as *Scenedesmus* and *Chlorella* are cultured under autotrophic or heterotrophic conditions in the presence and absence of herbicides.[117] The unicellular cyanobacterium *Aphanocapsa* has also been used extensively by the group at Konstanz (see Ref. 118).

Studies on intact cells or plants provide information on the relative potency of inhibitors and can usually pinpoint which step(s) is the target site. Additionally, they reflect factors such as metabolism and translocation of the herbicide. They do not provide, however, definitive values for inhibitor constants because of the factors previously described. Such values must be obtained from enzyme preparations, and preferably pure proteins.

4.5.2. Electron Microscopy

Electron microscopy of herbicide-treated tissues has often been carried out in order to show the intracellular damage caused by the inhibition of carotenogenesis[119] or to demonstrate the redistribution of the pigments after such treatment, for example, the increased presence of plastoglobuli containing phytoene.[120] Electron microscopy is most usefully carried out in conjunction with carotenoid analyses, especially or organelle fractions. In isolation, however, this technique cannot elucidate the primary mode of action of bleaching herbicides.

4.5.3. Cell-Free Systems

In order to study the molecular mechanisms of inhibitors and to prove directly that the target site is the enzyme in question, it is necessary to use the isolated enzyme, preferably in a pure form. This ideal is not yet possible with carotenogenic enzymes since, with the exception of phytoene synthase from *Capsicum*,[67] none of them has been purified (see Section 4.4.2). In addition, chloroplast extracts typically have very low carotenogenic activities and so are unsuitable for inhibition studies (see Ref. 66). It has been necessary, therefore, to compromise on the choice of tissue or organism for the preparation of cell extracts. The most promising *in vitro* system of an organism with oxygenic photosynthesis is from *Aphanocapsa* (*Synecococcus*).[118] Alternative extracts that have been used are from chromoplasts of *Narcissus* flowers[79] and *Capsicum* fruits.[57] A chloroplast preparation from *Chenopodium album* has also been used, although it is not as enzymatically active as the *Aphanocapsa* system.[121,122]

A second difficulty in the use of cell extracts is the availability of substrates. Enzymatic activities must be measured by the incorporation of radioactivity, preferably ^{14}C, from substrate into product, and such ^{14}C-labeled compounds are difficult to obtain (see Ref. 66). One approach has been to use a coupled enzyme assay in which ^{14}C-labeled phytoene is synthesized *in situ* from [2-^{14}C]mevalonic acid with a cell extract of the phytoene-accumulating mutant of *Phycomyces*. Once the ^{14}C-labeled phytoene has been formed, the extract under investigation is added, with and without herbicide, and the incubation continued in order to estimate desaturation of the radiolabeled phytoene.[123] Presenting the lipophilic substrate to the enzyme in this way is successful and avoids the use of surfactants. It is vital to exhaustively purify the products of the reaction in order to unequivocally demonstrate *in vitro* carotenogenesis.[109-111]

Providing that these constraints are recognized, such cell-free systems are most valuable in studying the molecular interactions between enzyme and inhibitor. Not only can inhibition constants be established but also the type of enzyme inhibition that occurs can be determined by the use of a Lineweaver–Burk plot. A second plot of enzyme concentration versus V_{max} determines whether the inhibition is reversible or irreversible.[118,121,122] A typical comparison of K_i and an *in vivo* I_{50} value is shown in Figure 4.6 for S-3422 (see Figure 4.7 for structure of S-3422). In this example, the two values are closely similar, indicating that there is no intracellular activation of the inhibitor nor any barrier to its uptake by the cells. This has generally proved to be the case, but CGA 22867 (see Figure 4.7) shows an I_{50} much lower than K_i, suggesting that the molecule is concentrated within the cell, perhaps at the target site, or that it is metabolized *in vivo* to a more inhibitory derivative.[124]

4.5.4. Quantitative Structure–Activity Relationships

Quantitative structure–activity relationship (QSAR) studies can be carried out on intact cells or plants or with cell extracts. The purpose of such investigations is to gain information on the functional moieties of a potential inhibitor that are necessary for activity. Typically, a series of analogues of a compound are synthesized and compared with respect to their inhibitory activity. To date, analogues of pyridazinones, difunones, phenoxybenzamides, and phenoxynicotinamides have been tested in this manner, with much of the work being carried out by the Konstanz group (see Ref. 118). All of these series inhibit phytoene desaturase (see Section 4.6.1.2), and representative structures are given in Figure 4.7.

4.5.4.1. Substituted Pyridazinones

Substituted pyridazinones can be categorized into five groups by cluster analysis, of which one is of analogues which inhibit phytoene desaturase.[118,125] A comparison of seven different phenylpyridazinones found that an *m*-trifluoro substituent at the phenyl ring and a methyl-

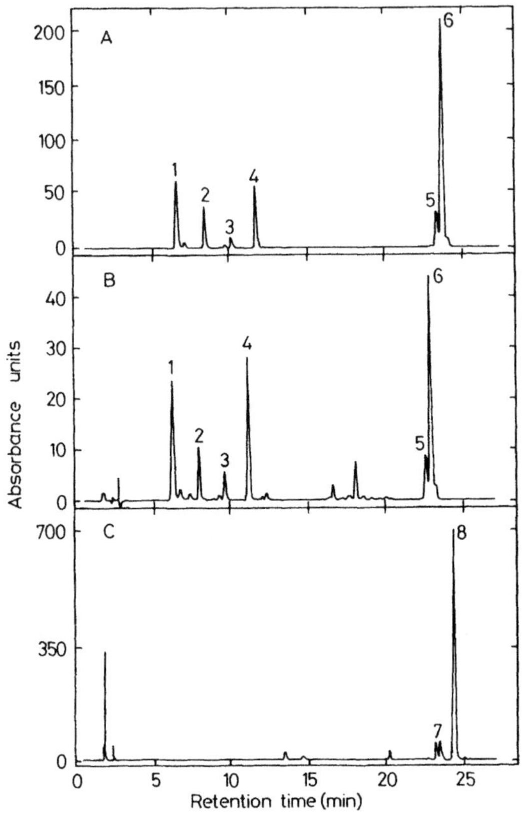

Figure 4.6. Separation by HPLC of plant carotenoids extracted from untreated (A) and diflufenican-treated (B, C) carrot cells. Traces A and B were monitored at 450 nm; trace C was monitored at 285 nm. 1, All-*E* neoxanthin; 2, violaxanthin; 3, antheraxanthin; 4, lutein; 5, α-carotene; 6, β-carotene; 7, phytofluenes; 8, 15-*Z* phytoene. Data kindly provided by Dr. K. Pallett.

substituted 4-amino group were favorable for herbicidal activity.[126] A more recent study has demonstrated that phenylpyridazinones require ligands which stabilize a resonance structure, leading to a positive partial charge on N-atom 2 and a negative charge on the oxygen at C-3, for effective inhibition of carotenogenesis.[127]

QSAR data on substituted 2-phenylpyridazinones have shown that the electronic properties of substituents strongly influence activity but can be countered by lipophilicity.[128] Data from whole cells and cell extracts of *Aphanocapsa* correlated well,[129,130] with the *in vitro* results showing a linear relationship between lipophilicity and inhibition of phytoene desaturase in accordance with the QSAR equation for intact cells.[131]

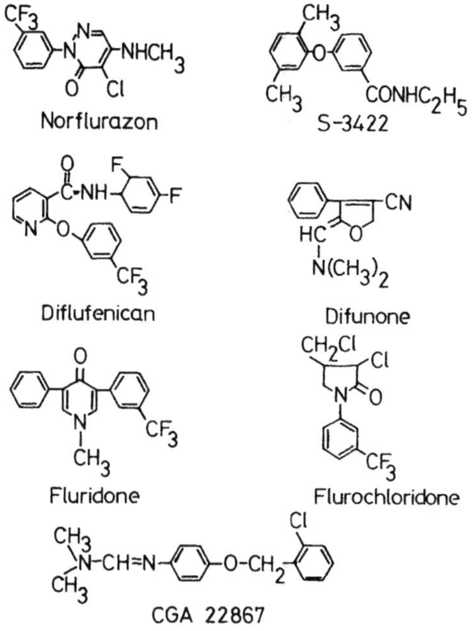

Figure 4.7. Inhibitors of phytoene desaturase.

4.5.4.2. Difunone Derivatives

The essential structural units of difunone (Figure 4.7) are the phenyl ring at C-3, a cyano group at C-4, and an oxo function at position 5.[132] Inhibition of phytoene desaturase increases with the insertion of additional groups *meta* to the phenyl ring, for example, $-SCH_3$ and $-CF_3$.[129,132]

4.5.4.3. Phenoxybenzamides

Phenoxyphenyl derivatives are herbicidal but do not have a single mode of action. Four clusters have been resolved according to their mode of action, one of which is the inhibition of carotenogenesis.[34,133] The benzamide structure of the substituted diphenyl ethers is essential for inhibition of carotenoid formation,[133] and so compounds such as S-3422 are termed phenoxybenzamides. The 3-phenoxy group can be replaced by $-CF_3$ to yield increased inhibitory activity. Substituents which reduce the positive charge on the carbonyl carbon enhance the effect. It is interesting to note that introduction of an $-NO_2$ group at C-6 alters the mode of action from that of a carotenoid biosynthesis inhibitor to that of a peroxidative herbicide.[130] Phenoxybenzamides are only inhibitory with a *meta* configuration, and substitutions at the phenoxy ring are more effective if they increase lipophilicity.[130] Lengthening the alkyl chain is also advantageous, with C_5 being optimal, but compounds with branched chains are less inhibitory than the equivalent straight-chain compounds.[131,134]

4.5.4.4. Phenoxynicotinamides

Studies with phenoxynicotinamide analogues have been reported.[118] It should be remembered that the lower potency of the commercial herbicide under laboratory conditions may be outweighed by other considerations such as selectivity, toxicity, and persistence in the soil, factors which can only be assessed by field trials.

4.6. TARGET SITES: THEORY VERSUS PRACTICE

The inhibition of carotenoid biosynthesis in plants could be caused by several effects. The most direct, and obvious, is the inhibition of enzyme synthesis or activity, and most studies have assumed this to be the case. There are, however, other possibilities that need to be considered, although they are difficult to show experimentally because of our ignorance of carotenogenesis and its regulation and of chloroplast development (see Sections 4.4.2 and 4.4.3). Alternative target sites include the unavailability of cofactors and/or substrates for the enzymes, disruption of the essential membrane organization within thylakoids, inhibition of the expression of regulatory genes, and the prevention of uptake of the nascent proteins into the chloroplast, perhaps by inhibition of a processing protease.

In this section, the available evidence for the modes of action of current bleaching herbicides with single, multiple, or unknown target sites is discussed. Previous reviews on the subject can be found elsewhere.[118,122,135,136]

4.6.1. Enzymes

4.6.1.1. Phytoene Synthase

In theory, phytoene synthase is a most suitable target site for herbicides. It is known to catalyze a two-step reaction (Figure 4.2), which has a different mechanism from that of the analogous reaction of squalene synthase[137] and should, therefore, be a carotenoid-specific target site. In addition, the protein is likely to be easily accessible to small, hydrophilic molecules as it is either stromal or loosely bound to the chloroplast membranes (see Section 4.4.2.1). In practice, however, there are no reports of a phytoene synthase-specific herbicide. Norflurazon, which is usually reported to be an inhibitor of phytoene desaturase (see Section 4.6.1.2), does inhibit phytoene synthase in cell extracts of the fungus *Phycomyces blakesleeanus*[138] and, more pertinently, in wheat seedlings, although only at low concentrations ($<0.05\ \mu M$).[139] The effect in wheat is partly reversed by several structurally related compounds such as desmethyl norflurazon[140] and by molecules which are inhibitors of kaurene synthetase.[141]

4.6.1.2. Phytoene Desaturase

The accumulation of phytoene is by far the most frequently reported consequence of treating plant and algal cells with commercial bleaching herbicide, and only representative references are given here. Phenylpyridazinones such as metflurazon[142-144] and norflurazon[107,120,145-148] as well as fluridone,[147] difunone,[126,149] methoxybenzophenone,[150] fluorochloridone,[151] haloxydine,[107] CGA 22867,[130] phenoxybenzamides,[134,152] and phenoxynicotinamides[116,124] (for structures, see Figure 4.7) have been shown to result in the inhibition of colored carotenoid synthesis and the buildup of massive amounts of phytoene and, to a lesser extent, phytofluene. The significance of the accumulation of hydroxy derivatives and geometric isomers of these carotenes under such conditions (see Section 4.5.1) has yet to be established.

The usual assumption has been that this accumulation of phytoene is a result of the inhibition of phytoene desaturase, but without the availability of a pure, chloroplastidic enzyme for enzyme–inhibitor studies, this suggestion cannot yet be proven unequivocally. The same effect could be caused by a lack of cofactors or substrates, or because of membrane disruption within the chloroplast. To some extent, these alternative hypotheses can be eliminated in the light of *in vitro* studies, using the cell extracts described in Section 4.5.3. In these experiments, phytoene accumulation has been demonstrated on addition of herbicides to cell extracts, which contain plentiful amounts of cofactors and substrate. These data imply that phytoene desaturase is inhibited directly by such inhibitors of carotenogenesis. Indeed, *in vitro* studies have also shown several compounds to be noncompetitive inhibitors,[118,153] although it is not known whether they all bind to the same site on the enzyme. There is no evidence to suggest that enzyme synthesis is inhibited by these compounds, and the close agreement of K_i and I_{50} values, for example, for S-3422 with *Aphanocapsa* (Figure 4.6) and norflurazon with daffodil,[154] argues strongly in support of a single target site, phytoene desaturase itself.

Sandmann and Boger have suggested that the phenylpyridazinones have two modes of action on phytoene desaturase.[118] Firstly, the compound mimics a carotene molecule and acts as an allosteric regulator. Secondly, an irreversible binding of the $—CF_3$ derivative occurs at the active site. Phenylpyridazinones such as Sandoz 9785 inhibit fatty acid desaturation in plants,[155] which may reflect the potential of these compounds to inhibit a range of desaturases (see also Chapter 3).

Fluorochloridone has also been shown to be effective *in vivo* and *in vitro*,[118,151] as has difunone and its analogues.[129]

4.6.1.3. ζ-Carotene Desaturase

Several herbicides have been reported to cause the accumulation of ζ-carotene both *in vivo* and *in vitro* and are referred to as inhibitors of ζ-carotene desaturase (Figure 4.8).

Pyrichlor,[156] dichlormate,[156] methoxyphenone,[150] and haloxydine[107] all cause the buildup of ζ-carotene, phytofluene, and phytoene in different plant seedlings, in a light-dependent fashion. *In vitro* studies have yet to be reported with these compounds. More recently, the experimental herbicide WL 110547 has been shown to cause ζ-carotene accumulation, especially at low light intensities.[157] Tests with several substituted 6-methylpyrimidines such as J852 on etiolated barley have shown them to be more inhibitory than dichloromate and to cause the massive accumulation of ζ-carotene. Simultaneous application of fluridone diminished ζ-carotene accumulation and instead caused the buildup of phytoene, suggesting separate, enzymatic target sites for these two compounds.[107] Studies with J852 have now been extended to *Triticum medicago*, *Aphanocapsa*, and *Chenopodium*.[121] In all cases, ζ-carotene was accumulated, together with some phytoene. The dihydropyrone LS 80707 also leads to the presence of ζ-carotene in *Aphanocapsa*.[124]

Inhibitor studies with LS 80707 and its isomer LS 80717 (containing *E*-groups at C-5 and C-6 of the pyrone ring) have revealed that LS 80717 does not inhibit ζ-carotene desaturase at the same concentration as LS 80707.[124] Another dihydropyrone, LS 780091, binds irreversibly and noncompetitively at the active site of the enzyme.[135]

Amitrole (Figure 4.9) also causes the accumulation of ζ-carotene in intact plants[156] and *Aphanocapsa*,[153] but a comparison of I_{50} values for *in vivo* and *in vitro* systems of the latter suggests that direct action on the enzyme is not the prime mode of action of this inhibitor. It is noteworthy

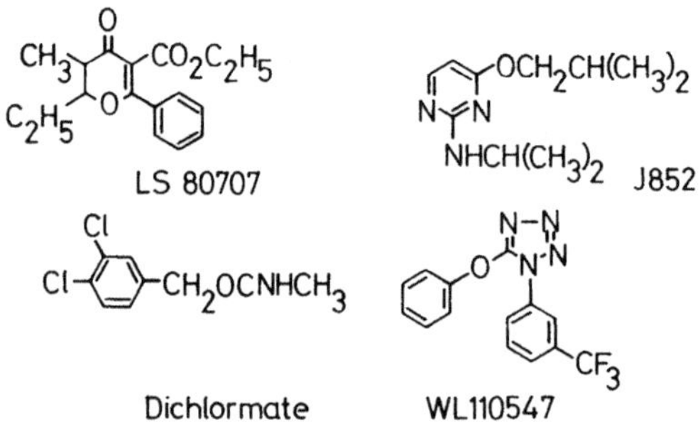

Figure 4.8. Inhibitors of ζ-carotene desaturase.

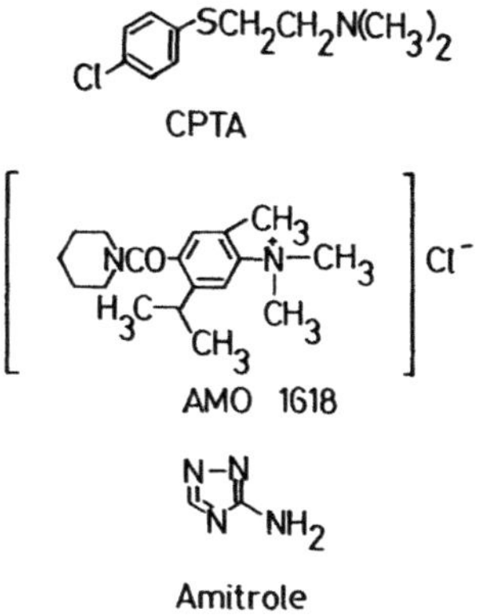

Figure 4.9. Inhibitors with multiple or uncertain target sites.

that amitrole has also been reported to inhibit cyclization[158] and phytoene desaturase.[164] Its precise mode of action, therefore, remains to be established.

4.6.1.4. Lycopene Cyclase

The accumulation of lycopene, at the expense of β-carotene formation, on treatment of plants with the substituted triethylamine CPTA (Figure 4.9) has been demonstrated in several cases, including citrus fruits[159] and *Aphanocapsa* cells and cell extracts.[153] In *Aphanocapsa, in vivo* and *in vitro* I_{50} values for lycopene cyclase were 45 and 30 μM, respectively, and a noncompetitive effect on the enzyme was shown.[153] However, in citrus fruits a range of onium compounds related to CPTA regulate the formation of all-*E* carotenoids, in that not only is cyclization blocked, but also the formation of lycopene is stimulated. The latter effect requires protein synthesis.[160] Indeed, the tertiary amine MPTA [2-(4-methylphenoxy)triethylamine] causes gene expression and translation of [poly(A)$^+$]RNA on 80S ribosomes.[161] Therefore, CPTA and related compounds may have multiple effects in plants.

AMO 1618 (Figure 4.9), an inhibitor of squalene cyclization[162] and kaurene synthetase,[163] is also an inhibitor of lycopene cyclization when used at high concentrations with the *Aphanocapsa* cell-free system.[124]

Finally, a group of phenoxytriethylamines, originally synthesized as potential uncouplers of oxidative phosphorylation, preferentially inhibit lycopene cyclization.[164]

4.6.1.5. Inhibitors of Xanthophyll Formation

Several hydroxylation reactions in xanthophyll biosynthesis are catalyzed by mixed-function oxidases and are therefore potential target sites. Although several inhibitors of similar reactions in sterol[165] and gibberellin[166] biosynthesis have been developed, there are no reports of herbicides which inhibit hydroxylations in carotenogenesis. Since the bioregulators tetcyclasis and LAB 11682, albeit at high concentrations, do inhibit the *in vitro* formation of β-cryptoxanthin, it should be possible to find other compounds which prevent xanthophyll formation. The resultant accumulation of carotenes in the plant would lead to impaired light harvesting, rather than preventing the photoprotective effect of carotenoids.

4.6.2. Other Target Sites

There is no experimental evidence to demonstrate that other reactions in plant carotenogenesis, whether they are related directly to enzyme activity or not, are target sites of the current range of bleaching herbicides. It is feasible that compounds such as amitrole and CPTA may act at multiple sites or on reactions which cannot be assayed at the present time.

4.7. FUTURE PROSPECTS

Our inadequate understanding of the properties of carotenogenic enzymes (see Section 4.4.2) precludes the rational design of herbicides which bind to the enzymes in question. It is important, therefore, to pursue a goal of purifying and characterizing these proteins, especially phytoene synthase, phytoene desaturase, and lycopene cyclase. In order for such investigations to be the most pertinent, photosynthetic, higher plant tissues should be used, although it is debatable at the present time whether the properties of the chloroplastidic enzymes are significantly different from those of the enzymes from nonphotosynthetic tissues or other carotenogenic systems. Ideally, the pure protein should be crystallized in order to establish its three-dimensional structure. Inevitably, this objective is some way into the future and will be technically difficult as the enzymes will be detergent solubilized. However, there are now reports of similar, solubilized enzymes being crystallized. The production of monoclonal and polyclonal antibodies to the pure enzymes will also be beneficial.

Alternative experimental approaches should also be pursued. Studies with radioactive inhibitors will enable competitive binding studies to be carried out. This approach would be facilitated by the identification of the carotenogenic enzyme on gels and the availability of herbicide-resistant mutants. One such mutant, from *Anacystis*, is resistant to norflurazon, fluorochloridone, and fluridone, in decreasing order of resistance.[168] Identification of the enzymes would be simplified if photosensitive, radioactive derivatives of inhibitors or substrates were available.

The isolation of carotenoid genes is now making progress, with the isolation of the *crtI* gene from *Rhodobacter*[85] and the cloning of the *al-3* and *al-1* genes from *Neurospora.*[169,170] Although such heterologous probes may cause problems in the screening of higher plant gene libraries, they should be used for this purpose. The small genome of *Arabidopsis* is an argument in favor of selecting this plant for the initial studies, with later work using the gene from this plant to probe gene banks from weed or crop species. Alternatively, oligonucleotide probes, synthesized from microsequence data of a pure plant enzyme, could be used for screening purposes. The availability of antibodies also raises the possibility of using an expression library for immunoscreening.

If the genes can be expressed in bacteria, then it will be possible to obtain large quantities of the proteins. As discussed earlier, such proteins are unlikely to be enzymatically active, but they may be amenable to reconstitution experiments. This would greatly facilitate studies on the properties of the enzymes and provide ample quantities of enzyme for routine *in vitro* screening of novel inhibitors.

The feasibility of targeting other sites within the chloroplast or cytoplasm is poor at the present time. Considerably more must be understood about chloroplast development and the integration of carotenogenic enzymes, and the carotenoids themselves, into the chloroplast. It may then be possible to inhibit such processes specifically.

Transgenic plants, resistant to bleaching herbicides, are a future possibility. Integration of a resistant gene into a crop plant would make it simpler to obtain selectivity of herbicides to weeds rather than crops.

We are at the beginning of an exciting era in plant carotenogenesis, which will lead to the purification of the enzymes involved and their genes. This, in turn, should provide the basis for elucidating chloroplast development with respect to carotenogenesis. It is hoped that such knowledge will enable future herbicides to be designed on a rational basis, rather than being obtained by routine screening of synthetic chemicals.

ACKNOWLEDGMENTS

I would like to thank Dr. K. Pallett of Rhone-Poulenc Agriculture for providing the data shown in Figure 4.5 prior to publication and Drs. G. Sandmann, R. E. Wilkinson, and C. Dahlin for providing preprints of publications.

REFERENCES

1. B. C. L. Weedon, in: *Carotenoids* (O. Isler, ed.), p. 29, Birkhauser, Basel (1971).
2. T. W. Goodwin, *The Biochemistry of the Carotenoids*, Vol. 1, Chapman and Hall, London (1980).

3. T. W. Goodwin, *The Biochemistry of the Carotenoids*, Vol. 2, Chapman and Hall, London (1983).
4. T. W. Goodwin, *Annu. Rev. Nutr. 6*, 273 (1986).
5. Commission on Biochemical Nomenclature, *Biochemistry 10*, 4827 (1971).
6. Commission on Biochemical Nomenclature, *Biochemistry 14*, 1803 (1975).
7. O. Straub, *Key to Carotenoids*, Birkhauser, Basel (1987).
8. G. Britton, *Nat. Prod. Rep. 6*, 359 (1989).
9. G. Britton and T. W. Goodwin, *Carotenoid Chemistry and Biochemistry*, Pergamon Press, Oxford (1982).
10. X.-S. Tang and K. Satoh, *FEBS Lett. 179*, 64 (1985).
11. J. P. Thornber, *Annu. Rev. Plant Physiol. 26*, 127 (1975).
12. H. K. Lichtenthaler, U. Prenzel, and G. Kuhn, *Z. Naturforsch., C 37*, 10 (1982).
13. D. Siefermann-Harms, *Biochim. Biophys. Acta 811*, 325 (1985).
14. J. P. Thornber, in: *Encyclopedia Plant Physiology, New Series*, p. 19, Springer-Verlag, Berlin (1986).
15. R. Cogdell, in: *Plant Pigments* (T. W. Goodwin, ed.), p. 183, Academic Press, London (1988).
16. G. F. Peter, O. Machold, and J. P. Thornber, in: *Plant Membranes, Structure, Assembly and Function* (J. L. Harwood and T. J. Walton, eds.), p. 17, The Biochemical Society, London (1988).
17. R. Douce, B. Holtz, and A. A. Benson, *J. Biol. Chem. 248*, 7215 (1973).
18. H. K. Lichtenthaler, U. Prenzel, R. Douce, and J. Joyard, *Biochim. Biophys. Acta 641*, 99 (1981).
19. M. J. Fishwick and A. J. Wright, *Phytochemistry 19*, 55 (1980).
20. M. Neuberger, E. P. Journet, R. Bligny, J.-P. Carde, and R. Douce, *Arch. Biochem. Biophys. 217*, 312 (1982).
21. H. W. Gardner, J. A. Miernyk, D. D. Christianson, and U. Khero, *Physiol. Plant. 69*, 541 (1987).
22. J. Joyard, Thèse de Doctorat d'Etat, Grenoble (1979).
23. C. Costes, C. Burghoffer, E. Carrayol, G. Ducet, and M. Diano, *Plant Sci. Lett. 6*, 253 (1976).
24. J. M. Whately and F. R. Whatley, *New Phytol. 106*, 667 (1987).
25. T. W. Goodwin, *Biochem. J. 70*, 612 (1958).
26. C. Lutz, *Z. Pflanzenphysiol. 104*, 43 (1981).
27. S. Liaaen-Jensen, *Pure Appl. Chem. 57*, 645 (1985).
28. J. Burczyk, *Phytochemistry 26*, 113 (1987).
29. N. I. Krinsky, in: *Carotenoids* (O. Isler, ed.), p. 669, Birkhauser, Basel (1971).
30. C. S. Foote, in: *Free Radicals in Biology* (W. A. Pryor, ed.), p. 85, Academic Press, New York (1976).
31. J. P. Knox and A. D. Dodge, *Phytochemistry 24*, 889 (1985).
32. C. S. Foote, Y. C. Chang, and R. W. Denny, *J. Am. Chem. Soc. 92*, 5216 (1970).
33. D. Gust, T. A. Moore, R. V. Bensasson, P. Mathis, E. J. Land, C. Chachaty, A. L. Moore, P. A. Liddel, and G. A. Neweth, *J. Am. Chem. Soc. 107*, 3631 (1985).
34. K. J. Kunert and A. D. Dodge, in: *Target Sites of Herbicide Action* (P. Boger and G. Sandmann, eds.), p. 45, CRC Press, Boca Raton, Florida (1989).
35. S. W. Jeffery, R. Douce, and A. A. Benson, *Proc. Natl. Acad. Sci. U.S.A. 71*, 807 (1974).
36. J. M. Anderson, J. C. Waldron, and S. W. Thorne, *FEBS Lett. 92*, 227 (1978).
37. A. N. D. Larkum and J. Barrett, in: *Advances in Botanical Research* (H. W. Woolhouse, ed.), Vol. 10, p. 1, Academic Press, New York (1983).
38. L. Axelsson, B. Klockare, H. Ryberg, and A. S. Sandelius, in: *Photosynthesis V. Chloroplast Development* (G. Akoyunoglou, ed.), p. 285, Balaban International Series, Philadelphia (1981).

39. C. Dahlin, H. Ryberg, and L. Axelsson, *Physiol. Plant 59*, 562 (1983).
40. B. V. Milborrow, in: *Plant Growth Substances 1988* (R. P. Pharis and S. B. Rood, eds.), p. 241, Springer, Berlin (1990).
41. S. L. Spurgeon and J. W. Porter, in: *The Biosynthesis of Isoprenoid Compounds* (J. W. Porter and S. L. Spurgeon, eds.), Vol. 2, p. 1, Wiley, New York (1983).
42. B. L. Jones and J. W. Porter, *CRC Crit. Rev. Plant Sci. 3*, 294 (1986).
43. G. Britton, in: *Plant Pigments* (T. W. Goodwin, ed.), p. 133, Academic Press, London (1988).
44. H. Kleinig, *Annu. Rev. Plant Physiol. Plant Mol. Biol. 40*, 39 (1989).
45. B. Camara, *Plant Physiol. 74*, 112 (1984).
46. B. Camara, O. Dogbo, F. Bardat, and R. Moneger, *C. R. Acad. Sci., Ser. 3, 301*, 749 (1985).
47. J. C. B. McDermott, G. Britton, and T. W. Goodwin, *Biochem. J. 134*, 1115 (1973).
48. F. Lutke-Brinkhaus, B. Liedvogel, K. Kreuz, and H. Kleinig, *Planta 156*, 176 (1982).
49. S. C. Kushwaha, G. Suzue, C. Subbarayan, and J. W. Porter, *J. Biol. Chem. 245*, 4708 (1970).
50. B. Camara, C. Payan, A. Escoffier and R. Moneger, *C. R. Acad. Sci., Ser. D, 291*, 306 (1980).
51. G. Britton and R. Powls, *Phytochemistry 16*, 1253 (1977).
52. B. Camara and R. Moneger, *Physiol. Veg. 20*, 757 (1982).
53. P. Beyer, G. Weiss, and H. Kleinig, *Eur. J. Biochem. 153*, 341 (1985).
54. G. Britton, W. J. S. Lockley, N. J. Patel, and T. W. Goodwin, *FEBS Lett. 79*, 281 (1977).
55. G. Britton, W. J. S. Lockley, R. Powls, T. W. Goodwin, and L. M. Heyes, *Nature 268*, 81 (1977).
56. H. M. Hill, S. K. Calderwood, and L. J. Rogers, *Phytochemistry 10*, 2051 (1971).
57. B. Camara, Thèse de Doctorat d'Etat, Université Pierre et Marie Curie, Paris (1981).
58. B. Maudinas, M. L. Bucholtz, C. Papastephanou, S. S. Katiyar, A. V. Briedis, and J. W. Porter, *Arch. Biochem. Biophys. 180*, 354 (1977).
59. G. Britton and T. W. Goodwin, *Phytochemistry 8*, 2257 (1969).
60. G. Britton, *Pure Appl. Chem. 47*, 223 (1976).
61. H. Yamamoto, *Methods Enzymol. 110*, 303 (1985).
62. D. Siefermann-Harms, in: *Lipids and Lipid Polymers in Higher Plants* (M. Tevini and H. K. Lichtenthaler, eds.), p. 218, Springer, Berlin (1977).
63. H. Y. Yamamoto, *Pure Appl. Chem. 51*, 639 (1979).
64. K. H. Grumbach, *Z. Naturforsch., C 39*, 455 (1983).
65. C. Costes, C. Burghoffer, J. Joyard, M. Block, and R. Douce, *FEBS Lett. 103*, 17 (1979).
66. P. M. Bramley, *Adv. Lip. Res. 21*, 243 (1985).
67. O. Dogbo, A. Laferriere, A. D'Harlingue, and B. Camara, *Proc. Natl. Acad. Sci. U.S.A. 85*, 7054 (1988).
68. S. C. Kushwaha, C. Subbarayan, D. A. Baler, and J. W. Porter, *J. Biol. Chem. 244*, 3635 (1969).
69. B. Maudinas, M. L. Bucholtz, C. Papastephanou, S. S. Katiyar, A. V. Briedis, and J. W. Porter, *Biochem. Biophys. Res. Commun. 66*, 430 (1975).
70. J. W. Porter and S. L. Spurgeon, *Pure Appl. Chem. 51*, 609 (1979).
71. S. L. Spurgeon, N. Sathyamoorthy, and J. W. Porter, *Arch. Biochem. Biophys. 230*, 446 (1984).
72. O. Dogbo, F. Bardat, A. Laferriere, J. Quennemet, J. Brangeon, and B. Camara, *Plant Sci. 49*, 89 (1987).
73. J. E. Graebe, *Phytochemistry 7*, 2003 (1968).
74. K. Kreuz, P. Beyer, and H. Kleinig, *Planta 154*, 66 (1982).
75. T. W. Goodwin, *Biochem. Soc. Trans. 11*, 473 (1983).
76. A. A. Qureshi, N. Qureshi, M. Kim, and J. W. Porter, *Arch. Biochem. Biophys. 162*, 117 (1974).
77. J. M. Clough and G. Pattenden, *J. Chem. Soc., Chem. Commun.*, 616 (1979).
78. J. M. Clough and G. Pattenden, *J. Chem. Soc., Perkin Trans. 1*, 3011 (1983).

79. P. Beyer, M. Mayer, and H. Kleinig, *Eur. J. Biochem. 184,* 141 (1989).
80. G. Mackinney, C. M. Rick, and J. A. Jenkins, *Proc. Natl. Acad. Sci. U.S.A. 42,* 404 (1956).
81. I. C. Anderson and D. S. Robertson, *Plant Physiol. 35,* 531 (1960).
82. P. M. Bramley and A. Mackenzie, *Curr. Top. Cell. Regul. 29,* 291 (1988).
83. M. D. de la Guardia, C. M. G. Aragon, F. J. Murillo, and E. Cerda-Olmedo, *Proc. Natl. Acad. Sci. U.S.A. 68,* 2012 (1971).
84. J. T. O. Kirk and R. Tilney-Bassett, *The Plastids,* W. H. Freeman and Co., San Francisco (1978).
85. G. A. Armstrong, M. Alberti, F. Leach, and J. E. Hearst, *Mol. Gen. Genet. 216,* 254 (1989).
86. G. E. Bartley and P. A. Scolnik, *J. Biol. Chem. 264,* 13109 (1989).
87. A. Schmidt, G. Sandmann, G. A. Armstrong, J. E. Hearst, and P. Boger, *Eur. J. Biochem, 184,* 375 (1989).
88. A.-H. Etemadi, *Adv. Lip. Res. 21,* 281 (1985).
89. D. Siefermann and H. Y. Yamamoto, *Arch. Biochem. Biophys. 171,* 70 (1975).
90. G. Sandmann and P. M. Bramley, *Biochim. Biophys. Acta 96,* 645 (1985).
91. W. Rau, in: *Pigments in Plants* (F.-C. Czygan, ed.), p. 80, Fischer, Stuttgart (1980).
92. H. Mohr, in: *Chloroplast Biogenesis* (N. R. Baker and J. Barber, eds.), p. 305, Elsevier, Amsterdam (1984).
93. J. H. Argyroudi-Akoyunoglou and G. Akoyunoglou, *FEBS Lett. 104,* 78 (1979).
94. K. O. Burkey, *Photosynth. Res. 10,* 37 (1986).
95. H. K. Lichtenthaler, C. Buschmann, and M. Doll, *Photosynth. Res. 2,* 115 (1981).
96. K. H. Grumbach and H. K. Lichtenthaler, *Photochem. Photobiol. 35,* 209 (1982).
97. M. J. Buggy, G. Britton, and T. W. Goodwin, *Biochem. J. 114,* 641 (1969).
98. D. Schulze-Siebert and G. Schultz, *Plant Physiol. 84,* 1233 (1987).
99. K. Kreuz and H. Kleinig, *Planta 153,* 578 (1981).
100. J. C. Gray, *Adv. Bot. Res. 14,* 25 (1987).
101. K. H. Grumbach and G. Britton, in: *Advances in Photosynthesis Research* (C. Sybesma, ed.), Vol. IV, p. 69, Martinus Nijhoff, Dordrecht (1984).
102. N. H. Chua and G. W. Schmidt, *Proc. Natl. Acad. Sci. U.S.A. 75,* 6110 (1978).
103. C. Robinson and R. J. Ellis, *Eur. J. Biochem. 142,* 343 (1984).
104. S. Smelkens, C. Baurene, J. Hageman, K. Keegstra, and P. Weisbeck, *Cell 46,* 365 (1986).
105. P. Eckes, J. Schell, and L. Willmitzer, *Mol. Gen. Genet. 199,* 216 (1985).
106. P. A. Scolnik, in: *UCLA Symposium: Molecular and Cellular Biology, New Series (Plant Gene Systems and Their Biology),* p. 383, Alan R. Liss, New York (1987).
107. S. M. Ridley, in: *Carotenoid Chemistry and Biochemistry* (G. Britton and T. W. Goodwin, eds.), p. 353, Pergamon, Oxford (1982).
108. G. Britton, in: *Regulation of Chloroplast Differentiation* (G. Akoyunoglou and H. Senger, eds.), p. 125, Alan R. Liss, New York (1986).
109. B. H. Davies, in: *Chemistry and Biochemistry of Plant Pigments* (T. W. Goodwin, ed.), p. 38, Academic Press, London (1976).
110. G. Britton, *Methods Enzymol. 111,* 113 (1985).
111. T. W. Goodwin and G. Britton, in: *Plant Pigments* (T. W. Goodwin, ed.), p. 61, Academic Press, London (1988).
112. S. J. Schwartz and J. H. Von Elbe, *J. Liq. Chromatogr. 5,* 43 (1982).
113. R. F. Taylor, *Adv. Chromatogr. 22,* 157 (1983).
114. G. Britton, P. Barry, and A. J. Young, *Proc. Br. Crop Prot. Conf.—Weeds,* p. 1015 (1987).
115. D. J. Mayonado, K. K. Hatzios, D. M. Orcutt, and H. P. Wilson, *Pestic. Biochem. Physiol. 35,* 138 (1989).
116. P. Barry and K. Pallett, *Z. Naturforsch. C 45,* 492 (1990).
117. G. Sandmann, K.-J. Kunert, and P. Boger, *Z. Naturforsch., C 34,* 1044 (1979).
118. G. Sandmann and P. Boger, in: *Target Sites of Herbicide Action* (P. Boger and G. Sandmann, eds.), p. 25, CRC Press, Boca Raton, Florida (1989).

119. L. Axelsson, C. Dahlin, and H. Ryberg, *Physiol. Plant. 55,* 111 (1982).
120. C. Dahlin and H. Ryberg, *Physiol. Plant. 68,* 39 (1986).
121. H. Kostal, Thesis, University of Konstanz (1986).
122. G. Sandmann and P. Boger, *Proc. Br. Crop Prot. Conf.—Weeds,* p. 139 (1987).
123. G. Sandmann and P. M. Bramley, *Planta 164,* 259 (1985).
124. G. Sandmann, P. M. Bramley, and P. Boger, *J. Pestic. Sci. 10,* 19 (1985).
125. J. B. St. John, U. Schirmer, F. R. Ruttig, and H. Bleiholder, in: *Pesticide Synthesis through Rational Approaches* (P. S. Magee, G. K. Kohn, and J. J. Menn, eds.), p. 145, American Chemical Society, Washington, D.C. (1984).
126. D. Urbach, M. Suchanka, and W. Urbach, *Z. Naturforsch., C 31,* 652 (1976).
127. G. Sandmann, K.-J. Kunert, and P. Boger, *Pestic. Biochem. Physiol. 15,* 288 (1981).
128. G. Sandmann and P. Boger, *Z. Naturforsch., C 37,* 1092 (1982).
129. I. E. Clarke, P. M. Bramley, G. Sandmann, and P. Boger, in: *Biochemistry and Metabolism of Plant Lipids* (J. F. G. M. Wintermans and P. J. C. Kuiper, eds.), p. 549, Elsevier, Amsterdam (1982).
130. G. Sandmann, I. E. Clarke, P. M. Bramley, and P. Boger, *Z. Naturforsch, C 39,* 443 (1984).
131. G. Sandmann, in: *Proceedings of the 6th IUPAC Congress on Pesticide Chemistry* (R. Greenhalgh and T. R. Roberts, eds.), p. 43, Blackwell Scientific, Oxford (1986).
132. R. Mengel, R. Sehring, A. Boehler-Kohler, and P. Boger, 5th International Congress on Pesticide Chemistry, Kyoto, Abstract IIc-5 (1982).
133. K.-J. Kunert, G. Sandmann, and P. Boger, *Rev. Weed Sci. 3,* 35 (1987).
134. I. E. Clarke, G. Sandmann, P. M. Bramley, and P. Boger, *Pestic. Biochem. Physiol. 23,* 335 (1985).
135. G. Sandmann and P. Boger, in: *Proceedings of the British Crop. Prot. Conf.—Weeds,* p. 139 (1987).
136. G. Sandmann and P. Boger, in: *Biochemical Responses Induced by Herbicides* (D. E. Moreland, J. B. St. John, and F. D. Hess, eds.), p. 111, American Chemical Society, Washington, D.C. (1982).
137. D. E. Gregonis and H. C. Rilling, *Biochem. Biophys. Res. Commun. 54,* 449 (1973).
138. G. Sandmann, P. M. Bramley, and P. Boger, *Pestic. Biochem. Physiol. 14,* 185 (1980).
139. R. E. Wilkinson, *Pestic. Biochem. Physiol. 23,* 370 (1985).
140. R. E. Wilkinson, *Pestic. Biochem. Physiol. 28,* 381 (1987).
141. R. E. Wilkinson, in: *Crop Safeners for Herbicides* (K. K. Hatzios and R. E. Hoagland, eds.), p. 221, Academic Press, New York (1989).
142. P. G. Bartels and C. McCullough, *Biochem. Biophys. Res. Commun. 48,* 16 (1972).
143. A. Ben-Aziz and E. Koren, *Plant Physiol. 54,* 956 (1974).
144. H. W. Kummel and L. H. Grimme, *Z. Naturforsch., C30,* 333 (1975).
145. A. D. Pardo and J. A. Schiff, *Can. J. Bot. 58,* 25 (1980).
146. A. J. Vaisberg and J. A. Schiff, *Plant Physiol. 57,* 260 (1976).
147. P. G. Bartels and C. W. Watson, *Weed Sci. 26,* 198 (1978).
148. G. Sandmann and P. Boger, *Weed Sci. 31,* 338 (1983).
149. K.-J. Kunert and P. Boger, *Weed Sci. 26,* 292 (1978).
150. Y. Fujh, J. Kurokawa, Y. Inoue, I. Yamaguchi, and T. Misato, *J. Pestic. Sci. 2,* 431 (1977).
151. M. M. Lay and A. M. Niland, *Pestic. Biochem. Physiol. 19,* 337 (1983).
152. R. Lambert and P. Boger, *Pestic. Biochem. Physiol. 20,* 183 (1983).
153. V. Zoebelin, Thesis, University of Konstanz (1985).
154. M. P. Mayer, D. L. Bartlett, P. Beyer, and H. Kleinig, *Pestic. Biochem. Physiol. 34,* 111 (1989).
155. J. B. St. John and J. L. Hilton, *Weed Sci. 24,* 579 (1976).
156. E. R. Burns, G. A. Buchanan, and M. C. Carter, *Plant Physiol. 47,* 144 (1971).
157. M. W. Kerr and D. P. Whitaker, *Proc. Br. Crop Prot. Conf.—Weeds,* p. 1005 (1987).

158. W. Rudiger, J. Benz, U. Lempart, S. Schoch, and D. Steffens, *Z. Pflanzenphysiol. 80,* 131 (1976).
159. H. Yokoyama, W.-J. Hsu, S. M. Poling, and E. Hayan, in: *Carotenoid Chemistry and Biochemistry* (G. Britton and T. W. Goodwin, eds.), p. 371, Pergamon Press, Oxford (1982).
160. W.-J. Hsu, H. Yokoyama, and C. W. Coggins, *Phytochemistry 11,* 2985 (1972).
161. H. Yokoyama, W.-J. Hsu, E. Hayman, and S. Poling, *Recent Adv. Phytochem. 18,* 231 (1984).
162. T. J. Douglas and L. G. Paleg, *Plant Physiol. 49,* 417 (1972).
163. J. D. Anderson and T. C. Moore, *Plant Physiol. 42,* 1527 (1967).
164. B. J. Wright, A. Baillie, K. Wright, J. R. Dowsett, and T. M. Sharpe, *Phytochemistry 19,* 61 (1980).
165. P. Benveniste, *Annu. Rev. Biochem. 37,* 275 (1986).
166. J. E. Graebe, in: *Plant Growth Substances* (P. F. Wareing, ed.), p. 71, Academic Press, New York (1982).
167. H. Linden, G. Sandmann, D. Chamovitz, J. Hirschberg, and P. Buger, *Pest. Biochem. Physiol. 36,* 46 (1990).
168. P. Boger, unpublished results (1989).
169. M. A. Nelson, G. Morelli, A. Carottoli, N. Romano, and G. Macino, *Mol. Cell. Biol. 9,* 1271 (1989).
170. T. J. Schmidhauser, F. R. Lauter, V. E. A. Russo, and C. Yanofsky, *Mol. Cell. Biol. 10,* 5064 (1990).

5

Other Primary Target Sites for Herbicides

KEN E. PALLETT

The target sites for the majority of commercially important herbicides are centered upon photosynthesis and biosynthesis of amino acids, lipids, and carotenoids and have been covered in the previous four chapters. However, a significant number of successful herbicides have target sites outside the scope of Chapters 1–4, and these target sites, where known, will be considered in this chapter.

These target sites will include respiration, chlorophyll biosynthesis, isoprenoid biosynthesis, microtubule biosynthesis and function, cellulose biosynthesis, protein and nucleic acid biosynthesis, folic acid biosynthesis, and hormone biosynthesis and function. Only herbicides with proven direct activity on these target sites will be reviewed in this chapter.

Reference will be made to several herbicides which can be classed as "multisite" inhibitors. Many of these herbicides have clearly established primary target sites. Examples of such herbicides are the hydroxybenzonitriles and dinitrophenols, which inhibit and uncouple photosynthetic electron transport (see Chapter 1). However, they are clearly mitochondrial electron transport inhibitory uncouplers (Section 5.1), which is accepted as contributing to their overall mode of action, and their inclusion here is justified.

KEN E. PALLETT • Department of Plant Biochemistry, Rhone-Poulenc Agriculture, Ongar, Essex CM5 0HW, United Kingdom.

Also, in addition to the effects of known commercial herbicides on the target sites, reference will be made to some noncommercial herbicide or inhibitor target sites which indicate the potential of those sites as targets for future commercialized products.

5.1. RESPIRATION

Figure 5.1 represents a simplified scheme summarizing the events in the process of respiration, which provides energy, in the form of ATP, necessary for the biosynthetic reactions and other energy-requiring events in cells. This process is universal, occurring in all organisms, and interference with it would be an effective target site resulting in the inhibition of growth and development. However, this universal nature does not make it a desirable choice as a target site for future herbicide design. Plants do possess some significant variations in the respiratory process such as the cyanide-resistant alternative oxidase.[1] Indeed, several herbicides which interfere with the respiratory chain also inhibit the alternative oxidase.[2]

Herbicide interference with respiration has been extensively reviewed.[3-5] However, several recent studies have been published which will form the major contribution to this section.

The majority of herbicides which interfere with the respiratory process do so by inhibiting and/or uncoupling the respiratory electron transport chain. Many of these compounds were discovered over 10 years ago; however, many are still commercially important. The reader is referred to previous reviews[3-5] for details on these compounds, examples of which are shown in Figure 5.2.

Although there are several mechanisms by which respiratory electron transport can be inhibited, the compounds in Figure 5.2 are classed as inhibitory uncouplers.[3-5] This activity is not restricted to mitochondrial electron transport as photosystem II (PSII) electron transport and phosphorylation are also affected by dinitrophenols and hydroxybenzonitriles (Refs. 3-5 and Chapter 1). In isolated mitochondrial preparations, lower concentrations of herbicides uncouple oxidative phosphorylation, that is, promote electron transport, without ATP synthesis, whereas higher concentrations inhibit electron transport. Fedtke calculated that uncoupling is 2-10 times more sensitive for a range of phenolic/acidic compounds.[4] A similar difference has also been reported for dinoterb (Figure 5.2), which uncouples oxidative phosphorylation at 0.05–0.15 mM and inhibits respiration at 0.5–1.0 mM.[6]

The mechanism(s) by which these effects are manifested has yet to be fully elucidated; however, the most favored proposal is a protonophore

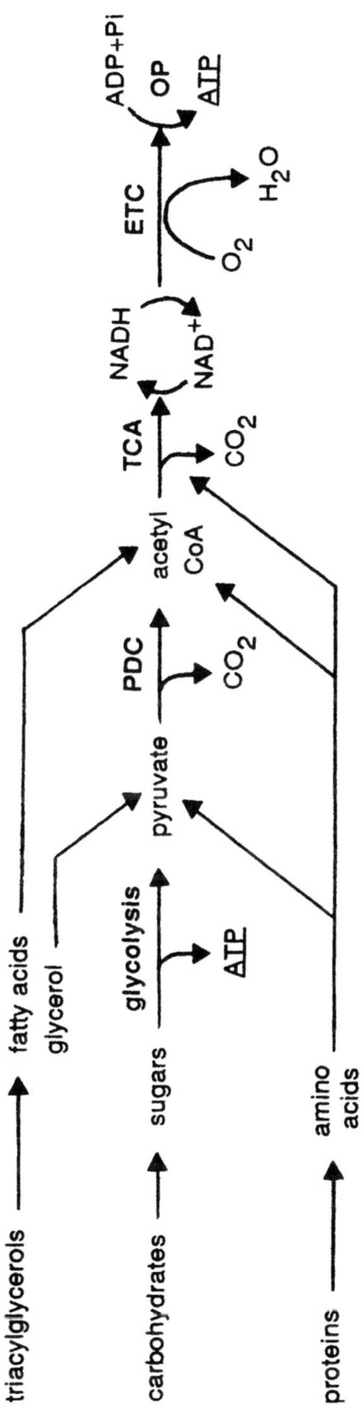

Figure 5.1. Summary of the events during respiration of triacylglycerols, carbohydrates, and proteins. PDC, Pyruvate dehydrogenase complex; TCA, tricarboxylic acid or Krebs cycle; ETC, mitochondrial electron transport; OP, oxidative phosphorylation.

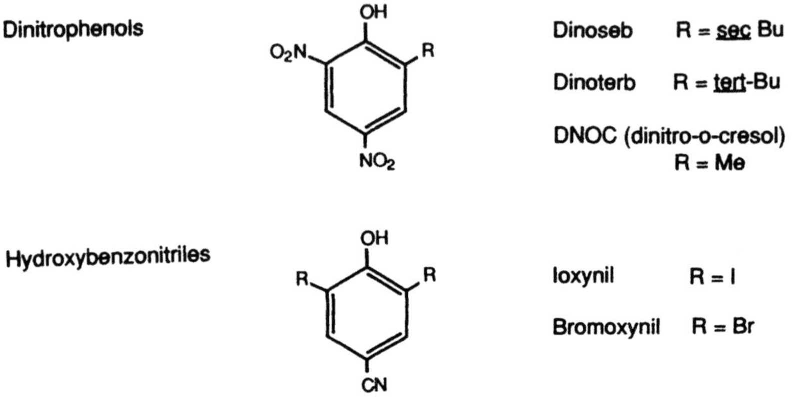

Figure 5.2. Examples of herbicides which have a primary action as inhibitory uncouplers.

action in accordance with the chemiosmotic theory (for a recent review, see Ref. 7). The herbicides in question contain a labile phenolic hydrogen and, in the dissociated form, would readily take up a protein in acidic conditions, giving a lipophilic moiety capable of traversing the inner mitochondrial membrane. Therefore, they could readily interfere with or abolish the proton gradient set up across the inner mitochondrial membrane during electron transport, which is necessary for oxidative phosphorylation.[7]

Recently, some *N*-phenylcarbamates have been shown to be very potent uncouplers.[8] The relatively acidic —NH— group of the carbamate function is proposed to be involved in the proton transfer across the membrane. The close analogy between 2,2-dichloroethyl *N*-(3,4-dichlorophenyl)carbamate and the proposed antimitotic *N*-phenylcarbamates such as barban and chloropropham (for structures, see Figure 5.15) may suggest a mitochondrial action for this herbicide. However, no uncoupling activity has been reported, although inhibition of O_2 uptake in isolated mitochondria and reduced ATP levels in excised hypocotyls has been shown.[9,10] In addition, respiratory inhibition *in vivo* has been reported but was concluded to be a consequence of inhibition elsewhere (see Ref. 11).

The protonophore mechanism of respiratory uncoupling has been questioned on the basis of results obtained with the use of affinity-labeled uncouplers[12–14] (see Ref. 5). Such studies have led to a proposal of an uncoupling binding site associated with ATPase.[5,12–15]

A further proposed reason for the uncoupling and also the less sensitive inhibitory action of these herbicides has arisen mainly from the work of Moreland and colleagues.[3,16] The lipophilic nature of these compounds when in their associated forms will allow partitioning into the mitochondrial inner membrane. Such partitioning may well produce alterations in the fluidity and permeability properties of the membrane, thus altering the conformation of respiratory chain components.[3,16] Chlorpropham, ioxynil, and dinoseb all perturb the inner mitochondrial membrane, as determined by cation permeability and swelling techniques (see Refs. 3 and 16 for original references).

It is not inconceivable that such permeability effects are a result of an initial direct effect of herbicide binding to ATPase or an electron transport chain component and/or a protonophore action. However, most, if not all, the herbicides discussed in this section can be classed as multisite inhibitors as evidenced by their inclusion in previous chapters, particularly Chapter 1, and elsewhere in this chapter. Such a membrane partitioning would not be specific to the inner mitochondrial membrane but would occur in other

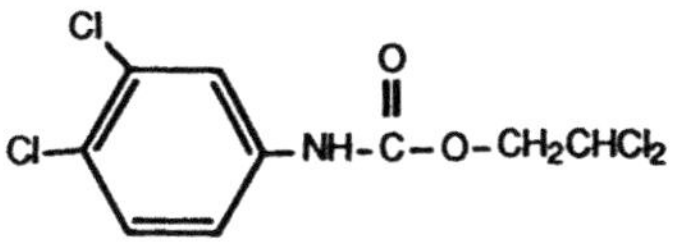

2,2-dichloroethyl-N-(3,4-dichloro phenyl) carbamate

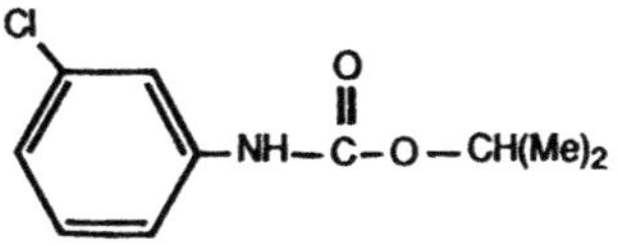

Chlorpropham

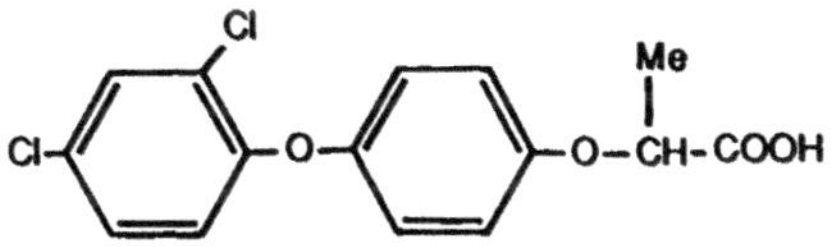

Diclofop

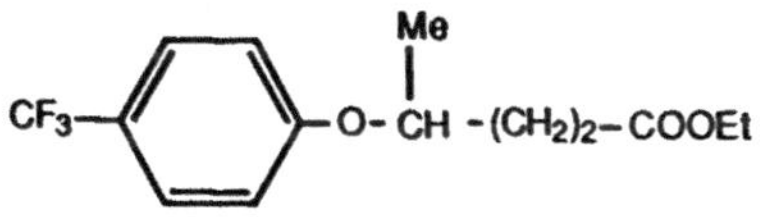

Difenopenten-ethyl

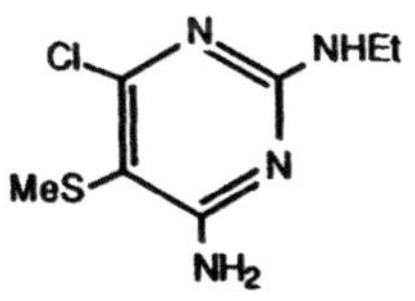

UK J72J

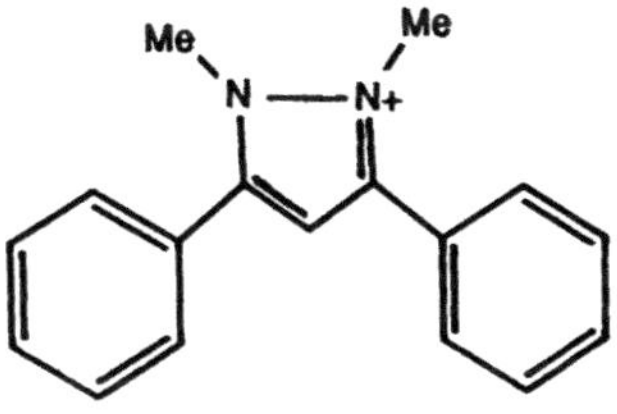

Difenzoquat

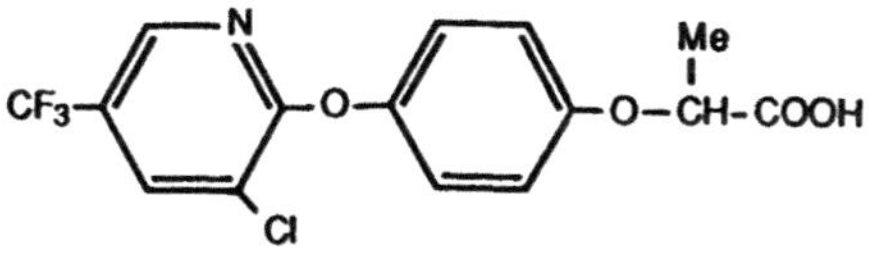

Haloxyfop

Figure 5.3. Examples of other herbicides which interfere with respiration and mitochondrial membranes. Diclofop and haloxyfop are the active free acids of their parent methyl esters (see Chapter 3).

cell membranes, causing indirect effects on any metabolic processes requiring specific membrane configuration/compartmentalization in the cell.

Nonspecific disruption of membrane permeability was proposed for difenopenten-ethyl {ethyl 4-[(4-trifluoromethyl)phenoxyl-2-pentanoate}, which uncoupled and inhibited respiration in corn and soybean mitochondria.[17] This was considered to be a secondary effect of this aryloxyphenoxypropionate analogue (see Chapter 3) due to the relatively high *in vitro* concentration. A similar conclusion was made for another aryloxyphenoxypropionate, diclofop, which uncoupled state 4 respiration, inhibited state 3 respiration, and caused swelling of isolated wheat and wild oat mitochondria.[18] The inhibition of respiration was proposed to be due to an effect prior to the electron transport chain, that is, an effect on succinate or malate dehydrogenase. A potent inhibition of state 3 respiration was also reported for the herbicide UK J72J (2-ethylamino-4-amino-5-thiomethyl-6-chloropyrimidine), with succinate as substrate.[19,20] Difenzoquat also inhibits state 3 respiration in isolated wheat and wild oat mitochondria; however, this was not thought to be a significant contribution to the chlorotic contact symptoms observed with this translocated herbicide.[21,22] Difenzoquat also possesses a weak bipyridinium-type PSI electron acceptor activity which causes chlorosis.[21,22]

No details on the mechanism of respiratory inhibition by diclofop, UK J72J, and difenzoquat have been reported; however, another aryloxyphenoxypropionate, haloxyfop, has been shown to inhibit both pyruvate dehydrogenase (PDC) and α-ketoglutarate dehydrogenase (αKGDC).[23] Haloxyfop was previously shown to inhibit respiration in corn root apices and isolated corn mitochondria.[24] It is a reversible noncompetitive inhibitor of PDC with respect to pyruvate, coenzyme A (CoA), and NAD^+, with K_i values ranging from 1 to 10 m*M*. It is an uncompetitive inhibitor of αKGDC with respect to α-ketoglutarate and a noncompetitive one with respect to CoA and NAD^+ with K_i values of approximately 1 m*M*.[23] These high K_i values and the nonselective nature of this inhibition (similar sensitivities with tolerant soybean and susceptible corn enzymes) suggest that these inhibitions are likely to be secondary, occurring only if high concentrations of herbicide accumulate in the mitochondrion.[23,24] The recent discovery that acetyl-CoA carboxylase is the primary target site of aryloxyphenoxypropionates (see Chapter 3) is further evidence that the action on PDC and αKGDC is secondary. However, it can be speculated that there may be a link, namely, carboxylation/decarboxylation, between the inhibitory mechanisms for the three enzymes and that this is common to diclofop, haloxyfop, difenopenten-ethyl, and other aryloxyphenoxypropionate herbicides.

Inhibition of PDC has recently been shown to be an effective primary target site for herbicides.[25] A series of acylphosphinates were synthesized

R = Me acetylmethyl phosphinate

R = OMe mono-methyl phosphonate

R = H acetyl phosphinate

pyruvate

Figure 5.4. The structures of pyruvate and the mechanism-based inhibitors of PDC.[25]

as mechanism-based inhibitors of the enzyme (Figure 5.4). The sodium monomethyl acetylphosphonate was a good competitive inhibitor, whereas the sodium acetylmethyl- and sodium acetylphosphinates were potent time-dependent inhibitors causing enzyme inactivation.[25] Inhibition of respiration with pyruvate as substrate was observed in isolated rat liver mitochondria. Evidence that PDC inhibition was phytotoxic was provided by an accumulation of pyruvate and alanine (the amination product of pyruvate) in treated plants.[25] In a separate study, the alanine analogue, 1-aminoethylphosphinate, was shown to be herbicidal due to its deamination to acetylphosphinate, which was shown to cause time-dependent inhibition following first-order and saturation kinetics requiring thiamine pyrophosphate.[26]

5.2. CHLOROPHYLL BIOSYNTHESIS

The crucial role of chlorophyll in plant metabolism suggests that inhibition of its synthesis would be a highly effective herbicide mode of action.[27] The first specific precursor for the formation of chlorophylls and other tetrapyrroles is α-aminolevulinic acid (ALA). Two pathways to ALA are thought to exist in plants.[28] The first, the Shemin pathway, involves a pyridoxal phosphate-requiring condensation of succinyl-CoA and glycine and may be located in the mitochondrion. The second, the so-called C_5 pathway, is plastid located and involves the transformation of glutamate via three enzymes and a specific tRNA intermediate (Figure 5.5 and Refs. 28–30). However, this second pathway has been proposed as the only source of ALA in plants.[27] Condensation of two ALA molecules, catalyzed by ALA dehydratase, gives rise to porphobilinogen, the first pyrrole compound in the pathway (Figure 5.5 and Ref. 28).

Four porphobilinogen molecules sequentially condense to ultimately form the tetrapyrrole ring system. The pathway appears to be common for chlorophylls and hemes until protoporphyrin IX, the branch-point intermediate before insertion of Mg^{2+} or Fe^{2+}, leading to chlorophylls and hemes, respectively. An outline of this pathway is shown in Figure 5.6.

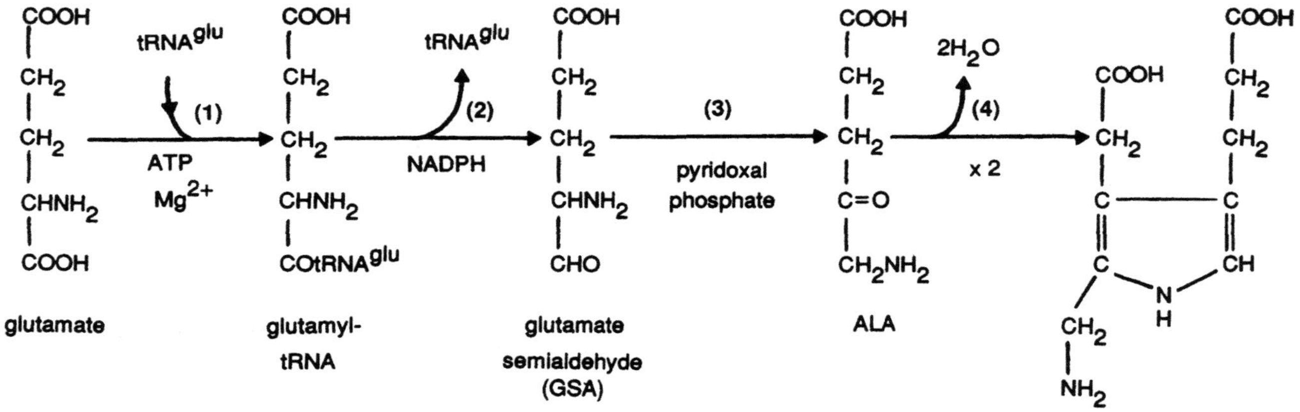

Figure 5.5. Proposed biosynthetic sequence of ALA synthesis from glutamate to the condensation of two ALA molecules, forming porphobilinogen. The enzymes catalyzing the steps in the pathway are (1) glutamyl-tRNA synthetase, (2) glutamyl-tRNA dehydrogenase, (3) ALA synthase (GSA aminotransferase), and (4) ALA dehydratase.

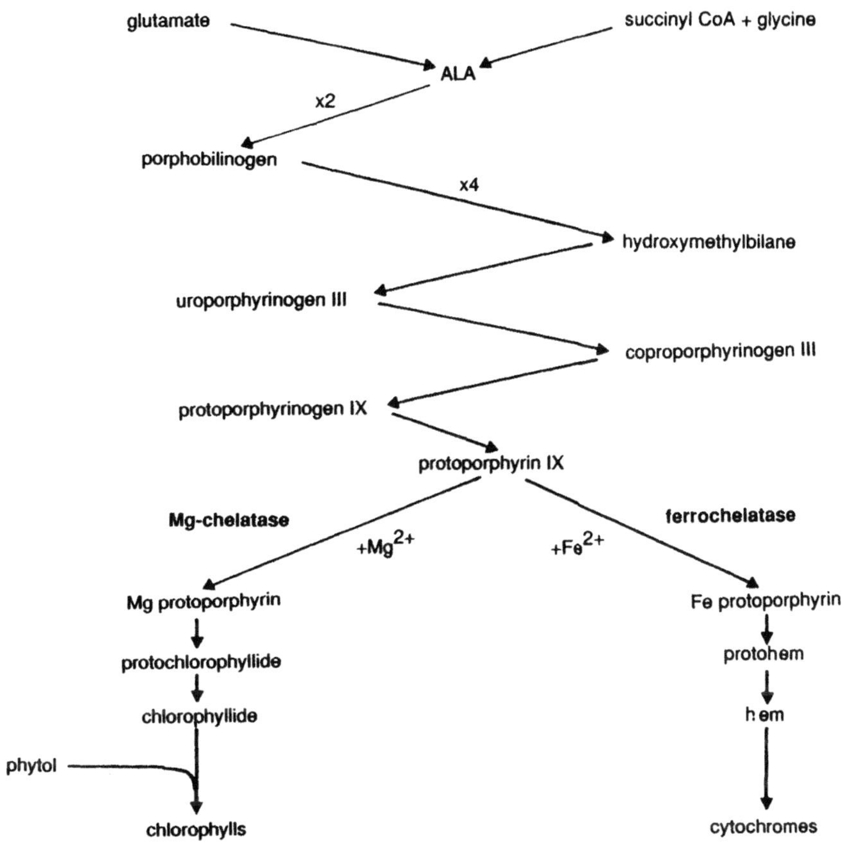

Figure 5.6. Outline of tetrapyrrole biosynthesis from ALA. For further detail, see Ref. 27.

5.2.1. Inhibition of ALA Synthesis

Gabaculine (3-amino-2,3-dihydrobenzoic acid) is a potent mechanism-based inhibitor of certain pyridoxal phosphate-linked enzymes.[27,29,30–34] In plants, it is a potent inhibitor of chlorophyll biosynthesis; the inhibition can be reversed by ALA.[27,35–38] Gabaculine and another transaminase inhibitor, 4-amino-5-fluoropentanoic acid (AFPA), are also reported to inhibit the biosynthesis of the linear tetrapyrrole chromophore of phytochrome.[27,39–41]

Leaves of gabaculine-treated plants appear pale yellow, and pigment analysis shows a specific effect only on chlorophyll and phytochrome.[27,38,40] Detailed analysis of photosynthetic pigments in gabaculine-treated plants shows a greater effect on chlorophyll biosynthesis in leaves which develop after treatment (Table 5.1). Leaf expansion continues in the absence of chlorophyll biosynthesis, and after seven days, once expansion is complete,

Table 5.1. Pigment Analysis of the Third Leaf of *Abutilon theophrasti* Two and Seven Days after Treatment of Two-Leaf Plants with Gabaculin

	μg Pigment/g F. wt.			
	Untreated		500 g/ha Gabaculin	
Pigment	2 days	7 days	2 days	7 days
α and β-Carotene	151.5	160.0	105.0	6.1
Lutein/zeaxanthin	290.6	265.6	314.6	43.0
Violaxanthin	117.0	116.9	32.9	19.0
9 *Z* Neoxanthin	55.9	53.8	33.4	3.8
All *E* neoxanthin	—	—	2.7	2.8
Antheraxanthin	35.1	28.8	42.4	12.9
Total carotenoids	650.2	625.1	530.5 (85)	87.5 (14)
Chlorophyll a	1945.4	1786.3	1025.2	72.8
Chlorophyll b	655.4	617.9	301.1	14.7
Total chlorophylls	2600.8	2404.2	1326.3 (55)	87.5 (3)

[a] Pallet and Barry, unpublished data.
[b] Analysis carried out as reported previously.[42,43]
[c] Figures in parentheses represent percentage of untreated.

the carotenoid profile resembles that of etiolated tissue rather than light-grown tissue.[43] This is an indication of the gabaculine effect on phytochrome synthesis, which is proposed to regulate carotenoid biosynthesis in the light.[44]

Gabaculine was first shown to inhibit γ-aminobutyric acid-α-ketoglutaric acid transaminase (GABA-T) by forming a covalent adduct with the pyridoxal phosphate cofactor of the enzyme.[27,32,34] Following binding, aromatization of the inhibitor occurs, with proton removal forming a tightly bound *m*-carboxyphenyl pyridoxamine.[27,34] In contrast, an enamine mechanism for enzyme inactivation by AFPA has been suggested.[45]

Knowledge of the inhibition of ALA synthesis by gabaculine provoked Flint to attempt to design some further inhibitors of ALA synthesis and evaluate their potential as herbicides. Some proved to be potent inhibitors and active herbicides.[36] To date, no commercialization of ALA synthesis inhibitors has occurred, but these inhibitors have been successfully used to probe and gain information on the biosynthesis of ALA and tetrapyrroles in plants.

Gabaculine-treated chlorophyll-deficient leaves respire normally, and there is normal cytochrome *c* oxidase in isolated mitochondria, providing evidence for the C_5 pathway in plastids and the Shemin pathway in mitochondria and indicating the specific effect of gabaculine on the plastid process.[35,38]

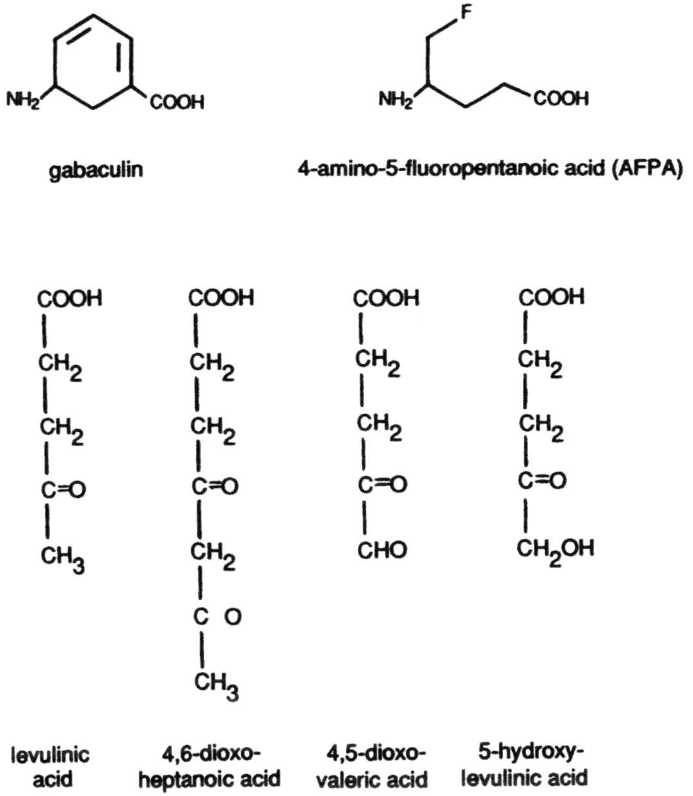

Figure 5.7. The structures of some ALA synthase and ALA dehydratase inhibitors.[27,28]

5.2.2. Inhibition of ALA Dehydratase

A number of ALA analogues are known as competitive inhibitors of ALA dehydratase (Figure 5.7). These contain the succinyl residue of ALA and are believed to form Schiff bases with the enzyme, causing ALA to accumulate in treated tissues.[28] Potentially, such a target for herbicides should be effective, resulting in inhibition of all tetrapyrrole synthesis. However, toxicological problems may be associated with the use of such compounds as herbicides owing to the universal occurrence of ALA dehydratase.

5.2.3. Inhibition of Protoporphyrinogen Oxidase

The *p*-nitrodiphenyl ethers (NDPEs) are a group of broad-spectrum herbicides, introduced in 1962, whose mechanism of action has been the subject of extensive research. Much of the earlier research has been recently reviewed.[46-50]

Table 5.2 shows examples of the major NDPE herbicides, and Figure 5.8 shows examples of other classes of miscellaneous herbicides whose mode of action has been considered identical to that of NDPEs based on their similar symptomology. The typical symptoms of these compounds are similar to those of paraquat. They cause a light-dependent, rapid chlorosis and necrosis due to a rapid induction of peroxidative degradation of cellular constituents, particularly membrane lipids.[46-50] Several ultrastructural studies have revealed that the NDPEs and other compounds (oxadiazon, LS 82-556, M&B 39279) show distinct differences in the sequence of ultrastructural damage compared to that induced by paraquat.[49,51-53]

Table 5.2. Examples of Major NDPE Herbicides

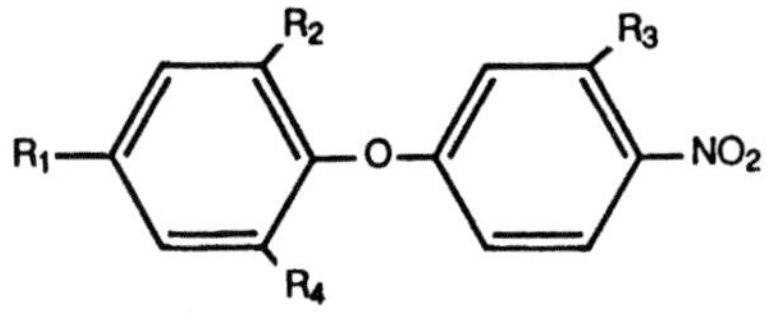

SUBSTITUENT				HERBICIDE NAME
R1	R2	R3	R4	
Cl	Cl	H	H	Nitrofen
Cl	Cl	H	Cl	Chlornitrofen (CNP)
CF_3	Cl	H	H	Nitrofluorfen
CF_3	Cl	COOH	H	Acifluorfen
CF_3	Cl	COONa	H	Acifluorfen-sodium
CF_3	Cl	$COOCH_3$	H	Acifluorfen-methyl
Cl	Cl	$COOCH_3$	H	Bifenox
Cl	Cl	OCH_3	H	Chlomethoxyfen
CF_3	Cl	OC_2H_5	H	Oxyfluorfen
CF_3	NO_2	H	H	Fluorodifen
CF_3	Cl	$CONHSO_2CH_3$	H	Fomesafen

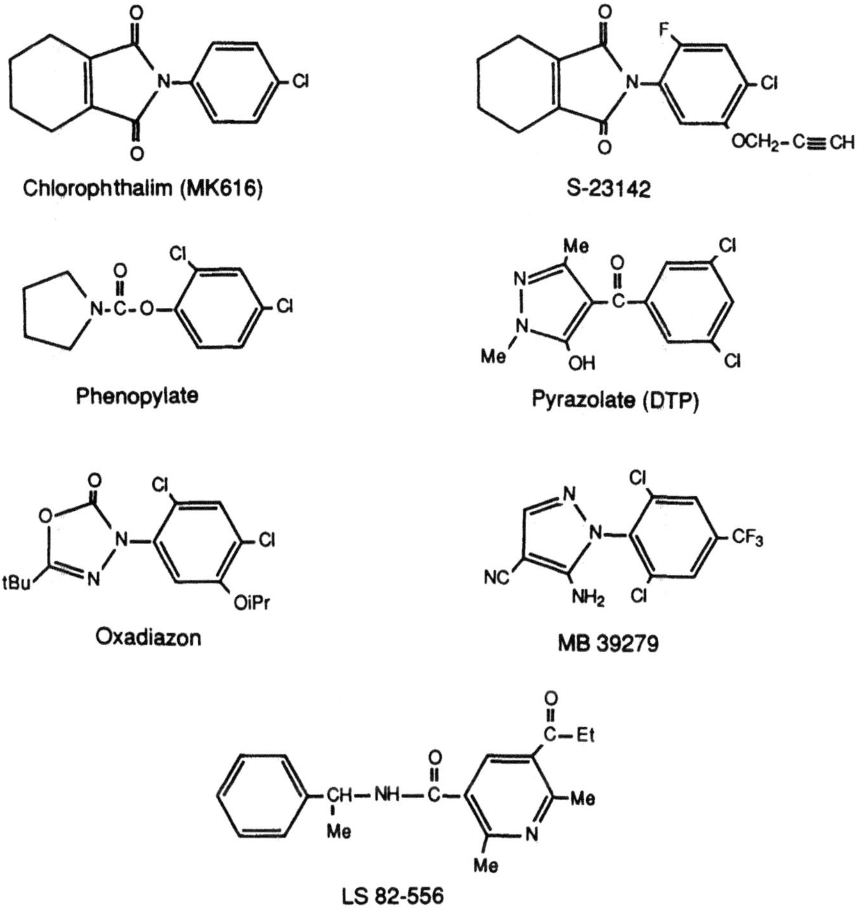

Figure 5.8. Examples of NDPE-type herbicides.

Several modes of action have been proposed for some individual NDPEs—for example, inhibition of PSII (nitrofen) and inhibition of carotenogenesis (non-*p*-nitro DPEs). However, such effects cannot be attributed to all NDPEs. The question of a role for photosynthesis in NDPE action is controversial. Oxyfluorfen has been proposed to be activated by probable reduction by ferredoxin to a *p*-nitro derivative which is proposed to be capable to inducing lipid peroxidation.[54] However, this mechanism cannot be applied to other NDPEs nor for acifluorfen methyl analogues where *p*-nitro is replaced by Cl or H.[55] In addition, NDPEs are active in chlorophyll-free and photosynthetically incapable plant tissues.[56–60] Direct irradiation of nitrofen and chlomethoxyfen in solution can lead to anionic radicals which may be capable of inducing lipid peroxidation.[61] However, this possible mechanism of activation cannot be applied to other NDPE and related compounds.

Carotenoids, or caroteno-proteins have been proposed as photoreceptors mediating NDPE phytotoxicity.[46,49,56] However, action spectra of acifluorfen methyl activity showed two peaks at 450 and 670 nm suggesting both carotenoid and chlorophyll involvement.[62] A study with S-23142 showed major peaks at 550 and 650 nm with only a minor peak at 450 nm indicating chlorophyll or a similar component as the major photoreceptor.[63]

In contrast to this, a study with nonchlorophyllous soybean cell cultures showed maximum injury at 350–450 nm with lesser activity at 450–750 nm.[64,65] Treated cells in darkness accumulated fluorescent pigments with spectral characteristics of protoporphyrin IX. The action spectrum of this tetrapyrrole was similar to that necessary for acifluorfen-induced damage and it was proposed that this herbicide caused abnormal accumulation of protoporphyrin IX which in turn induced lethal photooxidative reactions.[64,65] Accumulation of tetrapyrroles was subsequently reported in other plant tissues, i.e., acifluorfen[66,67]; oxyfluorfen, chlorophthalim, oxadiazon, and LS82-5566[68]; oxadiazon, fluorodifen, oxyfluorfen, nitrofen, and acifluorfen.[69,70] Inhibitors of tetrapyrrole biosynthesis such as gabaculin and 4,6-dioxoheptonoic acid (Figure 5.7) greatly reduced or prevented photooxidative damage of NDPEs and oxadiazon.[65,69,70] Therefore, after the efforts of several research groups and the reporting of a great deal of often conflicting data, a photoreceptor for NDPEs and related compounds was identified. The mechanism by which protoporphyrin IX accumulated was the subject of several proposals. Mg^{2+} chelatase was proposed as the target site.[70] Also, the conversion of protochlorophyllide to chlorophyllide (chllide) was proposed by Kouji *et al.*[66] An accumulation of chllide would feedback inhibit Mg^{2+} chelatase, and this, coupled to a stimulation of ALA synthesis by NDPE, was proposed to cause the rapid accumulation of protoporphyrin IX.[66]

However, Matringe, Scalla, and co-workers were the first to show that the primary target site was in fact protoporphyrinogen oxidase. Acifluorfen methyl was reported to strongly inhibit this enzyme with an LC_{50} of 4 nM and was thus some 5000 times more inhibitory than another protoporphyrinogen oxidase inhibitor, bilirubin.[71] Acifluorfen methyl only weakly inhibited Mg^{2+} or Fe^{2+} chelatases with an LC_{50} of 80–100 μM.[71] Matringe and co-workers have also reported that oxadiazon, M&B 39279, and LS 82-556 strongly inhibited protoporphyrinogen oxidase, and a scheme has been proposed to explain the accumulation of the product of the reaction rather than the substrate (Figure 5.9 and Ref. 72).

In treated cells, accumulated protoporphyrinogen IX is proposed to diffuse from its site of synthesis and is subjected to nonenzymatic oxidation to protoporphyrin IX. The abnormal cell location, beyond the reach of

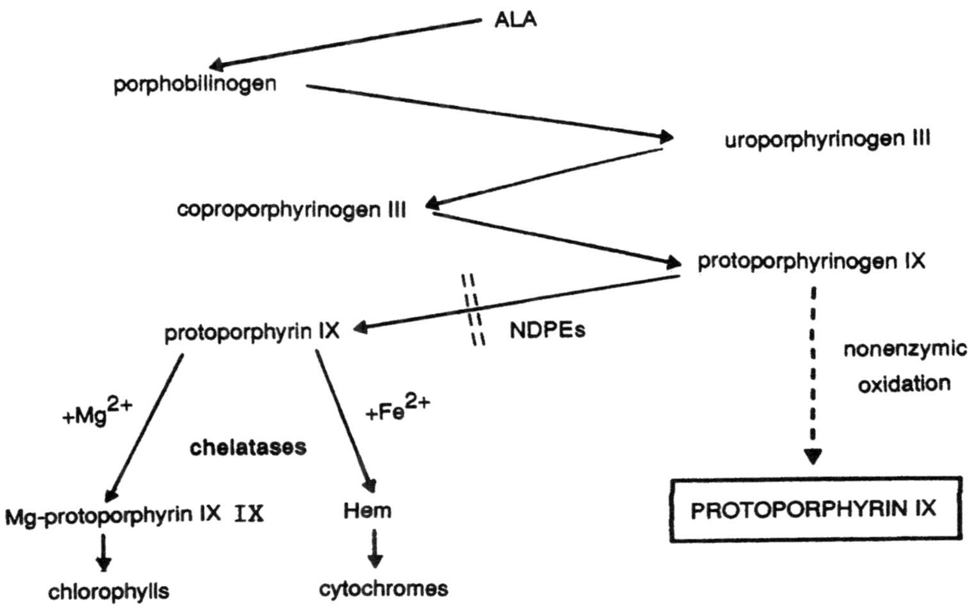

Figure 5.9. Postulated mechanism of NDPE-type herbicides (based on Ref. 72).

chelatases, leads to accumulation and gives rise to lethal photooxidative reactions due to the photodynamic nature of protoporphyrin IX.[72]

The mechanism of inhibition of protoporphyrinogen oxidase by NDPEs and related structures has yet to be elucidated. However, mammalian yeast and plant enzymes appear similarly sensitive,[71,72] indicating cause for concern with respect to toxicological problems.

5.2.4. Photodynamic Herbicides

Rebeiz and co-workers have attempted to develop the use of ALA and chlorophyll biosynthesis modulators as photodynamic herbicides.[73-75] This novel approach makes use of the fact that the biosynthesis of ALA appears to be the rate-limiting step in tetrapyrrole biosynthesis. The application of ALA circumvents this rate-limiting step, and treated plant tissues accumulate tetrapyrroles, which can act as photodynamic herbicides and induce phytotoxicity similar to that now established for NDPEs. The susceptibility of treated plants is dependent on the nature of the tetrapyrroles that accumulate.[73-75]

The dual application of ALA with one or more chlorophyll biosynthesis modulators can result in selective herbicide action due to the four different greening patterns within the multibranched chlorophyll pathway that occur in plants.[73-76] The chlorophyll biosynthesis modulators are classified into three major groups depending on their mode of action: enhancers of ALA conversion to tetrapyrroles (e.g., 2-pyridine aldehyde, picolinic acid, 4,4′-dipyridyl, phenanthridine, 2,2′-pyridylamine); inducers of tetrapyrrole accumulation (e.g., 1,10-phenanthroline, 2,2′-dipyridyl); and inhibitors of monovinyl protochlorophyllide accumulation (1,7- and 4,7-phenanthroline, 2,3-dipyridyl, 2,4-dipyridyl).[74,75]

5.3. ISOPRENOID BIOSYNTHESIS

Figure 5.10 represents an outline of the isoprenoid pathway in plants and highlights the importance of this pathway in providing molecules essential for the metabolism of plant cells. Many of the isoprenoids (terpenoids) derived from the pathway are unique to plants, and therefore their biosynthesis would be a good herbicide target. Indeed, this has already been seen in Chapter 4 with the inhibition of carotenoid biosynthesis from phytoene by several classes of successful herbicides. In this section, the evidence for herbicide inhibition of other sites within the isoprenoid pathway is reviewed.

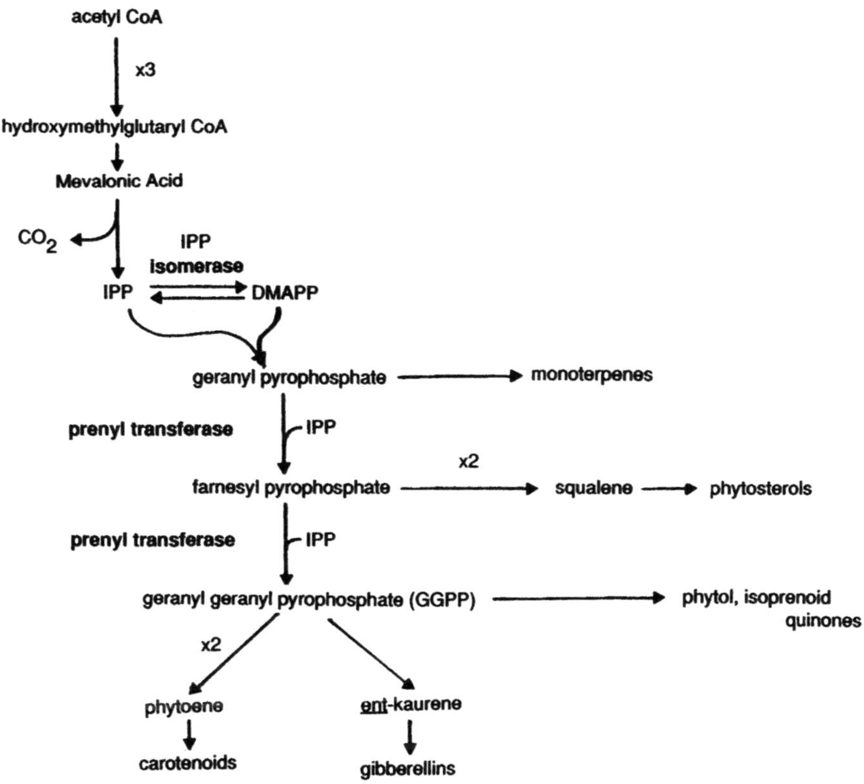

Figure 5.10. Outline of the isoprenoid pathway. IPP, Isopentenyl pyrophosphate; DMAPP, dimethylallyl pyrophosphate.

5.3.1. Inhibition of Isopentenyl Pyrophosphate Isomerase and Prenyl Transferase

Dimethazone or clomazone, 2-[(2-chlorophenyl)methyl]-4,4-dimethyl-3-isoxazolidinone, was introduced in 1984 as a selective preemergence/

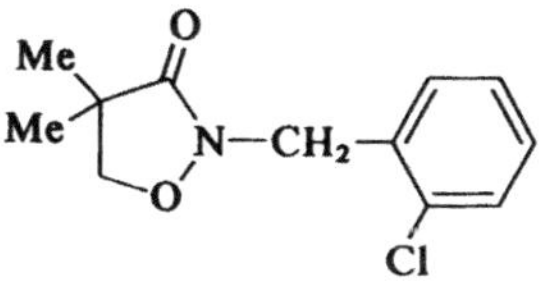

Dimethazone or clomazone

preplant-incorporated herbicide which causes characteristic bleaching symptoms in susceptible plants.[77,78] These symptoms provoked comparative mode of action studies with the phytoene dehydrogenase inhibitor norflurazon (see Chapter 4). However, although both compounds caused similar reductions in both chlorophylls and carotenoids, phytoene did not accumulate in dimethazone-treated leaves.[79,80] In addition, dimethazone did not inhibit phytoene and δ-carotene dehydrogenases or lycopene cyclase in a cell-free carotenogenic *Aphanocapsa* system.[81]

Retardation of the Shibata shift (a measure of the conversion of chlorophyllide to chlorophyll) suggested inhibition of phytol biosynthesis.[78,80] Gibberellin biosynthesis was also suggested as a target site for dimethazone as etiolated seedling growth was inhibited,[80] growth and bleaching was not affected in a gibberellin-deficient maize mutant,[82] and growth inhibition of pea seedlings was reversed by GA_3.[83]

Thus, inhibition of isoprenoid biosynthesis at an early stage, that is, prior to geranylgeranyl pyrophosphate (GGPP) synthesis, seemed a likely target.[83] Dimethazone was shown to affect the conversion of isopentenyl pyrophosphate (IPP) to GGPP (Figure 5.10), and it is proposed to inhibit IPP isomerase and/or prenyl transferase.[81] Interestingly, it appears that polymerization of IPP in the plastid is more susceptible than the extraplastidic process leading to sterols.[81] No inhibition by dimethazone of IPP isomerase, prenyl transferase, and phytoene synthase was detected in isolated plastids from mustard seedlings or in chromoplasts from daffodil flowers.[200] Therefore, the mode of action of this herbicide remains uncertain.

5.3.2. Phytosterol Biosynthesis

Ergosterol is the major sterol in fungi and plays essential roles in membrane structure and function, and its depletion is the target site of fungicides.[5] These inhibit the cytochrome P450-dependent monooxygenase responsible for the removal of the C-14α-methyl group in ergosterol biosynthesis.[5,84,85] γ-Ketotriazole has herbicidal properties and is an analogue of the triazole fungicides and also of the growth retardant paclobutrazol (Figure 5.11 and Ref. 85). The herbicidal properties are reduced shoot height with necrosis and chlorosis of leaf tissue.[85] As these symptoms develop, normal sterols (e.g., sitosterol, stigmasterol, campesterol) decrease and abnormal C-14α-sterols (e.g., obtusifoliol, 14α-methyl-Δ^8-cholesterol, 14α-methyl-Δ^8-ergosterol) accumulate, indicating inhibition of the C-14α-demethylase which catalyzes the C-14α-demethylation of obtusifoliol (Ref. 85).

The γ-ketotriazole did not inhibit C-14α-demethylation in fungi or *ent*-kaurene oxidase, the cytochrome P450 monooxygenase target site of

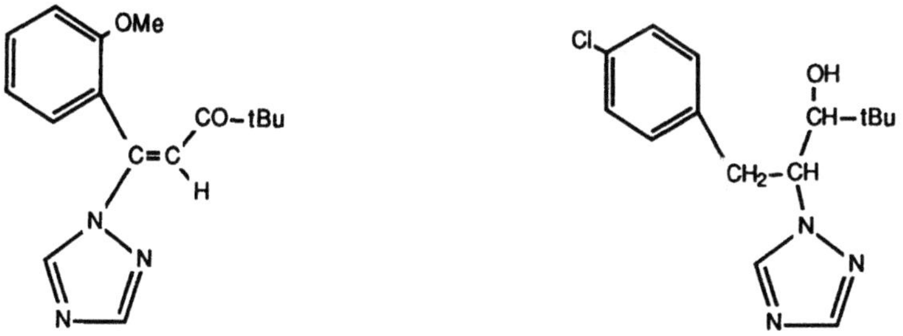

γ-keto triazole Paclobutrazol

Figure 5.11. The structures of γ-ketotriazole and paclobutrazol.

paclobutrazol.[85] Paclobutrazol did inhibit sterol biosynthesis in both plants and fungi as did the triazole fungicide triadimenol.[86,87] Therefore, it appears that there are differences in cytochrome P450-dependent monooxygenases[86,88] and that these enzymes can be exploited as a herbicide target site as C-14α-demethyl sterols do not adequately support efficient metabolic function.[86,87]

5.3.3. Gibberellin Biosynthesis

Gibberellin biosynthesis is a proven target site for growth retardants such as chlormequat chloride (CCC), ancymidol, chlorphonium chloride, and daminozide.[5,89] However, gibberellin biosynthesis has also been proposed as a target site for the members of the thiocarbamate and chloroacetamide classes of herbicides (Figure 5.12).

These two groups have yet to have a specfic primary target site ascribed to them; however, both types of herbicides are reported to inhibit lipid biosynthesis (see Chapter 3). Gibberellin biosynthesis is inhibited by the thiocarbamate herbicide EPTC at a step before *ent*-kaurene.[90,91] In a cell-free preparation from *Sorghum* seedlings, the conversion of [^{14}C]mevalonic acid (MVA) to kaurene was reduced and [^{14}C]-GGPP accumulated.

Recently, EPTC has been shown to inhibit the conversion of acetate and pyruvate to acetyl-CoA.[92] However, this cannot directly explain the inhibition of *ent*-kaurene synthesis from MVA. It seems likely that the thiocarbamate effect on gibberellin biosynthesis is a secondary target. This was the conclusion from a study with diallate, which inhibited *ent*-kaurene synthetase only at high concentrations.[93] Gibberellins may have a role in

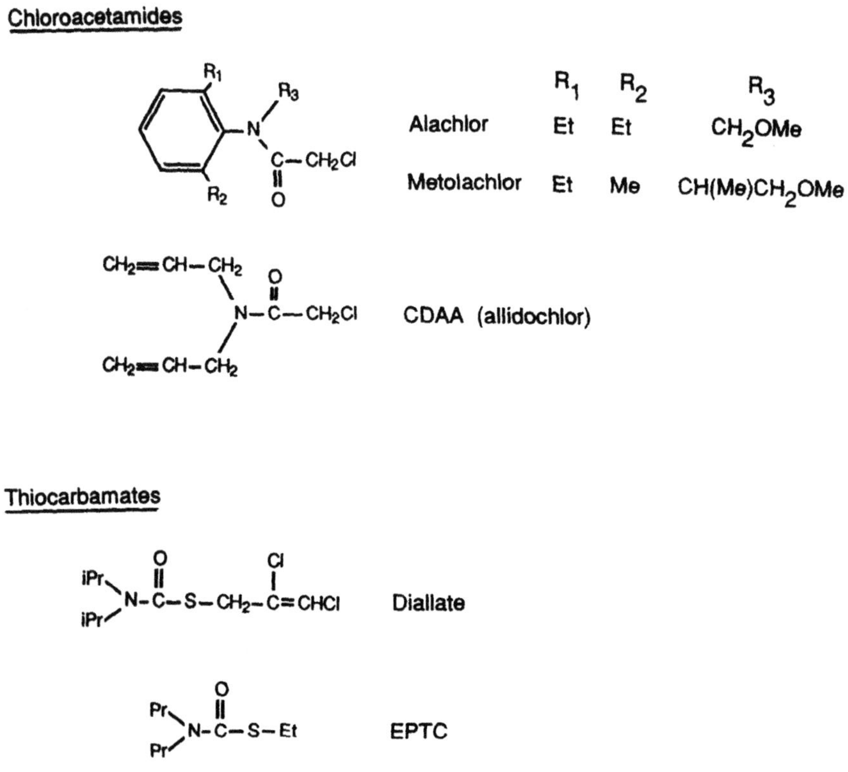

Figure 5.12. Examples of some chloroacetamide and thiocarbamate herbicides.

the overall action of thiocarbamates as gibberellin-deficient mutants of maize are more sensitive to EPTC than normal maize.[94]

There are also several reports implicating gibberellin biosynthesis as a target for chloroacetamides.[95-98] Gibberellic acid reversed growth inhibition caused by metolachlor; both metolachlor and alachlor inhibited gibberellin biosynthesis and CDAA inhibited *ent*-kaurene oxidation in a cell-free preparation from etiolated *Sorghum* seedlings. The oxidation of *ent*-kaurenal to *ent*-kaurenoic acid was the major site of inhibition, with possibly the oxidation of *ent*-kaurenol to *ent*-kaurenal as a secondary site (Figure 5.13) However, as with the thiocarbamates, this effect on gibberellins is unlikely to be the primary target site for chloroacetamides. Recently, inhibition of acetate incorporation into acyl lipids was reported as the process

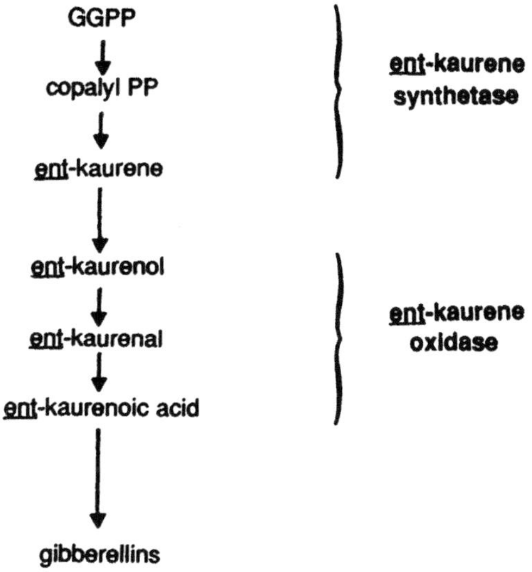

Figure 5.13. A scheme outlining the conversion of geranylgeranyl pyrophosphate (GGPP) to *ent*-kaurene. *ent*-Kaurene oxidase is the target site of growth retardants such as paclobutrazol and ancymidol.[5,87,88]

most sensitive to chloroacetamides in *Scenedesmus.*[99] It is proposed that these herbicides inhibit at the level of palmitate elongation and desaturation of oleate.[100] It is conceivable that the basis of this and other effects of the chloroacetamides is an interference with coenzyme A (CoA) synthesis or function, possibly via binding to its thiol group.[101]. Chloroacetamides are reported to inhibit the activation of *p*-coumaric acid to *p*-coumaryl-CoA in lignin and anthocyanin biosynthesis.[102–104]

5.3.4. Phytol Biosynthesis

Any inhibition of the isoprenoid pathway before GGPP (Figure 5.10), as occurs, for example, with dimethazone, will reduce phytol levels; however, the conversion of GGPP to phytol has also been implicated as a specific site for herbicides. EPTC was reported to inhibit the reduction of the side chain of chlorophyll to phytol.[105] Aminotriazole was also proposed to specifically inhibit the reactions involved in reducing the geranylgeraniol side chain. Treated wheat seedlings contained chlorophyll with geranylgeraniol or dihydrogeranylgeraniol replacing phytol.[106] However, this is almost certainly a secondary site for this herbicide, its primary target being carotenoid biosynthesis (Ref. 107 and Chapter 4).

5.4. MICROTUBULE ASSEMBLY AND FUNCTION

5.4.1. Formation and Function of Microtubules

The structure, properties, and function of microtubules have been extensively reviewed in recent years (see, e.g., Refs. 4, 107, and 109), and only a brief summary will be given here.

Microtubules (MTs) are filamentous subcellular structures composed mainly of the heterodimeric protein tubulin ($M_r \sim 100{,}000$) organized into specific functional arrays in higher plants: interphase array MTs (largely cortical MTs), preprophase band (PPB) MTs, spindle MTs, and cytokinetic MTs (cell plate MTs). They are organized into a complex microtubule cycle and are involved with structural differentiation (Figure 5.14). Specifically, they participate in several processes during the cell cycle, including chromosome migration, vesicle transport, and orientation of cellulose microfibril deposition in the developing cell wall.[4,108,109] The four types of MT arrays form at specific times during the cell cycle (Figure 5.14). These control the polymerization and depolymerization of MTs at different locations in the cell. There appears to be a pool of free α- and β-tubulin within the cytoplasm which is reversibly polymerized to MTs under the control of microtubule

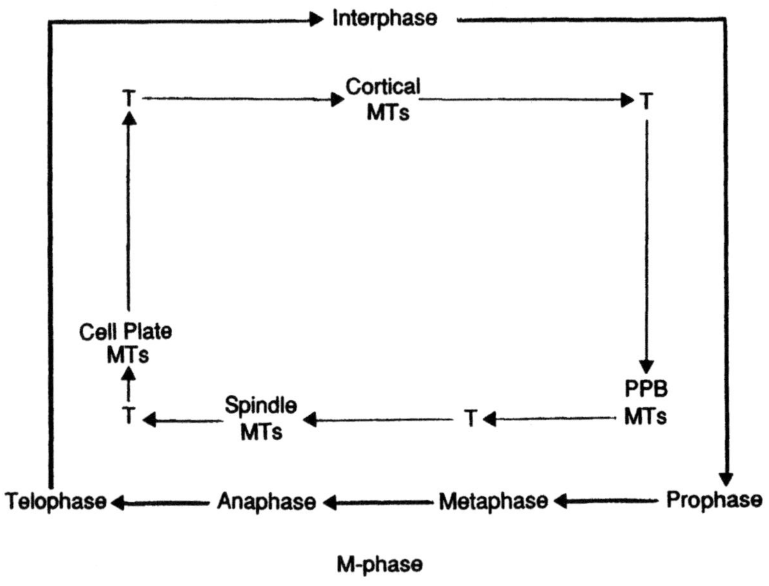

Figure 5.14. Scheme of the microtubule (MT) cycle within the cell cycle. The mitotic (M) phase of the cell cycle is exaggerated with respect to the length of the interphase. The polymerization of tubulin (T) is under the control of specific microtubule organizing centers (MTOCs) at each stage of the MT cycle.[4,108,109]

organizing centers (MTOCs). The proportion of tubulin assembled as MTs can vary from 0 to 90% throughout the cell cycle.[108]

In addition to the specific MTOCs controlling each type of MT polymerization/depolymerization, other factors have been shown to influence this process:

(a) the cations Ca^{2+} and Mg^{2+}. *In vitro* Ca^{2+} induces depolymerization whereas Mg^{2+} induces MT assembly;
(b) calmodulin, the Ca^{2+}-binding protein;
(c) microtubule-associated proteins (MAPs), e.g., *tau*-factor (58–65 kDa) and high-molecular-weight proteins (180–345 kDa) which bind to tubulin and stimulate *in vitro* polymerization[108];
(d) GTP; at least one GTP binding site per tubulin molecule and GTP binding is necessary for polymerization.[4,108,109]

5.4.2. Herbicide Interference with Microtubule Structure and Function

It has been known for many years that herbicides belonging to the *N*-phenylcarbamate and dinitroaniline classes of herbicides are antimicrotubule agents (see, e.g., Refs. 3–5, 108, and 109). Figure 5.15 shows representatives of these two classes and other anti-MT herbicides. The literature on much of the early research on the mode of action of these compounds has been the subject of extensive reviews (see, e.g., Refs. 3–5).

The intention here is to summarize just the major findings of this research that provide evidence concerning the target site of these herbicides.

5.4.2.1. Morphological and Cytological Effects

Dinitroanilines produce gross morphological abnormalities in plants, particularly in the regions of high meristematic activity, for example, root tip swelling. MTs disappear and mitosis is disrupted during metaphase with chromosome aberration, polyploidization, and bi- and polynuclear cells.[4,110,111] Fedtke has categorized the specific morphological and cytological effects of various dinitroanilines, *N*-phenylcarbamates, and other anti-MT herbicides.[4] All these herbicides show similar morphological effects. However, the *N*-phenylcarbamates do appear to differ from the dinitroanilines; for example, Bartels and Hilton reported that, in propham-treated wheat and oats, MTs were still evident but appeared disorganized.[112] These authors also showed that pronamide caused loss of MTs in treated tissues and proposed a similar mode of action to that of dinitroanilines. Amiprophos-methyl also appears to cause similar morphological and cytological effects to those produced by dinitroanilines.[109,113] Chlorthal-dimethyl caused disruption of mitosis after prophase, with metaphase, anaphase, and

N-phenycarbamates

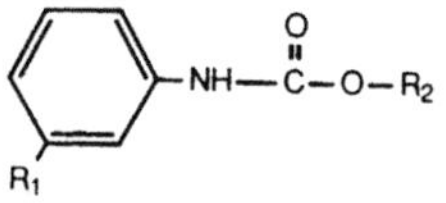

	R_1	R_2
Barban	Cl	$CH_2C{=}CCH_2Cl$
Carbetamide	H	CH(Me)CONHEt
Chlorpropham	Cl	iPr
Propham	H	iPr

Dinitroanilines

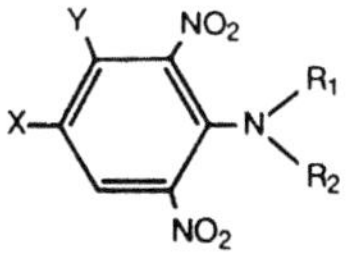

	X	Y	R_1	R_2
Dinitramine	CF_3	NH_2	Et	Et
Oryzalin	SO_2NH_2	H	Pr	Pr
Pendimethalin	Me	Me	H	$CH(Et)_2$
Trifluralin	CF_3	H	Pr	Pr

Other Compounds

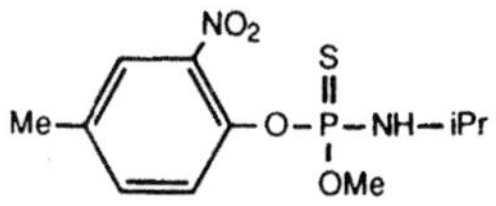

Amiprophos-methyl

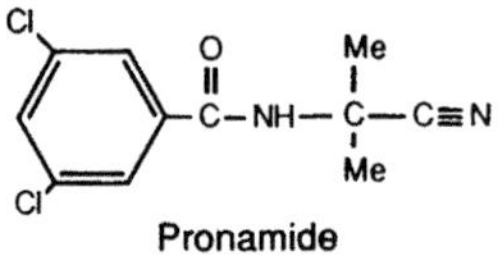

Pronamide
(Propyzamide)

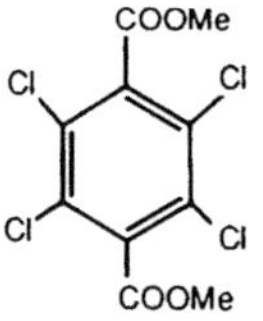

Chlorthal-dimethyl

Figure 5.15. Examples of herbicides which interfere with MT assembly or function.

telophase figures absent and condensed prophase figures predominant in treated oat roots.[114]

5.4.2.2. Mechanism of Herbicidal Anti-MT Activity

Cytological effects of dinitroanilines, pronamide, and amiprophos-methyl, manifested by an absence of MTs in treated tissues, suggest inhibition of tubulin synthesis or tubulin polymerization. Several studies with isolated pig brain tubulin have shown that these herbicides do not prevent tubulin polymerization (see, e.g., Ref. 112). It was concluded that trifluralin, oryzalin, pronamide, and propham did not act in the same manner as the anti-MT drug colchicine, which does bind to tubulin and prevent its polymerization. However, more recently, *in vitro* studies with plant and algal tubulin have shown differences between animal and plant tubulin with respect to the amino acid sequence of α-tubulin and in the binding of colchicine. Colchicine has only a low affinity for plant tubulin.[109,115,116]

Hess and Bayer were the first to report appreciable binding of a dinitroaniline herbicide, [^{14}C]trifluralin, to *Chlamydomonas* flagella tubulin.[117] Also [^{14}C]oryzalin was shown to bind to this flagella tubulin, and the complex did not polymerize.[111] More recently, oryzalin and amiprophos-methyl have been shown to bind to and prevent polymerization of tubulin isolated from *Rosa* cell cultures.[119] There is strong evidence suggesting that dinitroanilines and amiprophos-methyl directly disrupt MTs in plant cells although the mechanism(s) is still to be elucidated.[110] Corroborative evidence for this has arisen from the recent studies with an *Eleucine* biotype resistant to dinitroanilines. It is proposed that resistance is due to an alteration in α-tubulin structure.[118,119] Trifluralin-resistant *Eleucine* is also resistant to other dinitroanilines and amiprophos-methyl, confirming a similar target site for the latter.[118,119] The resistant *Eleucine* is classed as supersensitive to chlorpropham and propham but possessed normal sensitivity to pronamide.[119] This lack of cross-resistance to pronamide raises questions about previous assumptions of a common target site for dinitroanilines and this herbicide. Morphological and cytological effects are similar,[112,119] but electron microscopic and immunofluorescence studies reveal that pronamide, unlike trifluralin, causes cells to have shorter MTs. These are located only at the kinetochore region of the chromosomes, where they fail to allow mitosis to proceed normally and cause polymorphic nuclei.[120] It is proposed that the target site for pronamide is either a destabilization of MT production or a limitation of supply of protein(s) critical for MT development.

The presence of disorientated MTs in tissues treated with propham led to the proposal that its target site was located at the level of MTOCs.[4,112] The increased sensitivity to *N*-phenylcarbamates exhibited by dinitroaniline-resistant *Eleucine*[119] suggests different target sites for the two herbicide classes. This was the concensus view in previous reviews,[3-5] and little further information has since been published to provide details on any mechanism of *N*-phenylcarbamate action.

Chlorthal-methyl has been suggested as a disrupter of MT organization although no information on any mechanism has been presented.[114]

Therefore, it appears that the anti-MT herbicides fall into three classes:

(a) those that prevent polymerization of tubulin to MT by binding to the tubulin, for example, dinitroanilines and amiprophos-methyl;

(b) those that interfere with the MTOC, for example, *N*-phenylcarbamates and chlorthal-methyl;
(c) pronamide, which may destabilize MT production or decrease protein supply for MT development.

However, an indirect effect on MTs has been proposed for dinitroanilines and amiprophos-methyl mediated by effects on Ca^{2+} levels.[121] Dinitroanilines were shown to inhibit Ca^{2+} uptake by mitochondria, thus possibly increasing cytoplasmic Ca^{2+}, which would inhibit tubulin polymerization.[122] Propham and chlorpropham also inhibited Ca^{2+} uptake by mitochondria, but less so than the dinitroanilines.[122] Therefore, an indirect effect on MT asssembly and function cannot be excluded for the *N*-phenylcarbamates. A possible mechanism by which the *N*-phenylcarbamates could affect Ca^{2+} levels has already been discussed—namely, a direct or indirect effect on electron transport (Section 5.1). There are also several reports that dinitroanilines can inhibit or uncouple mitochondrial electron transport.[4,5,121]

In conclusion, it is unlikely that the mechanism of action of all the anti-MT herbicides is via a direct or indirect effect on Ca^{2+} levels. Similar morphological and cytological effects would be expected for all the anti-MT herbicides if they were mediated by an effect on Ca^{2+} levels. In addition, the importance of Ca^{2+} in metabolic control[123] would mean that any inhibition in Ca^{2+} compartmentalization would be manifested in many ways, and not specifically on MTs. A specific direct effect by these herbicides on MT structure and function is likely, the mechanism(s) of which is still to be elucidated. However, effects on processes which influence Ca^{2+} levels may contribute to the overall action.

5.5. CELLULOSE BIOSYNTHESIS

Cellulose is a homopolymer of β-(1 → 4)-linked glucose that plays a crucial structural role in plant cell walls.[124] This essential function of cellulose and its specificity to plants makes it a desirable herbicide target site.

Two herbicides, dichlobenil and chlorthiamid, which is degraded to dichlobenil in plants and soil, are thought to have cellulose biosynthesis as the primary target site[125] (Figure 5.16). Dichlobenil is a potent inhibitor of germination and of actively dividing meristematic cells.[126] Respiratory uncoupling has been proposed as a mode of action for dichlobenil.[5,127] Although dichlobenil itself only weakly uncouples both respiratory and photosynthetic electron transport, its hydroxylation products are potent uncouplers by virtue of their labile phenolic hydrogen (see Section 5.1 and

Figure 5.16. The structures of dichlobenil and chlorthiamid.

Refs. 125 and 127). However, although 3- and 4-hydroxyl-dichlobenil are significant metabolites in plants,[126] they are likely to be readily conjugated, and the respiratory uncoupling can be considered a secondary site of action.

The inhibition of [^{14}C]glucose incorporation into cellulose in azuki bean epicotyl sections was the first indication that polymer synthesis was a target site for dichlobenil.[128] Concentrations of dichlobenil below 1 μM caused specific inhibition of [^{14}C]glucose incorporation into cellulose. This effect was considered to be a direct action on polymer biosynthesis rather than an indirect one occurring via a disruption of MTs involved in cell wall synthesis.[128]

Further support for cell wall biosynthesis as a target site was provided by Umetsu *et al.*, who reported abnormal swelling of cultured soybean cells treated with dichlobenil.[129] Dichlobenil was also shown to be an effective and reversible inhibitor of cell wall formation in tobacco protoplasts.[130] It was considerably more potent than other cell wall inhibitors, such as coumarin, and it caused the formation of multinucleate protoplasts.[130] Inhibition of cellulose biosynthesis prevented cell wall formation and cytokinesis but nuclear division; that is, mitosis was unaffected. Further studies with tobacco protoplasts confirmed the specific and potent inhibition associated with the cellulose-synthesizing complex at the plasma membrane.[131]

Dichlobenil was reported to be more than 100 times more inhibitory than coumarin toward cellulose biosynthesis in two further cell wall synthesizing systems. In cotton fibers growing from *in vitro* cultured ovules, dichlobenil at a concentration of 10 μM caused 80% inhibition of cellulose biosynthesis within 2–3 h; the inhibition was reversed by removal of the herbicide.[132] Micromolar concentrations were effective within 10 min of incubation, indicating that cellulose biosynthesis is a likely primary target for this herbicide. The action of dichlobenil is more specific for the β-(1 → 4)-glucan synthesis of cellulose rather than noncellulosic β-(1 → 3)-glucan synthesis, which is less sensitive.[132] In tobacco cells, specific inhibition of cellulose biosynthesis was apparent at 0.5–5.0 μM dichlobenil; however, no inhibition of *in vitro* β-glucan synthesis in particulate membrane preparations was found.[132] This lack of *in vitro* effect may be explained by a disruption of the β-(1 → 4)-glucan-synthesizing complex during membrane isolation.[133]

More recently, with the use of [^{3}H]-2,6-dichlorophenylazide (DCPA), a photoaffinity-labeled analogue of dichlobenil, which also inhibits cellulose biosynthesis, an 18-kDa polypeptide has been identified as a possible receptor site for the herbicidal inhibition of cellulose biosynthesis.[134–136] This 18-kDa polypeptide is only loosely membrane associated, and it is considered unlikely to be the catalytic polypeptide of cellulose synthetase. It has been suggested that it may be a regulatory protein for β-glucan synthesis, and it has been proposed that dichlobenil binding leads to dissociation or a conformational change, thereby inhibiting cellulose biosynthesis.[134,135] More recently, DCPA was found to act as a reversible inhibitor of dichlobenil binding and cellulose biosynthesis in the dark; upon UV irradiation, the inhibition became irreversible. Furthermore, the biosynthesis of the 18-kDa polypeptide was correlated with the onset of cellulose biosynthesis in cotton fibers.[136]

Therefore, it appears that the target site for dichlobenil has been established as an 18-kDa polypeptide associated with cellulose biosynthesis. However, the exact function of this polypeptide and the mechanism of dichlobenil inhibition remain to be elucidated.

5.6. PROTEIN AND NUCLEIC ACID BIOSYNTHESIS

Inhibition of protein, RNA, and DNA synthesis by herbicides has been the subject of numerous reports and reviews (see, e.g., Refs. 3, 4, and 137–139). Inhibition of the biosynthesis of any of these macromolecules would lead to an effective and abrupt cessation of plant growth. However, with very few exceptions, the reports on herbicidal interference with protein and nucleic acid synthesis reflect secondary effects resulting from an effect at a target site elsewhere in the cell. For example, any interruption of the production of metabolic energy, as a result of respiratory uncoupling and/or inhibition, would lead to a rapid cessation of macromolecular synthesis. In several surveys of herbicide effects on ^{14}C-labeled precursor incorporation into protein and RNA in excised plant tissues and isolated mesophyll cells, dinitrophenols and hydroxybenzonitriles were the most potent inhibitors (Section 5.1; see, e.g., Refs. 138 and 140).

There are several examples of newly discovered herbicides being categorized as protein or nucleic acid inhibitors. In initial mode of action studies following discovery and commercialization, these processes have been identified as being the most sensitive with several herbicides whose primary target sites were not immediately obvious. For example, DNA synthesis was the most sensitive process affected by DPX-4189 (chlorsulfuron),[141] and protein and RNA synthesis were the most sensitive processes affected by HOE 39866 (glufosinate or phosphinothricin).[142] However, the primary target sites of these herbicides have been subsequently established as acetolactate synthesis and glutamine synthetase, respectively (Chapter 2).

However, there does appear to be at least one herbicide whose target site appears to be an inhibition of protein biosynthesis. The *R*-isomer of 2-(4-methyl-2,6-dinitroanilino)-*N*-methyl propionamide (MDMP; Figure 5.17) is a noncommercialized herbicide which has been shown to exert a direct effect on protein synthesis.[143] Protein synthesis was inhibited at the initiation stage in an *in vitro* translation system.[144] It was subsequently reported that MDMP interfered with the attachment of ribosomes to messenger RNA.[145] MDMP was also shown to specifically affect the larger eucaryotic (i.e., cytoplasmic) ribosomes and had no effect on the smaller procaryotic (i.e., bacterial and plastid) ribosomes.[146,147]

There are several examples of herbicides whose mode of action, and therefore primary target site, is unknown, but to which protein and nucleic acid synthesis has been identified as being sensitive. Endothal (Figure 5.17) at 11 μM caused a 63% inhibition of leucine incorporation in excised *Sesbania* segments within hours.[140] The target site for the translocated activity of difenzoquat (Figure 5.3) remains to be elucidated. However, in an investigation into the mechanism of crop tolerance of wheat cultivars, a greater sensitivity of thymidine incorporation into DNA was reported in apical meristem sections excised from the more susceptible cultivars.[148,149] This differential effect on DNA synthesis occurred at relatively high concentrations; however, the inhibition was apparent within 15 min.[148] It was concluded that, as this effect on DNA synthesis was the only identified difference in the activity of difenzoquat in tolerant and susceptible cultivars, it contributed to the basis of susceptibility.[148] However, there is no evidence that this is the target site in susceptible species, for example, wild oats, and

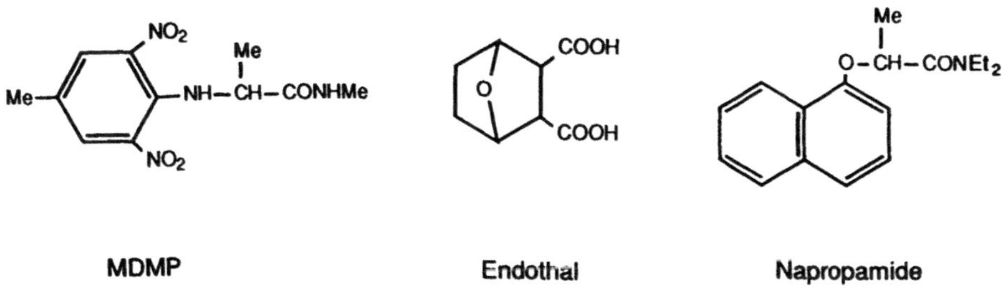

Figure 5.17. Herbicides which have been proposed to directly or indirectly inhibit protein synthesis.

the target site remains to be elucidated. In a recent investigation into the metabolic effects of napropamide (Figure 5.17), an inhibition of the cell cycle was proposed.[150] Cells were blocked at both the G_1 and G_2 stages of the cell cycle, and an inhibition of the synthesis or activity of cell-cycle-specific proteins was suggested.[150]

5.7. FOLIC ACID BIOSYNTHESIS

Folic acid plays the central role in one-carbon metabolism and is directly or indirectly involved in the biosynthesis of nucleic acid bases and amino acids.[151] In plants and microbes, the various folate derivatives involved in one-carbon metabolism originate from dihydropterin derivatives, via the enzymes 7,8-dihydropteroate and dihydrofolate synthase (Figure 5.18). 7,8-Dihydrofolate then undergoes reduction to 5,6,7,8-tetrahydrofolate (THF), which can become substituted with formyl, methylene, or methyl groups for utilization in one-carbon donor reactions.[151]

Asulam (Figure 5.19) is a benzenesulfonyl carbamate, and its mode of action was originally proposed to be similar to that of *N*-phenylcarbamates.[152] Onion root tip cells treated with asulam become arrested in metaphase, suggesting interference with MT assembly or function (Section 5.4 and Ref. 152).

However, asulam bears a structural resemblance to 4-aminobenzoic acid (Figure 5.19), one of the substrates of 7,8-dihydropteroate synthase. There are several reports that asulam interferes with the synthesis of folate derivatives in plants and that 4-aminobenzoic acid or folic acid can partially or completely overcome growth inhibition of plants or cell cultures treated with asulam.[153-156] Asulam is structurally related to sulfanilamide (Figure 5.19) and the sulfonamide drugs, which are known to inhibit 7,8-dihydropteroate synthase.[157,158] In cell-free extracts of wheat seedlings, asulam and sulfanilamide gave similar inhibition of 7,8-dihydropteroate synthesis.[159] The reversal of asulam-induced growth inhibition by 4-aminobenzoic acid indicated competitive inhibition of the enzyme.[153] Depletion of folate derivatives following asulam treatment is further evidence of a direct inhibition of folate synthesis.[159]

Therefore, it seems certain that the primary target site for asulam is an inhibition of 7,8-dihydropteroate synthesis. The mitotic interference[152] and also decreased protein and RNA synthesis[160] are a consequence of inhibited 7,8-dihydropteroate synthesis, which would reduce or prevent amino acid and nucleotide production. Support for this arises from a report that DCDT (Figure 5.19), an inhibitor of dihydrofolate reductase (Figure 5.18), causes mitotic interference.[161]

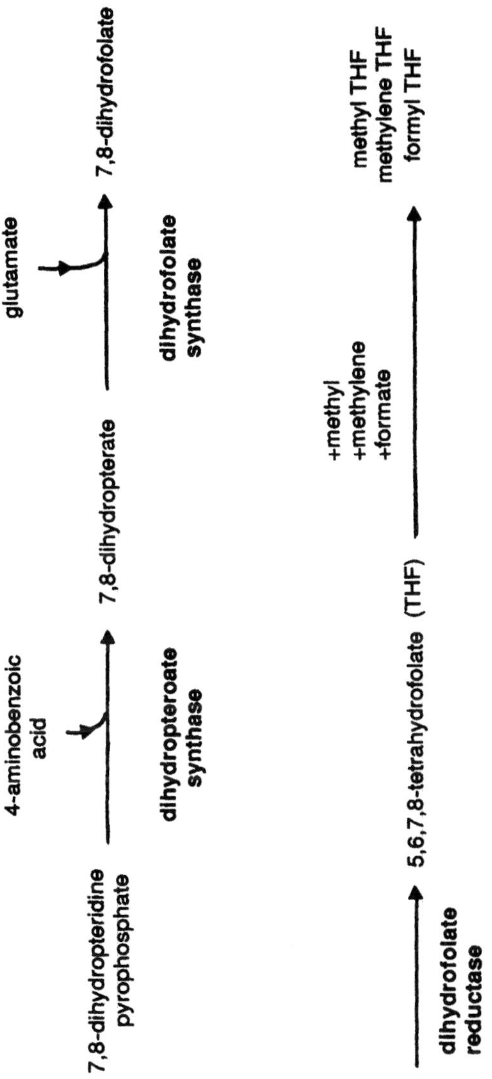

Figure 5.18. An outline of the synthesis of folate derivatives in plants.

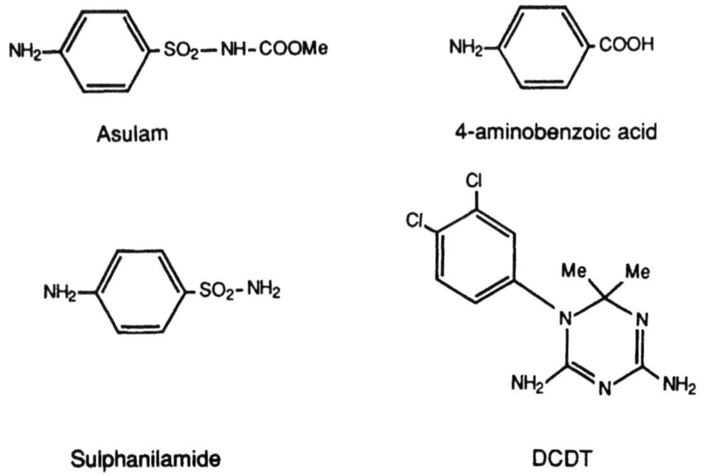

Figure 5.19. The structures of asulam, 4-aminobenzoic acid, sulfanilamide, and 4,6-diamino-1-(3,4-dichlorophenyl)-1,2-dihydro-2,2-dimethyl-1,3,5-triazine (DCDT).

5.8. AUXIN AND ANTI-AUXIN ACTIVITY

This section will deal with herbicides which either possess or antagonize auxin activity in auxin bioassays such as the oat coleoptile and bean hypocotyl elongation systems. The auxin herbicides, that is, herbicides that possess auxin activity, include the highly successful phenoxyacetic acid herbicides (2,4-D and MCPA) discovered over 40 years ago.[162] Their mode of action has been clearly established as acting as mimics of auxin, but they possess a high degree of metabolic stability, unlike endogenous auxin.[4,5,163] The second herbicide group comprises the so-called wild oat herbicides, for example, benzoylprop-ethyl, flamprop-isopropyl, and chlorfenprop-methyl.[164]

5.8.1. Auxin Herbicides

It is not the purpose of this section to repeat previous reviews on auxins and auxin herbicides.[4,5,163,165] A brief summary will be presented on the current views on the target site(s) for the auxin herbicides and recent developments concerning mode of action.

Figure 5.20 shows some of the auxin herbicides in current use. The phenoxy and benzoic acids and their related compounds have been in use

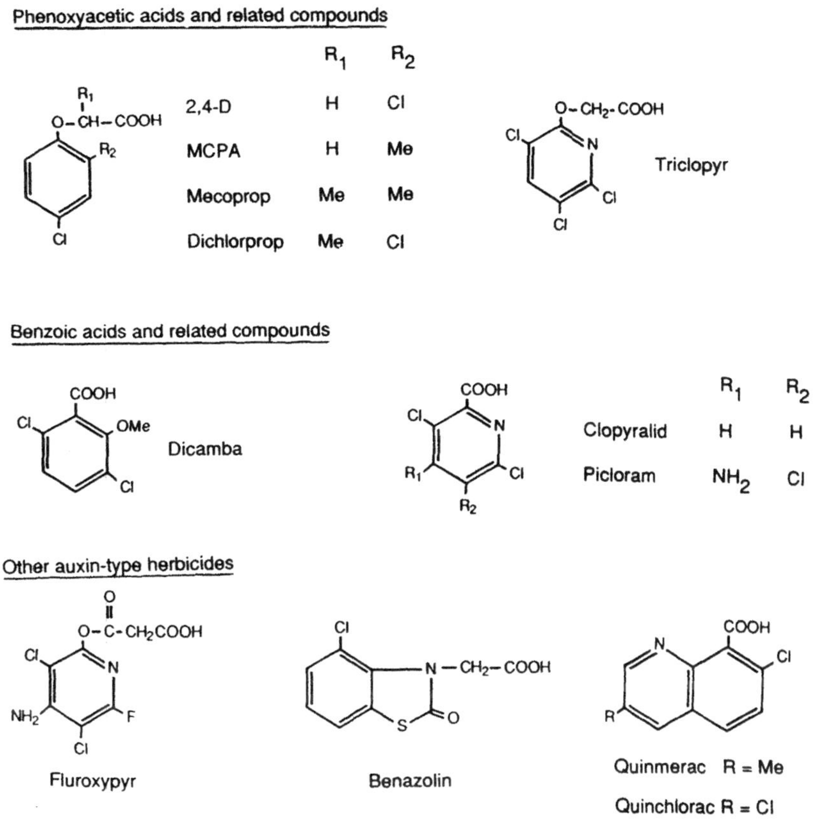

Figure 5.20. Examples of some auxin herbicides which display auxin activity in bioassays.[4,5,166-171]

for many years,[4,5] whereas fluroxypyr, quinchlorac, and quinmerac are more recently developed herbicides.[166-168]

5.8.1.1. Herbicidal Phytotoxicity

Anatomical and morphological studies in auxin herbicide-treated plants reveal a catalogue of symptom development including severe chloroplast damage leading to leaf chlorosis, altered stomatal function, stem tissue proliferation, root initiation in stem tissue, root tissue disintegration, and abnormal apical growth.[4,163,166] Fedtke[4] summarized the sequence of symptoms to be associated with three phases of herbicide action: (i) stimulation of metabolic processes, (ii) distribution leading to aberrant morphology, and (iii) tissue collapse and decay.

The reason for the phytotoxic action of the auxin herbicides is proposed to be an aberrant nucleic acid metabolism induced by a hormonal imbalance in treated tissues.[4,163]

5.8.1.2. Auxin Mode of Action

It is necessary to briefly consider the current view on the mode of action of auxins in order to identify the basis of phytotoxicity of the auxin herbicides. However, in spite of 40 years of extensive research, aided by the use of the auxin herbicides, there is no universal view on the auxin binding sites (receptors) and the events following binding.

Two types of auxin-induced responses are recognized following receptor binding—a rapid response within 10–20 min and a slower response after 1–2 h. The rapid response leads to an activation of plasmalemma ATPase, which induces proton efflux, resulting in cell wall acidification and loosening. This is the first detectable physiological effect of receptor site binding and has been extensively investigated with respect to auxin and herbicide binding.[172-176] This rapid response alone cannot explain the biochemical activity of auxins or the reason for auxin herbicide phytotoxicity. A second response occurs some 35–45 min after receptor binding and involves transcription and translation, that is, RNA and protein synthesis. Increased RNA synthesis has been reported within a few hours of auxin herbicide treatment (see Ref. 4 for original references). It was initially thought that the rapid response resulted from proton efflux induced by auxin binding directly to plasmalemma ATPase; however, it appears that ATPase is not an auxin receptor.[177,178]

In fact, it seems likely that both responses are induced by an as yet unknown second messenger which triggers the auxin responses. Figure 5.21 represents a scheme encompassing the currently held views on events following auxin receptor binding. It is proposed that the rapid physiological response of proton efflux is mediated by rapidly induced mRNAs.[175] The second response results from indirect regulation by auxin and includes induction or synthesis of mRNAs coding for cellulases, hydrolases, and polypeptides necessary for cell division, differentiation, and meristematic induction.[175]

5.8.1.3. Auxin Receptors/Binding Proteins

The existence of auxin receptors/binding proteins, the likely target sites of auxin herbicides, has been the subject of considerable research in recent years (for reviews, see Refs. 4, 173, and 179–181). There seems little

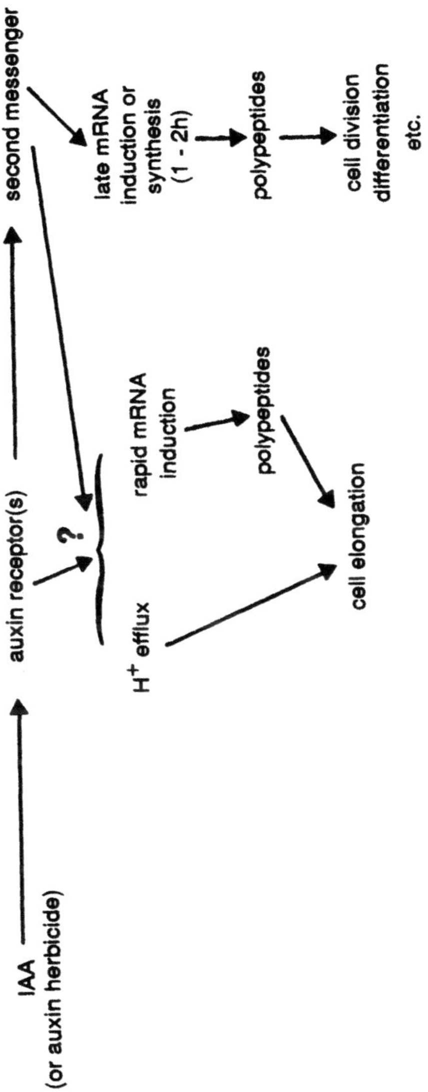

Figure 5.21. A scheme of the two responses resulting from auxin or auxin herbicide binding to the auxin receptor(s). Based on Ref. 175.

doubt that auxin binding sites are located on more than one type of membrane. Three membrane binding proteins have been clearly identified in maize coleoptiles: sites 1 (endoplasmic reticulum), 2 (plasmalemma), and 3 (tonoplast). Also, there appears to be an additional microsomal membrane binding site in maize leaf and coleoptile tissue. These binding sites and those from other species have been characterized with respect to substrate specificity, binding, and dissociation constants.[4,173,177,179,181] Furthermore, there are several reports of nonmembrane auxin binding sites, for example, ribulose-bisphosphate carboxylase (pea and bean leaves) and soluble cytoplasmic proteins from tobacco callus, soybean, and maize.[4,173,177,179,181]

However, there is no unequivocal evidence of a physiological role of auxin binding to any of these protein sites. Indeed, auxin has been shown to bind to bovine serum albumin (BSA) with substrate specificity and binding affinity similar to that of binding sites of plant origin.[182,183] Therefore, until binding to a plant auxin binding site can be linked directly to an auxin-induced response, the identity of the auxin receptor(s), and therefore the auxin herbicide target site, remains uncertain.

5.8.1.4. Structure–Activity Relationships of Auxins

A series of structure–activity proposals have been put forward in the 30 years of research into auxins. For many years, the most widely accepted suggestion was the charge separation theory of Porter and Thimann.[184,185] This proposes that an active auxin possesses a fractional positive change (indole nitrogen of IAA) located at a distance of 0.55 nm from the negatively charged carboxyl group. The auxin activity of phenoxyacetic acids, benzoic acids, and their derivatives can be accommodated within this theory.[184,185] However, this theory was unable to reconcile the contrasting auxin activities of enantiomeric pairs and inactive and active auxins.[186] Further theories have been proposed which take into account receptor configurations necessary to accommodate the diverse range of auxin structures, including the auxin herbicides.[186–189] Lehmann proposed a three-point attachment theory,[186] whereas both the models of Kaethner[188] and Rakhaminova *et al.*[187] suggest that auxin activity is dependent on a conformational change of the receptor induced by binding. Katekar proposed a site complementary to the IAA molecule, with IAA binding in an extended planar conformation.[189] These theories have been reviewed recently,[5,173,181] and it appears that none can wholly accommodate the diverse nature of auxin herbicides. If a universal receptor site exists, a further model is necessary.

In conclusion, the auxin herbicides are clearly auxin mimics; however, despite many years of research, their target site(s) and detailed mechanisms of action remain to be clearly identified.

5.8.2. Auxin Inhibitor Herbicides

Auxin inhibitor herbicides include the so-called wild oat herbicides shown in Figure 5.22.[164] Their classification as auxin inhibitors is based on their inhibition of auxin-induced responses in auxin bioassays and their antagonism of auxin herbicides.[191-195] This anti-auxin activity of diclofop-methyl is almost certainly secondary in importance to its inhibition of acetyl-CoA carboxylase (see Chapter 3).

The wild oat herbicides are esters which are hydrolyzed to the free acids necessary for activity and translocation.[164] The auxin activity of diclofop-methyl is proposed to be due to the ester[193]; however, hydrolysis to the free acid is rapid, and any physiological response by treated tissues in periods greater than 20–60 min may in fact be due to the acid. The wild oat herbicides also possess an asymmetric carbon atom, and where separate enantiomers have been available, it appears that auxin activity is restricted to the *R*-configuration.[4,192]

The symptomology induced by these herbicides reveals inhibited cell division and elongation[4,164,190-194]; however, the anti-auxin activity has only been investigated in detail for chlorfenprop-methyl and diclofop-methyl.[192,193] It was concluded that chlorfenprop did not act competitively at an auxin binding site but interacted with a cell membrane, altering its integrity.[192] A specific receptor for chlorfenprop seems likely to take into account the specific activity of the L(−)-enantiomer. Binding to this site subsequently interferes with auxin binding to its receptor(s).

Growth inhibition by diclofop-methyl was only partially overcome by IAA or dicamba, and it was concluded that diclofop-methyl and auxin were not competing for the same site.[5,193] It is possible that the anti-auxin activity of diclofop-methyl and the other wild oat herbicides results from an interaction with a common membrane site which alters auxin binding to its receptor(s). However, the nature of this site is unknown. It may not be a specific site but rather a nonspecific interaction of herbicide and cell membrane. A swelling of isolated mitochondria has been reported (see Section 5.1 and Ref. 18). Also, benzoylprop-ethyl and flamprop-isopropyl alter membrane fluidity of isolated rat liver mitochondria,[195] indicating the potential of these herbicides to disrupt membrane integrity and therefore any process requiring unimpaired membrane integrity.

In conclusion, it seems likely that the wild oat herbicides bind to a membrane site(s) and that this binding alters the binding of auxin to its receptor site(s).

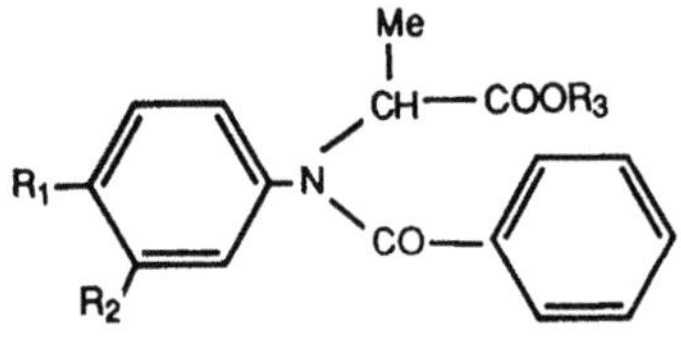

	R_1	R_2	R_3
Benzoylprop-ethyl	Cl	Cl	Et
Flamprop-methyl	F	Cl	Me
Flamprop-isopropyl	F	Cl	iPr

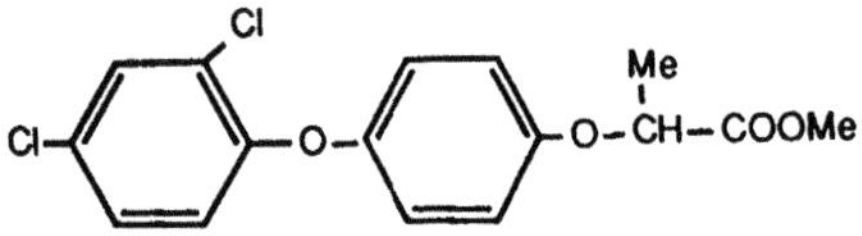

Diclofop-methyl

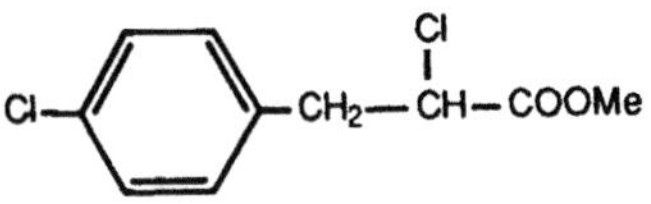

Chlorfenprop-methyl

Figure 5.22. The wild oat herbicides which possess anti-auxin activity.

5.8.3. Auxin Transport Inhibitors

A brief consideration should be made of naptalam, which possesses both herbicidal and plant growth regulatory activity.[5] Naptalam is proposed

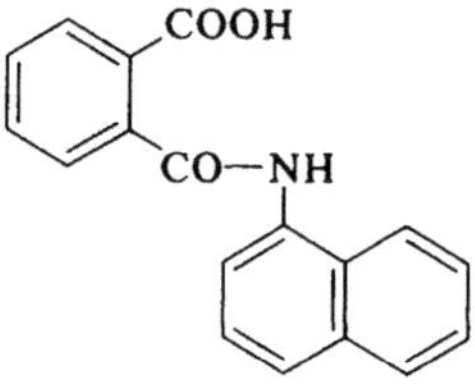

Naptalam

to bind to a phytotropin binding site, leading to an inhibition of auxin efflux from the basal ends of cells and therefore basipetal auxin transport.[5] The common structural requirements for auxin transport inhibitors have been characterized and a model of a binding site has been proposed by Katekar and co-workers.[189,196–199]

Naptalam seems to specifically inhibit auxin transport, and it must be concluded that different sites exist for auxin transport and other auxin-induced responses.

5.9. CONCLUDING REMARKS

This chapter has included consideration of a large number of herbicidal molecules with effects on several diverse areas of plant metabolism. In many instances conclusive evidence that these molecules affect a particular target site, plus the mechanism of their interactions, are still unavailable. This is due essentially to an absence of detailed biochemical knowledge of the target sites and processes. It can be speculated that as our knowledge of plant metabolism increases so the mechanisms of herbicidal inhibition will be discovered. Indeed, the herbicides themselves are likely to have a role in increasing our biochemical knowledge by acting as a direct impetus for the research or as tools for its study. Evidence for this has been seen in Chapters 1–4 where the various herbicides included within the scope of these chapters have clearly contributed to our understanding of photosynthesis, amino acid, lipid and carotenoid metabolism. In addition, the use of the auxin-type herbicides is widespread in hormonal research and compounds such as gabaculin and dichlobenil have been used in research on chlorophyll and cellulose biosynthesis respectively.

REFERENCES

1. B. T. Storey, in: *The Biochemistry of Plants. A Comprehensive Treatise, Vol. 2, Metabolism and Respiration* (D. D. Davies, ed.), p. 125, Academic Press, New York (1980).
2. J. P. Blein, *Plant Sci. Lett. 19*, 65 (1980).
3. D. E. Moreland, *Annu. Rev. Plant Physiol. 31*, 597 (1980).
4. C. Fedtke, *Biochemistry and Physiology of Herbicide Action*, Springer-Verlag, Berlin (1982).
5. J. R. Corbett, K. Wright, and A. C. Baillie, *The Biochemical Mode of Action of Pesticides*, 2nd ed., Academic Press, London (1984).
6. O. Belbachir, M. Matringe, D. Chevallier, and M. Tissut *Pestic. Biochem. Physiol. 14*, 309 (1980).

7. R. Douce, R. Brouquisse, and E.-P. Journet, in: *The Biochemistry of Plants. A Comprehensive Treatise, Vol. 11, Biochemistry of Metabolism* (D. D. Davies, ed.), p. 177, Academic Press, New York (1987).
8. S. Mona, P. Ravanel, X. de Cherade, M. Bergon, J.-P. Calmon, and M. Tissut, *Pestic. Biochem. Physiol. 27,* 261 (1987).
9. P. D. Lotlikar, L. F. Remmert, and V. H. Freed, *Weed Sci. 16,* 161 (1968).
10. R. D. Gruenhagen and D. E. Moreland, *Weed Sci. 19,* 319 (1971).
11. R. C. Kirkwood, in: *Herbicides: Physiology, Biochemistry and Ecology* (L. J. Andus, ed.), Vol. 1, p. 443. Academic Press, New York (1976).
12. W. G. Hanstein, *Biochim. Biophys. Acta. 456,* 129 (1976).
13. W. G. Hanstein and R. Kiehl, *Biochem. Biophys. Res. Commun. 100,* 118 (1981).
14. Y. Hatefi, *Ann. N.Y. Acad. Sci. 346,* 434 (1980).
15. R. S. Criddle, R. F. Johnston, and R. J. Stack, in: *Current Topics in Bioenergetics* (D. R. Sanadi, ed.), Vol. 9, p. 89, Academic Press, New York (1979).
16. D. E. Moreland, S. C. Huber, and W. P. Novitzky, in: *Biochemical Responses Induced by Herbicides* (D. E. Moreland, J. B. St. John, and F. D. Hess, eds.), p. 79, American Chemical Society, Washington, D.C. (1982).
17. D. R. Gealy, R. A. Boydston, R. R. Klein, and D. E. Koeppe, *Pestic. Biochem. Physiol. 27,* 106 (1987).
18. A. S. Cohen and I. N. Morrison, *Pestic. Biochem. Physiol. 16,* 110 (1981).
19. C. Gauvrit and R. Scalla, *FEBS Lett. 158,* 222 (1983).
20. C. Gauvrit, *Phytochemistry 24,* 2181 (1985).
21. B. P. Halling and R. Behrens, *Weed Sci. 31,* 693 (1982).
22. K. E. Pallett, *Weed Res. 22,* 329 (1982).
23. H.-Y, Cho, J. M. Widholm, and F. W. Slife, *Plant Physiol. 87,* 334 (1988).
24. J. W. Gronwald, *Weed Sci. 34,* 196 (1986).
25. A. C. Baillie, K. Wright, B. J. Wright, and C. G. Earnshaw, *Pestic. Biochem. Physiol. 30,* 103 (1988).
26. B. Laber and N. Amrhein, *Biochem. J. 248,* 351 (1987).
27. L. J. Rogers and A. J. Smith, in: *Amino Acid Biosynthesis Inhibitors,* British Crop Protection Council Monograph No. 42 (L. G. Copping, J. Dalziel, and A. D. Dodge, eds.), p. 183 (1989).
28. W. Rudiger and S. Schoch, in: *Plant Pigments* (T. W. Goodwin, ed.), p. 1, Academic Press, London (1988).
29. J. L. Corriveau and S. I. Beale, *Plant Sci. 45,* 9 (1986).
30. Y. J. Avissar and S. I. Beale, *Plant Physiol. 89,* 852 (1989).
31. J. K. Hoober, A. Khan, D. E. Ash, S. Gough, and C. G. Kannangara, *Carlsberg Res. Commun. 53,* 11 (1988).
32. R. R. Rando, *Biochemistry 16,* 4604 (1977).
33. C. Kannangara and A. Schouboe, *Carlsberg Res. Commun. 50,* 179 (1985).
34. T. S. Soper and J. M. Manning, *J. Biol. Chem. 257,* 13930 (1982).
35. D. H. Flint, *Plant Physiol. 75, S*-170 (1984).
36. D. H. Flint, Abstracts, 27th Harden Conference, Wye College, England (1986).
37. D. P. Caiger, S. A. Pearson, A. J. Smith, and L. J. Rogers, *Plant Cell Environ. 9,* 495 (1986).
38. T. B. May, J. A. Guikema, R. L. Henry, M. K. Schuler, and P. P. Wong, *Plant Physiol. 84,* 1309 (1987).
39. G. Gardner and H. L. Gorton, *Plant Physiol. 77,* 540 (1985).
40. G. Gardner, H. L. Gorton, and S. A. Brown, *Plant Physiol. 87,* 8 (1988).
41. T. D. Elich and J. C. Lazarias, *Plant Physiol. 84,* 304 (1987).
42. P. Barry and K. E. Pallett, *Proc. EWRS Symp. Factors Affecting Herbicidal Activity and Selectivity,* p. 51 (1988)

43. P. Barry, Ph.D. thesis, University of Liverpool, U.K. (1988).
44. R. Oelmuller and H. Mohr, *Planta 164*, 390 (1985).
45. R. B. Silverman and B. J. Invergo, *Biochemistry 25*, 6817 (1986).
46. G. L. Orr and F. D. Hess, in: *Biochemical Responses Induced by Herbicides* (D. E. Moreland, J. B. St. John, and F. D. Hess, eds.), p. 131, American Chemical Society, Washington, D.C. (1982).
47. R. Lambert, G. Sandmann, and P. Boger, in: *Pesticide Chemistry: Human Welfare and the Environment 3* (J. Miyamoto, P. C. Kearney, S. Matsunaka, D. H. Hutson, and S. D. Murphy, eds.), p. 79, Pergamon Press, New York (1983).
48. P. Boger, *Z. Naturforsch., C 39*, 468 (1984).
49. S. O. Duke and W. H. Kenyon, *Z. Naturforsch., C 42*, 813 (1987).
50. K. J. Kunert, G. Sandmann, and P. Boger, *Rev. Weed Sci. 3*, 35 (1987).
51. W. H. Kenyon, S. O. Duke, and K. C. Vaughn, *Pestic. Biochem. Physiol. 24*, 240 (1985).
52. M. Matringe, J. L. Dufour, J. Lherminier and R. Scalla, *Pestic. Biochem. Physiol. 26*, 150 (1986).
53. P. M. Derrick, A. H. Cobb, and K. E. Pallett, *Pestic. Biochem. Physiol. 32*, 153 (1988).
54. D. J. Gillham and A. D. Dodge, *Pestic. Sci. 19*, 25 (1987).
55. M. P. Ensminger, F. D. Hess, and J. T. Bahr, *Pestic. Biochem. Physiol. 23*, 163 (1985).
56. S. O. Duke, K. C. Vaughn, and R. L. Meeusen, *Pestic. Biochem. Physiol. 21*, 368 (1984).
57. S. O. Duke and W. H. Kenyon, *Plant Physiol. 81*, 882 (1986).
58. B. P. Halling and G. R. Peters, *Plant Physiol. 84*, 1114 (1987).
59. M. Matringe and R. Scalla, *Pestic. Biochem. Physiol. 27*, 267 (1987).
60. J. R. Bowyer, B. J. Smith, P. Camilleri, and S. A. Lee, *Plant Physiol. 83*, 613 (1987).
61. N. Takahashi and R. P. Mason, *J. Pestic. Sci. 12*, 745 (1987).
62. M. P. Ensminger and F. D. Hess, *Plant Physiol. 77*, 503 (1985).
63. R. Sato, E. Nagaro, H. Oshio, K. Kamoshita, and M. Furuya, *Plant Physiol. 85*, 1146 (1987).
64. M. Matringe and R. Scalla, *Proceedings of the British Crop Protection Conference—Weeds*, Vol. 3, p. 981 (1987).
65. M. Matringe and R. Scalla, *Plant Physiol. 86*, 619 (1988).
66. H. Kouji, T. Masuda, and S. Matsunaka, *Pestic. Biochem. Physiol. 33*, 230 (1989).
67. D. A. Witkowski and B. P. Halling, *Plant Physiol. 87*, 632 (1988).
68. G. Sandmann and P. Boger, *Z. Naturforsch., C 43*, 699 (1988).
69. J. Lydon and S. O. Duke, *Pestic. Biochem. Physiol. 31*, 74 (1988).
70. S. O. Duke, J. Lydon, and R. N. Paul, *Weed Sci. 37*, 152 (1989).
71. M. Matringe, J.-M. Camadro, P. Labbe, and R. Scalla, *Biochem. J. 260*, 231 (1989).
72. M. Matringe, J.-M. Camadro, P. Labbe, and R. Scalla, *FEBS Lett. 245*, 35 (1989).
73. C. A. Rebeiz, A. Montazer-Zouhoor, H. J. Hopen, and S. M. Wu, *Enzyme Microb. Technol. 6*, 390 (1984).
74. C. A. Rebeiz, A. Montazer-Zouhoor, J. M. Mayasich, B. C. Tripathy, S. M. Wu, and C. C. Rebeiz, in: *Light Activated Pesticides* (J. R. Heitz and K. R. Downum, eds.), p. 295, American Chemical Society, Washington, D.C. (1987).
75. C. A. Rebeiz, A. Montazer-Zouhoor, J. M. Mayasich, B. C. Tripathy, S. M. Wu, and C. C. Rebeiz, *CRC Crit. Rev. Plant Sci. 6*, 385 (1988).
76. P. A. Castlefranco and S. I. Beale, in: *The Biochemistry of Plants. A Comprehensive Treatise, Vol. 8, Photosynthesis* (M. D. Hatch and N. K. Boardman, eds.), p. 376, Academic Press, New York (1981).
77. J. H. Chang and M. J. Konz, Abstracts 187th National Meeting of the American Chemical Society, Pesticide Chemistry Division, No. 22 (1984).

78. S. O. Duke and W. H. Kenyon, *Pestic. Biochem. Physiol. 25*, 11 (1986).
79. G. E. Darke and J. W. Boksak, *Abstr. Weed Sci. Soc. Am. 25*, 69 (1985).
80. S. O. Duke, W. H. Kenyon, and R. N. Paul, *Weed Sci. 33*, 786 (1986).
81. G. Sandmann and P. Boger, *Z. Naturforsch., C 42*, 803 (1987).
82. W. K. Vencill, K. K. Hatzios, and H. P. Wilson, *Abstr. Weed Sci. Soc. Am. 27*, 65 (1987).
83. G. Sandmann and P. Boger, *Z. Naturforsch., C 41*, 729 (1986).
84. J. D. Weete, in: *Ecology and Metabolism of Plant Lipids* (G. Fuller and W. D. Nes, eds.), p. 268, American Chemical Society, Washington, D.C. (1987).
85. R. S. Burden, C. S. James, D. T. Cooke, and N. H. Anderson, *Proceedings of the British Crop Protection Conference—Weeds*, Vol. 1, p. 171 (1987).
86. R. S. Burden, G. A. Carter, T. Clark, D. T. Cooke, S. J. Croker, A. H. B. Deas, P. Hedden, C. S. James, and J. R. Lenton, *Pestic. Sci. 21*, 253 (1987).
87. L. J. Goad, P. A. Haughan, and J. R. Lenton, in: *Plant Lipids: Targets for Manipulation* (N. J. Pinfield and A. K. Stobart, eds.), British Plant Growth Regulator Group Monograph 17, p. 91 (1988).
88. W. Rademacher, H. Fritsch, J. E. Graebe, H. Sauter, and J. Jung, *Pestic. Sci. 21*, 241 (1987).
89. J. W. Dicks, in: *Recent Developments in the Use of Plant Growth Regulators* (D. R. Clifford and J. R. Lenton, eds.), p. 1, British Plant Growth Regulator Group (1980).
90. R. E. Wilkinson, *Pestic. Biochem. Physiol. 19*, 321 (1983).
91. R. E. Wilkinson, in: *Pesticide Chemistry: Human Welfare and Environment 3* (J. Miyamoto, P. C. Kearney, S. Matsunaka, D. H. Hutson, and S. D. Murphy, eds.), p. 233, Pergamon Press, New York (1983).
92. R. E. Wilkinson and T. H. Oswald, *Pestic. Biochem. Physiol. 28*, 38 (1987).
93. R. E. Wilkinson, *Pestic. Biochem. Physiol. 25*, 93 (1986).
94. T. Komives, A. Hulesch, A. V. Komives, and F. Dutka, *Biochem. Physiol. Pflanz. 182*, 257 (1987).
95. R. E. Wilkinson, *Pestic. Biochem. Physiol. 32*, 25 (1988).
96. R. E. Wilkinson, *Pestic Biochem. Physiol. 17*, 177 (1982).
97. R. E. Wilkinson, *Pestic. Biochem. Physiol. 16*, 199 (1981).
98. R. E. Wilkinson, *Pestic. Biochem. Physiol. 23*, 19 (1985).
99. H. Weisshaar and P. Boger, *Pestic. Biochem. Physiol. 28*, 286 (1987).
100. H. Weisshaar, G. Retzlaff, and P. Boger, *Pestic. Biochem. Physiol. 32*, 212 (1988).
101. E. P. Fuerst, *Weed Technol. 1*, 270 (1987).
102. J. Lydon and S. O. Duke, *Pestic. Sci. 25*, 361 (1989).
103. W. T. Molin, E. J. Anderson, and C. A. Porter, *Pestic. Biochem. Physiol. 25*, 105 (1986).
104. T. C. Lee, W. T. Molin, and C. A. Porter, *Abstr. Weed Sci. Soc. Am. 26*, 81 (1986).
105. R. E. Wilkinson, *Pestic. Biochem. Physiol. 23*, 289 (1985).
106. W. Rudiger and J. Benz, *Z. Naturforsch., C 34*, 1055 (1979).
107. K. E. Pallett and P. Barry, *Z. Naturforsch. C 45*, 492 (1990).
108. B. E. S. Gunning and A. R. Hardham, *Ann. Revu. Plant Physiol. 33*, 651 (1982).
109. P. J. Dawson and C. W. Lloyd, in: *The Biochemistry of Plants. A Comprehensive Treatise, Vol. 12, Physiology of Metabolism* (D. D. Davies, ed.), p. 3, Academic Press, New York (1987).
110. L. C. Morejohn, T. E. Bureau, J. Mole-Bajer, and D. E. Fosket, *Planta 172*, 252 (1987).
111. S. D. Strachan and F. D. Hess, *Pestic. Biochem. Physiol. 20*, 141 (1983).
112. P. G. Bartels and J. L. Hilton, *Pestic. Biochem. Physiol. 3*, 462 (1973).
113. O. Kiermayer and C. Fedtke, *Protoplasma 92*, 163 (1977).
114. J. D. Holmsen and F. D. Hess, *Weed Sci. 32*, 732 (1984).
115. L. C. Morejohn and D. E. Fosket, *J. Cell Biol. 99*, 141 (1984).
116. L. C. Morejohn, T. E. Bureau, L. P. Tocchi, and D. E. Fosket, *Planta 170*, 230 (1987).
117. F. D. Hess and D. E. Bayer, *J. Cell Sci. 24*, 351 (1977).

118. K. C. Vaughn, *Pestic. Biochem. Physiol. 26*, 66 (1986).
119. K. C. Vaughn, M. D. Marks, and D. P. Weeks, *Plant Physiol. 83*, 956 (1987).
120. M. A. Vaughan and K. C. Vaughn, *Pestic. Biochem. Physiol. 28*, 182 (1987).
121. C. Hertel, H. Quader, D. G. Robinson, and D. Marme, *Planta 149*, 336 (1980).
122. C. Hertel and D. Marme, *Pestic. Biochem. Physiol. 19*, 282 (1983).
123. E. F. Allan and A. J. Trewavas, in: *The Biochemistry of Plants. A Comprehensive Treatise, Vol. 12, Physiology of Metabolism* (D. D. Davies, ed.), p. 117, Academic Press, New York (1987).
124. J. R. Colvin, in: *The Biochemistry of Plants. A Comprehensive Treatise, Vol. 3, Carbohydrates: Structure and Function* (J. Preiss, ed.), p. 543, Academic Press, New York (1980).
125. K. I. Benyon and A. N. Wright, *Residue Rev. 43*, 23 (1972).
126. F. M. Ashton and A. S. Crafts, *Mode of Action of Herbicides*, 2nd ed., p. 254, Wiley, New York (1981).
127. D. E. Moreland, G. G. Hussey, and F. S. Farmer, *Pestic. Biochem. Physiol. 4*, 356 (1974).
128. T. Hogetsa, H. Shibaoka, and M. Shimokoriyama, *Plant Cell Physiol. 15*, 389 (1984).
129. N. Umetsu, S. Satoh, and K. Matsoda, *Plant Cell Physiol. 17*, 1071 (1976).
130. Y. Meyer and W. Herth, *Planta 142*, 253 (1978).
131. D. W. Galbraith and B. A. Shields, *Physiol. Plant. 55*, 25 (1982).
132. D. Montezinos and D. P. Delmer, *Planta 148*, 305 (1980).
133. W. Blaschek, V. Semler, and G. Franz, *J. Plant Physiol. 120*, 457 (1985).
134. D. P. Delmer, G. Cooper, D. Alexander, J. Cooper, T. Hayashi, C. Nitsche, and M. Thelen, *J. Cell Sci. Suppl. 2*, 33 (1985).
135. D. P. Delmer, S. M. Read, and G. Cooper, *Plant Physiol. 84*, 415 (1987).
136. G. Cooper, D. P. Delmer, and S. M. Read, Abstracts, 196th National Meeting of the American Chemical Society, No. ORGN 101 (1988).
137. J. H. Cherry, in: *Herbicides: Physiology, Biochemistry and Ecology* (L. J. Audus, ed.), Vol. 1, p. 525, Academic Press, New York (1976).
138. F. M. Ashton, O. T. deVilliers, R. K. Glenn, and W. B. Duke, *Pestic. Biochem. Physiol. 7*, 122 (1977).
139. D. E. Moreland, S. S. Malhotra, R. D. Gruenhagen, and E. H. Shokraii, *Weed Sci. 17*, 556 (1969).
140. J. D. Mann, L. S. Jordan, and B. E. Day, *Plant Physiol. 40*, 840 (1965).
141. T. B. Ray, *Proceedings of the British Crop Protection Conference—Weeds*, Vol. 1, p. 7 (1980).
142. R. R. Bellinder, K. K. Hatzios, and H. P. Wilson, *Weed Sci. 33*, 779 (1985).
143. M. W. Kerr and R. J. Avery, *Biochem. J. 128*, 132P (1972).
144. D. P. Weeks and R. Baxter, *Biochemistry 11*, 3060 (1972).
145. R. Baxter and J. E. McGowan, *J. Exp. Bot. 27*, 525 (1976).
146. R. Baxter, V. C. Knell, H. J. Somerville, H. M. Swain, and D. P. Weeks, *Nature 243*, 139 (1973).
147. R. J. Ellis, *Phytochemistry 14*, 89 (1975).
148. K. E. Pallett and J. C. Caseley, *Pestic. Biochem. Physiol. 14*, 426 (1980).
149. K. E. Pallett, *Weed Res. 24*, 163.
150. J. M. DiTomaso, T. L. Rost, and F. M. Ashton, *Pestic. Biochem. Physiol. 31*, 166 (1988).
151. E. A. Cossins, in: *The Biochemistry of Plants. A Comprehensive Treatise, Vol. 11, Biochemistry of Metabolism* (D. D. Davies, ed.), p. 318, Academic Press, New York (1987).
152. R. B. Skerrett and T. A. Fretz, *Hort. Sci. 10*, 162 (1975).
153. J. L. Killmer, J. M. Widholm, and F. W. Slife, *Plant Sci. Lett. 19*, 203 (1980).
154. N. H. Stephen, G. T. Cook, and H. J. Duncan, *Ann. Appl. Biol. 96*, 227 (1980).
155. P. Veerasekaran, R. C. Kirkwood, and E. W. Parnell, *Pestic. Sci. 12*, 325 (1981).
156. N. A. Hewertson and H. A. Collin, *Weed Res. 24*, 79 (1984).
157. O. Okinaka and K. Iwai, *J. Vitaminol. 16*, 201 (1970).

158. G. H. Miller, P. H. Doukas, and J. K. Seydel, *J. Med. Chem. 15,* 700 (1972).
159. P. Veerasekaran, R. C. Kirkwood, and E. W. Parnell, *Pestic. Sci. 12,* 330 (1981).
160. P. Veerasekaran, R. C. Kirkwood, and W. W. Fletcher, *Weed Res. 17,* 85 (1977).
161. L. Rudenberg, G. E. Foley, and W. D. Winter, *Science 121,* 899 (1955).
162. C. Kirby, *The Hormone Weedkillers,* BCPC Publications, London (1980).
163. J. B. Pillmoor and J. K. Gaunt, in: *Progress in Pesticide Biochemistry,* Vol. 1 (D. H. Hutson and T. R. Roberts, eds.), p. 147, John Wiley and Sons, Chichester (1981).
164. K. E. Pallett, *Proceedings of the British Crop Protection Conference—Weeds,* Vol. 3, p. 843 (1980).
165. A. J. Trewavas, *Plant Cell Environ. 4,* 203 (1981).
166. G. E. Sanders and K. E. Pallett, *Ann. Appl. Biol. 111,* 385 (1987).
167. B. Wuerzer, R. Berghaus, M. Hagen, R. D. Kohler, and J. Markert, *Proceedings of the British Crop Protection Conference—Weeds,* Vol. 1, p. 63 (1985).
168. R. Berghaus and G. Relzlaff, *Proc. EWRS Symp. Factors Affecting Herbicidal Activity and Selectivity,* p. 81 (1988).
169. L. L. Whatley and F. W. Slife, *Weed Sci. 31,* 801 (1983).
170. L. R. Chang and C. L. Foy, *Pestic. Biochem. Physiol. 19,* 203 (1983).
171. P. J. Fitzsimons, P. Barnwell, and A. H. Cobb, *Proc. EWRS Symp. Factors Affecting Herbicidal Activity and Selectivity,* p. 63 (1988).
172. D. L. Rayle, *Planta 114,* 63 (1973).
173. P. H. Rubery, *Ann. Rev. Plant Physiol. 32,* 569 (1981).
174. M. L. Evans, *Crit. Rev. Plant Sci. 2,* 317 (1985).
175. A. Theologis, *Annu. Rev. Plant Physiol. 37,* 407 (1986).
176. P. J. Fitzsimons, P. R. Miller, and A. H. Cobb, *Proceedings of the British Crop Protection Conference—Weeds,* Vol. 1, p. 179 (1987).
177. K. R. Libbenga, A. C. Maan, P. C. G. van der Linde, and A. M. Mennes, in: *Hormones, Receptors and Cellular Interactions in Plants* (C. M. Chadwick and P. R. Garrod, eds.), p. 1, Cambridge University Press, Cambridge (1986).
178. R. Serrano, *Annu. Rev. Plant Physiol. 40,* 61 (1989).
179. M. A. Venis, *Biochem. Soc. Trans. 6,* 325 (1978).
180. J. W. Cross, *Plant Cell Environ. 8,* 351 (1985).
181. M. A. Venis, *Hormone Binding Sites in Plants* Longman, New York (1985).
182. M. A. Matlib, R. C. Kirkwood, and J. D. E. Patterson, *Weed Res. 11,* 190 (1971).
183. G. T. P. Murphy, *Plant Sci. Lett. 15,* 183 (1979).
184. W. L. Porter and K. V. Thimann, *Phytochemistry 4,* 229 (1965).
185. K. V. Thimann, in: *The Physiology of Plant Growth and Development* (M. B. Wilkins, ed.), p. 1, McGraw-Hill, London (1969).
186. P. A. F. Lehmann, *Chem. Biol. Interact. 20,* 239 (1978).
187. A. B. Rakhaminova, E. E. Khavin, and L. S. Yaguzhinskii, *Biokhimya 43,* 639 (1978).
188. T. M. Kaethner, *Nature 267,* 19 (1977).
189. G. F. Katekar, *Phytochemistry 18,* 223 (1979).
190. B. Jeffcoat and W. N. Harries, *Pestic. Sci. 4,* 891 (1973).
191. B. Jeffcoat and W. N. Harries, *Pestic. Sci. 6,* 283 (1975).
192. G. K. Andreev and N. Amrhein, *Physiol. Plant. 37,* 175 (1976).
193. M. A. Shimabukuro, R. H. Shimabukuro, W. S. Nord, and R. A. Hoerauf, *Pestic. Biochem. Physiol. 8,* 199 (1978).
194. I. N. Morrison, B. D. Hill, and L. G. Dushnicky, *Weed Res. 19,* 385 (1979).
195. C. Gauvrit, *Pestic. Biochem. Physiol. 21,* 377 (1984).
196. G. F. Katekar and A. E. Geissler, *Plant Physiol. 60,* 826 (1977).
197. G. F. Katekar and A. E. Geissler, *Plant Physiol. 66,* 1190 (1980).

198. G. F. Katekar, J.-F. Nave, and A. E. Geissler, *Plant Physiol. 68*, 1460 (1981).
199. G. F. Katekar, A. E. Geissler, C. H. L. Kennard, and G. Smith, *Phytochemistry 26*, 1257 (1987).
200. M. Luetzow, P. Beyer, and H. Kleinig, *Z. Naturforsch.*, *C 45*, 856 (1990).

6

Modifiers of Herbicide Action at Target Sites

KRITON K. HATZIOS

6.1. INTRODUCTION

The spectrum of weed control afforded by most of the currently available herbicides is rather limited. Attempts to improve the performance of single herbicides by using herbicide mixtures date back to the early beginnings of the development of chemical weed control.[1] Combinations of a graminicide and a broadleaf herbicide have been particularly successful and are recommended for use in most agronomic crops.[2] More recently, many of the newer herbicides with high potency and a single target site (e.g., sulfonylureas and imididazolinones) are mixed with several old and cheap herbicides for broad-spectrum, season-long weed control.[3,4]

Another major reason for the increased use of herbicide mixtures in recent years is to delay the appearance of herbicide-resistant weed biotypes. Repeated use of a single herbicide possessing long-lasting activity and acting on a single target site imposes a strong selective pressure on sensitive weeds which may give rise to the development of herbicide-resistant biotypes.[5-8]

In addition to the use of two- or three-way tank mixtures of selected herbicides, herbicides are also used in combinations with other pesticides (e.g., insecticides, fungicides, nematocides) or with auxiliary chemicals such as synergists, safeners, extenders, and adjuvants.

KRITON K. HATZIOS • Department of Plant Pathology, Physiology and Weed Science, Virginia Polytechnic Institute and State University, Blacksburg, Virginia 24061-0330.

In many cases, the components of selected herbicide/herbicide or herbicide/nonherbicide mixtures act independently of each other, and the responses of target species to such combinations are predictable from the performance of each chemical applied individually. When this happens, the combined effect of the two chemicals is described as additive.[9,10] In some cases, however, the components of agrochemical mixtures may interact, and the biological activity of a herbicide may be modified by the prior, simultaneous, or sequential application of another herbicide or auxiliary chemical to the same target weed. When this happens, the response of the target species to the joint application of the two chemicals is not predictable from the performance of each chemical applied alone. If, as a result of such interactions, the biological activity of the agrochemical mixture is decreased, the mixture is described as antagonistic. When the biological activity of the mixture is enhanced, the mixture is described as synergistic.[9,10]

General or specific aspects of the interactions of herbicides with other herbicides or nonherbicidal agrochemicals have been reviewed previously.[4,9–18] The purpose of this chapter is to review and consolidate more recent contributions to our knowledge and understanding of the principles and mechanisms of action of chemical modifiers (synergists and antagonists) of herbicide action at target sites.

6.2. INTERACTIONS OF HERBICIDES AND MODIFIERS: FUNDAMENTALS

6.2.1. Definitions and Test Methods

The task of characterizing the joint action of herbicide/herbicide or herbicide/nonherbicide mixtures on plant tissues has challenged weed scientists, pesticide technologists, and statisticians for a long time. The terminology used in describing the performance of herbicide mixtures and the methodology applied in characterizing it have been reviewed extensively.[9–25] In general, three major types of joint action of herbicide mixtures on target plants are commonly acceptable and they are defined as follows[9]:

- *Additive effect*: cooperative action of two herbicides such that the observed response of a test plant to their joint application is equal to the response predicted to occur by an appropriate reference model.
- *Synergism*: cooperative action of two herbicides such that the observed response of a test plant to their joint application appears to be greater than the response predicted to occur by an appropriate reference model.
- *Antagonism*: cooperative action of two herbicides such that the observed response of a test plant to their joint application appears

to be less than the response predicted to occur by an appropriate reference model.

From the above discussion, it is evident that in order to test the observed response of a target plant to the application of a herbicide mixture and characterize it as additive, synergistic, or antagonistic, an "expected" or "predicted" response should be well defined. The calculation of a predicted response requires the selection of an appropriate reference model.[9,13] Such a selection is not difficult when only one component of a mixture is biologically active on a target plant. When both components of a herbicide mixture are biologically active, the definition and selection of a reference model is more complex.[13]

The controversy as to what constitutes an acceptable definition and calculation of a "predicted" response has not been resolved.[10] Nevertheless, two main reference models, known as the "additive dose model" (ADM) and the "multiplicative survival model" (MSM), are generally acceptable for the detection of interactions and characterization of the joint action of herbicide/herbicide or herbicide/nonherbicide mixtures.[9,13]

The ADM is valid for studying the interactions of herbicides with other herbicides or nonherbicides that exhibit similar action. Under the assumptions of the ADM, the growth responses of a test plant to the joint application of two components of a herbicide mixture are equated to the "sum" of the responses to each chemical tested separately.[13] Thus, the predicted dose-response relationships for a herbicide mixture using the ADM are such that if one component of the mixture is replaced at a constant proportion by the other, the predicted response remains unchanged.[9,10,13] Methods developed for the prediction of plant responses to herbicide mixtures using the ASM include the isobole method,[25] the ANOVA method,[9] and the two-parameter method.[24]

The MSM is appropriate for predicting plant responses to herbicide mixtures whose components act by different modes of action. Use of the MSM requires that experimental observations be expressed as percentages or proportions of a hypothetical maximum measure of activity.[9,13] Under these assumptions, the MSM equates the proportion of the observed response of a test plant to the action of a herbicide mixture to the "product" of the corresponding proportions observed when the two components of the mixture are administered separately.[13] Methods developed for the prediction of plant response to herbicide mixtures using the MSM include Colby's equation,[21] the MSM equation developed by Morse,[13] and the regression method of Nash.[14] Recently, Flint *et al.*[22] devised a method for the statistical treatment of Colby's equation.

Departures from the selected reference model, following the comparison of observed responses to those predicted by the model, or rejection of the null hypothesis of "no interactions" would be indicative of significant antagonistic or synergistic interactions.[9,13] Further discussion of the ADM and MSM models as well as applications, comparisons, and criticisms of the specific methods used for the prediction of plant responses to the application of herbicide mixtures are available in the literature.[9–25]

6.2.2. Mechanisms of Synergism and Antagonism

Chemical modifiers of the action of selected herbicides can interact with the respective herbicides either outside or inside the target or nontarget plant species. Interactions may also occur before application while herbicides are mixed with other herbicides or auxiliary chemicals in the spray tank. Chemical or physical incompatibility of herbicide mixtures has been reviewed elsewhere[26–28] and is outside the scope of this chapter.

Synergism or antagonism of the activity of a given herbicide on a target plant brought about by the prior or simultaneous application of a modifier may be the result of changes in the amount of the herbicide reaching its target site in an active form. Alternatively, synergists or antagonists may also influence the toxicity of a herbicide by interacting with it at the site of its biochemical or physiological action within the plant cell.

In most of the documented cases, the synergism or antagonism of herbicidal activity by chemical modifiers appears to be the result of altered penetration, translocation, and biotransformation of a given herbicide caused by the prior or simultaneous application of the chemical modifier. The synergism of atrazine activity on grass weeds such as foxtails (*Setaria* sp.) by tridiphane results from a strong inhibition of the glutathione-mediated metabolism of atrazine caused by this synergist.[29] Similarly, the safener-mediated protection of large-seeded grass crops such as corn, sorghum, and rice from thiocarbamate and chloroacetanilide herbicide injury is due to an enhanced metabolic detoxification of these herbicides brought about by safener treatments.[30–34]

Investigations on the mechanisms underlying the synergism or antagonism of herbicide action by chemical modifiers at target sites are rather limited. Until recently, this was mainly due to our poor understanding of the exact mechanisms of action of most herbicides and their interactions with specific target sites. Recent advances in molecular biology, however, have aided tremendously studies in this area. Today, the mechanisms of action of several classes of herbicides are well characterized, and they are reviewed in detail in other chapters of this book. Examples of synergism or antagonism of herbicide action at target sites as well as discussions of the mechanisms involved in such interactions are presented in the remaining sections of this chapter.

6.3. SYNERGISTS OF HERBICIDE ACTION AT TARGET SITES

The status of synergizing herbicides has been reviewed recently by Gressel and Shaaltiel.[35,36] In most of the cases studied, herbicide synergism did not result from interactions between herbicides and synergists at target sites.[9,36] Inhibition of herbicide degradation in plants or in soils is currently recognized as the major mechanism of action of successful synergists of herbicide action.[36] The aim of this section is to discuss briefly the relatively few examples of herbicide synergism which appear to result from interactions of herbicides and synergists at the target sites.

6.3.1. Synergists of Herbicides at Single Target Sites

A herbicide and a synergist may act on the same target site, and each one may complement the action of the other by exerting its effect on this site. Such a mechanism for herbicide synergism would presume the existence of a target site (e.g., enzyme or protein) which possesses two or more active domains. Each domain should exhibit a differential affinity for the herbicide and the synergist so that they can both bind onto the site and inhibit its function. When a target site possesses only a single active domain which can be blocked either by the herbicide or by the modifier of herbicide activity, competitive antagonism rather than synergism would be more likely to occur.

Examples of herbicide synergism resulting from such a mechanism are provided by the reported cases of synergistic interactions of photosynthesis-inhibiting herbicides. Extensive studies conducted in the last decade have identified a 32-kDa protein (now known as D1 protein) as the target site of photosystem II (PSII)-inhibiting herbicides such as the *s*-triazines, triazinones, ureas, phenylpyridazinones, uracils, and biscarbamates.[37] D1 is a highly conserved plant protein, and its conserved portion is much larger than that which would be necessary for building high-affinity binding for a single family of herbicides.[37] Therefore, it has been suggested that the multiple families of photosynthesis-inhibiting herbicides have high affinity for different domains of the evolutionarily conserved portion of the D1 protein.[37] The finding that atrazine-resistant plant mutants are not resistant to diuron further supports the view that the binding regions of D1 for different chemical families do not overlap completely.[37,38] Mixtures of PSII herbicides which have been reported to be synergistic on crop or weed plants include metamitron and methabenzthiazuron on annual bluegrass (*Poa annua* L.),[39] lenacil and metamitron on cultivated oats (*Avena sativa* L.),[40] and pyrazon and lenacil on cultivated oats.[40]

In other cases, herbicide synergism may result from interactions of a herbicide and a synergist on a secondary rather than a primary target site. The enzyme phytase (EC 3.1.3.26) has been proposed as a likely target site for the synergistic action of atrazine or chloramben with phosphorus.[41] High, but not phytotoxic, levels of phosphorus enhanced the inhibition of phytase activity by atrazine and chloramben.[41] A severe inhibition of chloroplast protein synthesis caused by the joint application of atrazine and alachlor has been proposed as the basis for the observed synergism of these two herbicides on Japanese millet [*Echinochloa crus-galli* (L.) Beauv. var. *frumentacea* L.].[42]

6.3.2. Physiological Synergism

Herbicides acting on different target sites may synergize each other by producing complementary effects on the same physiological process. Each compound is unhindered in its ability to react with its receptor site and elicit a characteristic response, but the joint effect of the two chemicals is the result of their inhibitory actions on the same physiological process. At first glance, such a mechanism would seem more appropriate for explaining the additive rather than synergistic effects of two herbicides applied jointly to the same target plant. Published reports in the literature, however, demonstrate that such a mechanism could indeed explain the observed synergism of herbicides acting on the same physiological process at different sites.

Prosch and Weber[43] proposed that the synergism of dinitroaniline and chloroacetanilide herbicides on ivyleaf morning glory [*Ipomoea hederacea* (L.) Jacq.] could be the result of the differential effects of these herbicides on mitosis. Dinitroaniline herbicides such as trifluralin and oryzalin inhibit mitosis by binding to plant tubulin and inhibiting microtubule polymerization.[44] An acetanilide-induced inhibition of specific proteins in the spindle fibers of protein cells has been suggested,[43] but it has yet to be demonstrated conclusively.

Another example of physiological synergism has been reported by Bocion,[45] who demonstrated a very strong synergistic interaction between glyphosate and imidazolinone herbicides on several grass and broadleaf weeds. Both of these classes of herbicides act as inhibitors of amino acid biosynthesis in plants. Glyphosate blocks the biosynthesis of aromatic amino acids by inhibiting the activity of 5-enolpyruvylshikimate-3-phosphate (EPSP) synthase (EC 2.5.1.19), a key enzyme in the shikimic acid pathway of plants.[46] Imidazolinone herbicides block the synthesis of branched-chain amino acids by inhibiting the activity of the enzyme acetolactate synthase (ALS; EC 4.1.3.18).[47] Joint application of 250 g/ha of experimental or commercialized imidazolinone herbicides and 100 g/ha of glyphosate produced the same level of plant necrosis as that caused by individual application of 2 kg/ha of imidazolinone herbicides or 1 kg/ha of glyphosate.[45]

6.3.3. Other Synergists of Herbicidal Action

Apart from acting via a specific and direct mechanism, some herbicides can also exert their phytotoxic activity indirectly by enhancing the photogeneration of oxyradicals and reactive oxygen species.[48] The major toxic oxygen species include superoxide (O_2^-), hydrogen peroxide (H_2O_2), hydroxyl radical ($OH\cdot$), and singlet oxygen ($^1O_2$), and they are products of the reduction of molecular oxygen (O_2) or its excitation to the singlet state.[49] These oxyradicals and intermediates are highly reactive and can cause severe cellular damage and eventually death, unless they are scavenged by endogenous defensive systems which are present in the chloroplasts of higher plants.[49] The major defense systems affording protection to plants against toxic oxygen species include either enzymatic scavengers such as superoxide dismutases (EC 1.15.1.1), catalases (EC 1.11.1.6), and peroxidases (EC 1.11.1.7) or nonenzymatic substrates such as glutathione, ascorbic acid, hydroquinones, carotenes, and α-tocopherol.[49–51] Many of these enzymatic or substrate scavengers of reactive oxygen species function in a sequential order, and they are components of the so-called Asada–Halliwell photoscavenging pathway.[50,51]

Enhanced generation of toxic oxygen species has been implicated in the action of several chemically diverse classes of herbicides including the bipyridiniums, diphenyl ethers, oxadiazoles, triazines, and phenylureas.[48] All of these classes of herbicides require light for activity, and they facilitate the excess buildup of toxic oxygen species to the point that the Asada–Halliwell photoscavenging system becomes overloaded.[48] Therefore, it is reasonable to expect that any compound which reduces the activity or the levels of any component of the Asada–Halliwell photoscavenging pathway may be a potential synergist of the oxidant-generating herbicides.

Gressel and his colleagues[35,36] put this biorational hypothesis to test and developed several synergists for the herbicide paraquat. They screened a number of copper- and zinc-chelating compounds as potential synergists of paraquat activity on *Asparagus sprengeri.* Diethyldithiocarbamate, the oxime of 4-octyl-2-acetylphenol, triethylenepentamine, di-2-ethylhexylphosphoric acid, and pyridine-2,6-dicarboxylic acid synergized significantly the activity of paraquat on *Asparagus sprengeri.*[36] Since chloroplast superoxide dismutase and ascorbate peroxidase are copper/zinc- and copper/iron-containing enzymes,[51] it was proposed that the basis for the observed synergism of paraquat by these chelators was probably due to the chelation of copper and zinc of these scavenging enzymes.[36]

Chemicals that conjugate with reduced glutathione have also been implicated as potential synergists of oxidant-generating herbicides. Dionigi and Dekker[52] reported recently that pretreatments with tridiphane synergized the damaging effect of paraquat on soybeans [*Glycine max* (L.) Merr.]. As mentioned earlier, tridiphane is a potent inhibitor of plant glutathione *S*-transferases (EC 2.5.1.18) and synergizes the activity of the herbicide atrazine on foxtails and panicoid grasses.[39] Tridiphane exerts its synergistic action after it forms a conjugate with reduced glutathione. It was proposed that the synergism of paraquat and tridiphane on soybeans results from the limited availability of glutathione needed for the quenching of oxyradicals in tridiphane-treated plants because of the formation of the glutathione-tridiphane conjugate.[36,52]

6.4. ANTAGONISTS OF HERBICIDE ACTION AT TARGET SITES

From a practical point of view, herbicide antagonism is generally less desirable than herbicide synergism.[4,9,10,36] This is particularly true when the antagonistic interactions between herbicides or between herbicides and nonherbicides result in reduced weed control. On selected occasions, however, herbicide antagonism may increase crop safety.[4,9,10] Indeed, exploitation of specific antagonistic interactions between herbicides and other agrochemicals has led to the commmercial development of herbicide safeners.[30,33,34] Herbicide safeners (also known as herbicide antidotes) protect crops against herbicide injury without protecting any weeds. Many of the practical and biological aspects of their use and action have been reviewed in a recent book on this topic.[32]

A significant decrease in the amount of a given herbicide that reaches its site of action caused by an antagonist-mediated reduction of herbicide uptake and transport, sequestration, or an enhancement of herbicide detoxification is currently viewed as the major mechanism of action of herbicide antagonists including safeners.[9,10,32] Nevertheless, interactions between herbicides and antagonists at target sites do occur, and they may result from the following mechanisms: (a) competition for binding to the same target site, (b) circumvention of the target site, and (c) compensation of the target site. The principles involved and specific examples illustrating these mechanisms will be discussed next.

6.4.1. Competition for Binding to the Same Target Site

Competitive binding for an active site occurs when an antagonist acts reversibly on the same target site as the herbicide; this mechanism has been referred to also as the "lock and key" mechanism.[53,54] By reacting with the receptor site, the antagonist prevents the binding of a given herbicide to the same receptor site and the formation of a herbicide–receptor complex that may be necessary for the expression of herbicidal activity. Although the concept of receptors for natural or synthetic plant growth regulators is not as advanced in plants as it is in mammalian systems, a number of metabolic enzymes or membrane-bound proteins have been identified recently as target sites for specific herbicides.[55]

6.4.1.1. Antagonists of Auxin-Type Herbicides

The first example of herbicide antagonism due to competitive binding for an active site was reported in 1953 when Hoffman[53] demonstrated that tomato plants treated with 2,4,6-T, an inactive structural analogue of 2,4-D, were not injured by low doses of 2,4-D vapors under greenhouse conditions. A more recent example of competitive antagonism between two auxin-type herbicides was provided by Hall and Soni.[56] They showed that the phytotoxic effects caused by picloram rates of up to 250 g/ha on rapeseed (*Brassica napus* L.) could be antidoted successfully by pretreating rapeseed with 500 g/ha of clopyralid, a chemical analogue of picloram. Hall and Soni[56] showed that clopyralid pretreatment significantly reduced ethylene evolution induced by picloram but did not influence the uptake, translocation, or metabolism of [^{14}C]picloram in rapeseed. On the basis of these results, they postulated that the antagonism of picloram by clopyralid on rapeseed may be due to an interaction of the two herbicides at a common receptor site. The biochemical nature of such a receptor is currently unknown. The subject of binding proteins and receptors for natural or synthetic auxins has been reviewed recently by Libbenga *et al.*[57]

6.4.1.2. Antagonists of Acetolactate Synthase-Inhibiting Herbicides

Sulfonylurea, imidazolinone, and triazolopyrimidine herbicides are potent inhibitors of the enzyme acetolactate synthase (ALS) and of the biosynthesis of branched-chain amino acids.[58] An example of competitive antagonism between two herbicides with this mode of action was provided recently by Green,[4] who reported a strong antidoting of the postemergence activity of 16 g/ha of chlorsulfuron on grain sorghum by a simultaneous postemergence application of 500 g/ha of DPX-D5293 [*N*-(aminocarbonyl)-2-chlorobenzenesulfonamide], an inactive structural analogue of chlorsulfuron. Competitive binding between chlorsulfuron and DPX-D5293 for a single receptor site of ALS offers a likely explanation for the observed antagonism between these two herbicides on grain sorghum. However, specific data on the binding affinity of these two sulfonylureas for ALS were not provided.[4]

In other studies, it was shown that pretreatment with the safeners dichlormid and naphthalic anhydride (NA) increased significantly the activity of ALS in treated corn seedlings.[59,60] The *in vitro* inhibition of ALS by chlorsulfuron, however, was either unaltered[59] or increased[60] by pretreatment with the safeners. Rubin and Casida[59] proposed that the dichlormid-induced elevation in ALS activity *in vivo* is sufficient to offer partial protection to corn against injury from the herbicide chlorsulfuron. In similar studies on the interactions between safeners and imidazolinone herbicides, Barrett[61] failed to detect any measurable effects of the safeners NA, dichlormid, oxabetrinil, and flurazole on extractable ALS activity in shoots or roots of corn and sorghum seedlings.

6.4.1.3. Antagonists of Acetyl-CoA Carboxylase-Inhibiting Herbicides

Aryloxyphenoxypropionate and cyclohexanedione herbicides have been recently identified as potent inhibitors of the enzyme acetyl-CoA carboxylase (ACCase; EC 6.4.1.2).[62] A strong antagonism between these herbicides and certain broadleaf herbicides, including phenoxyalkanoic acids, diphenyl ethers, and bentazon, which are commonly mixed with them in the spray tank has been documented by several investigators.[10,63-71] The controversy as to whether the antagonism of postemergence-applied graminicides by specific broadleaf herbicides is related to biological reasons or chemical incompatibility reasons is still unresolved.[10]

The reduced activity of sethoxydim on grass weeds following tank mixing with the sodium salt of bentazon (Na-bentazon) is well documented.[68,69] This antagonism does not seem to result from interactions of the two herbicides at the target site of sethoxydim. Retzlaff *et al.*[72] reported that bentazon at 100 μM did not antagonize the effect of 10 μM sethoxydim on a crude preparation of ACCase from oat chloroplasts. Rendina *et al.*[73] have also shown that bentazon does not inhibit the ACCase activity in corn. Oxime ether safeners such as oxabetrinil and CGA-133205 did not affect the activity of ACCase isolated from grain sorghum leaves.[74] Dichlormid antagonized the effects of sethoxydim on lipid synthesis of isolated grain sorghum leaf protoplasts,[75] but the effects of this safener on the activity of ACCase have not been examined.

A bentazon-induced reduction in the foliar absorption of sethoxydim by goose grass (*Eleusine indica* L.) has been proposed by Rhodes and Coble[76] as a likely mechanism for the observed antagonism between these two herbicides. However, Retzlaff *et al.*[72] suggested that the site of the antagonistic interaction between bentazon and sethoxydim is located in biomembranes such as the chloroplast envelope or the plasmalemma. The effects of sethoxydim on membrane potential of plant cells have been reported recently by Weber and his colleagues.[77,78]

Thus, it appears likely that the antagonism of the activity of cyclohexanedione herbicides by bentazon and other broadleaf herbicides may result from interactions occurring at target sites other than the enzyme ACCase, which is currently accepted as the primary target site of these graminicides. Experimental data supporting such a postulation are rather limited with regard to the antagonism of sethoxydim by bentazon. However, accumulated evidence from studies on the antagonism of the aryloxyphenoxypropionate herbicide diclofop-methyl by 2,4-D on wild oats (*Avena fatua* L.) appears to strongly support this possibility.

The antagonism of diclofop-methyl(DM) by 2,4-D is reciprocal[79] and has been demonstrated in coleoptiles or whole plants of wild oat[63,64,80] as well as in corn and soybean tissue culture.[79] DM and 2,4-D are structural analogues with different biological activities. DM is phytotoxic to grasses but inactive against broadleaf weeds, whereas the activity of 2,4-D is just the opposite. While the effects of DM on wild oat are much reduced by 2,4-D, the activity of 2,4-D on broadleaf weeds is not affected by DM.[67] Other auxinlike herbicides such as dicamba and MCPA also antagonize the action of DM on wild oats.[64,79] The timing of the application of DM and auxin herbicides appears to be critical for their antagonistic interaction on wild oats. In oat coleoptiles, 2,4-D reverses the action of DM when applied within 2 h after exposure to DM,[84] whereas in whole plants, antagonism occurs even when auxin herbicides are applied within 8 h following the application of DM.[64]

The following mechanisms have been proposed to explain the antagonism of DM by 2,4-D on wild oats: a 2,4-D-induced reduction of DM absorption and translocation[66,81,82]; a 2,4-D-induced decrease in the rate of hydrolysis of DM to diclofop acid, the active form of this herbicide[66,82]; a 2,4-D-induced increase in the rate of herbicide detoxification by conjugation of diclofop to water-soluble metabolites[66,67]; a competitive antagonism between DM and 2,4-D on a common receptor site[67,83]; and a 2,4-D-induced reversal of the DM effects on transmembrane proton gradient.[84]

While earlier investigations on the antagonism of DM by 2,4-D on wild oats tended to favor an enhanced detoxification of DM induced by 2,4-D,[66,67,81,82] more recent studies appear to connect the mechanism of this antagonism with the site of action of DM.[83,84] Shimabukuro *et al.*[84] suggested that the reversal of the effects of DM on membrane potential, acidification, and growth by 2,4-D and other auxins is the most likely mechanism for the observed antagonism of DM by 2,4-D on wild oats. A similar explanation has been also suggested for the antagonism of DM by MCPA.[85] Lipid metabolism and ACCase do not appear to be involved in the antagonism of DM by 2,4-D.[84] Rendina *et al.*[73] showed that ACCase is insensitive to

2,4-D, whereas Nalewaja *et al.*[86] reported that exogenous applications of fatty acids and their esters enhanced rather than decreased the phytotoxicity of DM. Kocher *et al.*[87] have also shown that the safener HOE 70542, which protects wheat from fenoxaprop-ethyl injury, did not influence the degree of ACCase inhibition by fenoxaprop-ethyl.

6.4.2. Circumvention of the Target Site

Circumvention of the target site of a herbicide resulting from modifications of the binding site has been documented as a major mechanism for the development of weed biotypes which are resistant to specific classes of herbicides.[5–7,58] Atrazine-resistant smooth pigweed (*Amaranthus hybridus* L.), chlorsulfuron-resistant prickle lettuce (*Lactuca scariola* L.), and trifluralin-resistant goose grass (*Eleusine indica* L.) have all resulted from alterations in the target or binding sites of the herbicides.[58] Circumvention of the target site of herbicides in a biosynthetic pathway may also be involved in the action of herbicide antagonists. Specific examples illustrating this mechanism of herbicide antagonism are given below.

6.4.2.1. Antagonists of Photosynthesis-Inhibiting Herbicides

It is currently accepted that photosystem II (PSII)-inhibiting herbicides block electron transport by replacing the native quinone (termed Q_A) which acts as the second electron carrier on the reducing side of PSII.[37] Given this fact, it can be postulated that synthetic quinones may compete with the binding of PSII herbicides, in a sense circumventing the target site of these herbicides. Indeed, Vermaas and his colleagues[88,89] have shown that the quinone derivative 6-azido-5-decyl-2,3-dimethoxy-*p*-benzoquinone (6-azi-do-$Q_{10}C_{10}$) was a competitive inhibitor of atrazine binding to pea (*Pisum sativum* L.) thylakoids without inhibiting electron transfer through PSII. Oettmeier and Soll[90] reported also that plastoquinone-1 (PQ-1), a short-chain analogue of native plastoquinone, displaced competitively the herbicide diuron from plastoquinone-depleted thylakoid membranes of spinach (*Spinacia oleracea* L.). Further exploitation of these results, however, was hindered by the great differences observed in the affinities of diuron and PQ-1 for the binding site. The affinity of diuron for the binding site was higher by more than three orders of magnitude compared to that of PQ-1.[90]

6.4.2.2. Antagonists of the Herbicide Paraquat

A similar approach has been utilized in attempts to chemically regulate plant tolerance to the herbicide paraquat at the chloroplast level. Paraquat exerts its phytotoxicity by accepting photosynthetic electrons from photosystem I (PSI) and transferring these electrons to oxygen-producing toxic oxygen species which cause lipid peroxidation and membrane breakdown.[49] During this process, paraquat is reduced to a cationic radical which is quickly reoxidized by molecular oxygen. The toxic oxygen species produced were discussed earlier in this chapter.

Until recently, it was accepted that paraquat is reduced by the primary electron acceptor of PSI, which is also responsible for the reduction of $NADP^+$.[49] Recent advances in this area, however, suggest that electron flow after PSI is branched (see Ref. 91). One branch is involved in $NADP^+$ reduction whereas another branch is involved in paraquat- or ferredoxin-mediated O_2 reduction. The branch for reduction of O_2 offers a potential site for regulating paraquat toxicity since electron flow to paraquat could be inhibited without a block of $NADP^+$ reduction. An unidentified inhibitor from hemolyzed rabbit sera was found to inhibit the paraquat- or ferredoxin-mediated O_2 evolution without inhibiting $NADP^+$-dependent O_2 evolution.[92] Similar results were obtained with hemoglobin, which offered partial protection of pea chloroplasts against paraquat by inhibiting only the paraquat-mediated O_2 reduction.[93]

Attempts to antagonize the activity of paraquat with benzyl viologen and other pyridyl compounds were unsuccessful.[93,94] D-Penicillamine, a copper chelate with superoxide dismutating activity, however, partially protected flax (*Linum usitatissimum* L.) cotyledons from paraquat injury.[95] Preconditioning of plants to sublethal doses of stresses which facilitate oxygen toxicity, such as ferrous sulfate and sulfite, has been reported to induce tolerance against subsequent treatments with paraquat.[53,96] An enhancement of the endogenous protective mechanisms toward oxygen species appears to be the reason for the cross-tolerance of these plants to paraquat.[96] Ferrous sulfate applied as a spray with paraquat could protect wheat (*Triticum aestivum* L.) and oats from paraquat injury, but the antidote-to-herbicide ratio (100:1) was too high for commercial exploitation.[53]

6.4.3. Compensation of the Target Site

An ultimate proof of the exact characterization of the mechanism of action of a herbicide is often provided by the reversal of its biological activity following the exogenous application of precursors, substrates, or products of a biosynthetic pathway which is inhibited by the herbicide in question. Antagonists of antibiotics or fungicides whose action is based on compensation of the target site of these biocides have been reported.[97,98] Selected examples of herbicide antagonists compensating for the target sites of herbicides are discussed in the following sections.

6.4.3.1. Antagonists for Herbicides Inhibiting the Biosynthesis of Folic Acid

The antagonism of bactericidal sulfanilamides by *p*-aminobenzoic acid (PABA), a precursor of the vitamin folic acid, has been well documented in the pharmacological literature.[97] More recently, it was shown that the phytotoxic activity of asulam, a herbicidal sulfanilamide derivative, also results from an inhibition of the biosynthesis of the vitamin folic acid.[99–102] In particular, asulam inhibits competitively the enzyme 7,8-dihydropteroate synthase, which catalyzes the conversion of 2-amino-4-hydroxy-6-hydroxymethyl-7,8-dihydropteridine and PABA to dihydropteroic acid, leading to folate depletion.[99] A consequence of the depletion of folic acid derivatives is the buildup of intermediates of the *de novo* synthesis of purine nucleotides.[102] Such intermediates include 5′-phosphoribosyl glycineamide (GAR) or 5′-phosphoribosyl-5-amino-4-imidazole (AICAR). The accumulation of GAR and AICAR in asulam-treated pea seedlings has been reported.[102]

The reversal of the inhibitory action of asulam on wheat, wild oats, and carrots by exogenous applications of folic acid and *p*-aminobenzoic acid has been demonstrated by a number of investigators.[99–102] The antagonism of asulam by 2.4-D on bracken fern [*Pteridium aquilinum* (L.) Kuhn] appeared to be linked to an interaction of the two herbicides on protein synthesis rather than the biosynthesis of folic acid.[103]

The phytotoxicity caused by high concentrations of the fungicide sodium sulfanilate has also been linked to an inhibition of folic acid biosynthesis. The differential response of tolerant asparagus long bean (*Vigna sesquipedalis* Fruvirth) and of susceptible peanut (*Arachis hypogaea* L.) to sodium sulfanilate was attributed to the differential efficacy of this inhibitor on the biosynthesis of folic acid in these plants.[104,105] Exogenous applications of PABA reversed the phytotoxicity of sodium sulfanilate.[104]

6.4.3.2. Compensation-Type Antagonists for Herbicides Inhibiting Amino Acid Biosynthesis

During the last decade, the biosynthesis of amino acids in plants has been exploited successfully as a desirable target site for herbicides because of the low mammalian toxicity of inhibitors having such a mode of action.[106] Inhibitors of amino acid biosynthesis that are currently marketed as herbicides include glyphosate and sulfosate, which inhibit the biosynthesis of the aromatic amino acids phenylalanine, tryptophan, and tyrosine by blocking the activity of the enzyme EPSP synthase[46,106]; glufosinate and bialaphos, which inhibit the biosynthesis of glutamine by blocking the activity of glutamine synthetase (EC 6.3.1.2)[107,108]; and sulfonylurea, imidazolinone, and triazolopyrimidine derivatives which inhibit the biosynthesis of the branched-chain amino acids by blocking the activity of acetolactate synthase (also known as acetohydroacid synthase).[47,58]

The mode of action of these herbicides has been verified by the reversal of their inhibitory action by exogenous applications of the depleted amino acids.[109-114] The inhibitory action of glyphosate on plants could also be reversed with exogenous applications of the amino acids aspartate and glutamate and several organic acids of the tricarboxylic acid cycle such as α-ketoglutarate, succinate, or malate.[115] In contrast, exogenous applications of glutamine did not antagonize the herbicidal activity of bialaphos on Japanese barnyard millet (*Echinochloa utilis* Ohwi).[116] This result confirms that the rapid accumulation of intracellular ammonia resulting from the bialaphos-induced inhibition of glutamine synthetase and an associated disruption of chloroplast structure is the actual cause of plant cell death, rather than the depletion of glutamine.[108]

6.4.3.3. Compensation-Type Antagonists for Other Herbicides

In early studies on the mode of action of PSII-inhibiting herbicides such as substituted phenylureas and *s*-triazines, it was observed that the phytotoxic symptoms of these herbicides could be alleviated by the exogenous addition of sugars.[117,118] Similarly, purine derivatives, such as adenine or guanine, and pantothenic acid have been reported as compensation-type antagonists of the activity of the herbicides amitrole and dalapon, respectively.[119,120]

A number of studies have demonstrated that the phytotoxic action of trifluralin and of other dinitroaniline herbicides can be antidoted with the use of compensation-type antagonists (for a review, see Ref. 121). Ascorbic acid, cysteine, 2,3-d-mercaptoethanol, and D-α-tocopherol acetate have all been reported to reduce growth inhibitions induced by trifluralin or other dinitroaniline herbicides.[121,122] It is evident, however, that the antagonism of dinitroaniline herbicidal activity by these compounds does not seem to be related to the primary mechanism of action of these herbicides. As mentioned in Section 6.3.2, dinitroaniline herbicides act by interfering with the function and polymerization of plant microtubules.[44]

Additional studies have proposed that the selective phytotoxicity of trifluralin to young seedlings is determined partially by the amount of endogenous liipids available for trapping trifluralin and preventing it from reaching its site of action.[123,124] A wide variety of externally applied lipids have been reported to prevent the phytotoxic action of trifluralin under laboratory or field conditions.[123,124]

Finally, exogenous application of ascorbate or galactonolactone, a direct precursor of ascorbate in plants, has been reported to antagonize the activity of herbicides generating toxic oxygen species, such as paraquat and diphenyl ethers.[49] Ascorbate decreased the paraquat-induced ethane evolution and chlorophyll bleaching in thylakoid membranes of spinach,[125] whereas galactonolactone protected alfalfa (*Medicago sativa* L.) seedlings from the peroxidative action of the herbicide acifluorfen.[126] The free-radical scavenger α-tocopherol has also been reported to protect plants against injury caused by diphenyl ether herbicides.[127]

6.5. CONCLUDING REMARKS

Adverse interactions between herbicides and other herbicides or nonherbicide chemicals on higher plants have been well documented. The continuous discovery and development of new herbicides dictates the need for continuing studies of the potential interactions of herbicide mixtures. A better understanding of herbicide interactions will facilitate the efficient use of mixtures of old and new herbicides in future crop production.

Of the currently known interactions of herbicides with other agrochemicals on higher plants, only a few appear to result from the interference of a selected agrochemical with herbicidal activity at target sites. Examples of such interactions are the main topic of the discussion presented in this chapter. Further advances in our understanding of the mechanisms of action and metabolism of specific herbicides will facilitate our ability to predict the performance of herbicide mixtures.

Environmental and genetic factors that affect herbicide interactions should be identified and studied more in the future to ensure optimum performance of herbicide mixtures. Advances in computer modeling coupled with a better understanding of the physiological principles underlying the interactions of herbicides with other agrochemicals would play a critical role in predicting and evaluating the performance of future herbicide mixtures. The results of such efforts will undoubtedly provide us with additional herbicide synergists or safeners for commercial use.

ACKNOWLEDGMENTS

I express my sincere appreciation to Professor J. Gressel of the Weizmann Institute of Science, Rehovot, Israel, for supplying me with a preprint of his review on "Synergizing Herbicides" which has been accepted for publication in *Reviews in Weed Science.* I am also indebted to the many colleagues who supplied reprints of their work for inclusion in this publication.

REFERENCES

1. A. S. Crafts and A. W. Cleary, *Hilgardia 10,* 401 (1936).
2. T. J. Martin, *Pestic. Sci. 20,* 289 (1987).
3. H. W. Delvo, *Proc. North Cent. Weed Control Conf. 42,* 28 (1987).
4. J. M. Green, *Weed Technol. 3,* 217 (1989).
5. H. M. LeBaron and J. Gressel, *Herbicide Resistance in Plants,* John Wiley & Sons, New York (1982).
6. J. Gressel, *Proc. Br. Crop Prot. Conf.—Weeds,* p. 479 (1987).
7. J. Gressel, *Oxford Surv. Plant Mol. Cell Biol. 5,* 195 (1987).
8. J. Gressel and L. A. Segel, in: *Managing Resistance to Agrochemicals* (M. B. Green, H. M. LeBaron, and W. K. Moberg, eds.), ACS Symposium Series, No. 421, p. 430, American Chemical Society, Washington, D. C. (1990).
9. K. K. Hatzios and D. Penner, *Rev. Weed Sci. 1,* 1 (1985).
10. J. M. Green and S. B. Bailey, in: *Methods of Applying Herbicides* (C. G. McWhorter and G. B. Gebhardt, eds.), p. 37, Weed Science Society of America, Champaign, Illinois (1988).
11. R. E. Drury, *Proc. Northeast. Weed Sci. Soc. Suppl. 34,* 69 (1980).
12. R. E. Drury, *Weed Sci. 28,* 573 (1980).
13. P. M. Morse, *Weed Sci. 26,* 58 (1978).
14. R. G. Nash, *Weed Sci. 29,* 147 (1981).
15. A. R. Putnam and D. Penner, *Residue Rev. 50,* 73 (1974).
16. J. C. Streibig, *Weed Sci. 29,* 469 (1981).
17. J. C. Streibig, *Ber. Fachg. Herbol. 24,* 183 (1983).
18. P. Kudsk, K. E. Thonke, and J. C. Streibig, *Weed Res. 27,* 425 (1987).
19. I. O. Akobundu, R. D. Sweet, and W. B. Duke, *Weed Sci. 23,* 20 (1975).
20. T. A. Campbell, W. A. Genter, and L. L. Danielson, *Weed Sci. 29,* 378 (1981).
21. S. R. Colby, *Weeds 15,* 20 (1967).
22. J. L. Flint, P. I. Cornelius, and M. Barrett, *Weed Technol. 2,* 304 (1988).
23. D. P. Gowing, *Weeds 8,* 379 (1960).
24. F. H. A. Rummens, *Weed Sci. 23,* 4 (1975).
25. P. M. L. Tammes, *Neth. J. Plant Pathol. 70,* 73 (1964).
26. R. D. Houghton, in: *Pesticide Tank Mix Applications: First Conference* (J. F. Wright *et al.,* eds.), p. 3, American Society of Testing and Materials, Baltimore, Maryland (1982).
27. H. T. Johanson and H. E. Kaldon, in: *2nd International IUPAC Congress of Pesticide Chemistry* (A. S. Tahori, ed.), Vol. 5, p. 485, Gordon and Breach, London (1972).
28. N. R. Smith, in: *Pesticide Formulations and Application Systems: Third Symposium* (T. M. Kaneko and N. B. Akesson, eds.), p. 52, American Society of Testing and Materials, Ann Arbor, Michigan (1983).
29. G. L. Lamoureux and D. G. Rusness, *Pestic. Biochem. Physiol. 27,* 318 (1986).
30. O. L. Hoffman, *Weeds 10,* 322 (1962).
31. K. K. Hatzios, *Adv. Agron. 36,* 265 (1983).
32. K. K. Hatzios and R. E. Hoagland (eds.), *Crop Safeners for Herbicides: Development, Uses, and Mechanisms of Action,* Academic Press, San Diego (1988).
33. F. M. Pallos and J. E. Casida, *Chemistry and Action of Herbicide Antidotes,* Academic Press, New York (1978).
34. C. Parker, *Pestic. Sci. 14,* 40 (1983).
35. J. Gressel and Y. Shaaltiel, in: *Biotechnology in Crop Protection* (P. A. Hedin, J. J. Menn, and R. M. Hollingworth, eds.), ACS Symposium Series, No. 379, p. 4, American Chemical Society, Washington, D.C. (1985).
36. J. Gressel, *Rev. Weed Sci. 5,* 49 (1990).
37. L. Mets and A. Thiel, in: *Target Sites of Herbicide Action* (P. Boger and G. Sandmann, eds.), p. 1, CRC Press, Boca Raton, Florida (1989).

38. C. J. Arntzen, K. Pfister, and K. E. Steinback, in: *Herbicide Resistance in Plants* (H. M. LeBaron and J. Gressel, eds.), p. 185, John Wiley & Sons, New York (1982).
39. F. W. Kirkham and R. J. Harger, *Weed Res. 22,* 57 (1982).
40. J. C. Streibig, *Weed Res. 26,* 207 (1986).
41. D. Penner, *Weed Sci. 18,* 360 (1970).
42. I. O. Akobundu, W. B. Duke, R. D. Sweet, and P. L. Minotti, *Weed Sci. 23,* 43 (1975).
43. S. D. Prosch and J. B. Weber, *Proc. South. Weed Sci. Soc. 35,* 352 (1982).
44. L. C. Morejohn, T. E. Bureau, J. Mole-Bajer, A. S. Bajer, and D. E. Fosket, *Planta 172,* 252 (1987).
45. P. Bocion, European Patent EP 234, 379 (1986).
46. N. Amhrein, *Recent Adv. Phytochem. 20,* 83 (1986).
47. M. J. Muhitch, D. L. Shaner, and M. A. Stidham, *Plant Physiol. 83,* 451 (1987).
48. M. L. Salin, *Physiol. Plant. 72,* 681 (1987).
49. K. J. Kunert and A. D. Dodge, in: *Target Sites of Herbicide Action* (P. Boger and G. Sandmann, eds.), p. 46, CRC Press, Boca Raton, Florida (1989).
50. K. Asada, M.-A. Takahasji, K. Tanaka, and Y. Nakano, in: *Biochemical and Medical Aspects of Active Oxygen* (O. Hayaishi and K. Asada, eds.), p. 45, University Park Press, Baltimore (1977).
51. B. Halliwell, *Chloroplast Metabolism: The Structure and Function of Chloroplasts in Green Leaf Cells,* Oxford University Press, Oxford (1981).
52. C. P. Dionigi and J. H. Dekker, *Abstr. Weed Sci. Soc. Am. 28,* 201 (1988).
53. O. L. Hoffman, in: *Chemistry and Action of Herbicide Antidotes* (F. M. Pallos and J. E. Casida, eds.), p. 1, Academic Press, New York (1978).
54. D. H. McRae and J. H. Bonner, *Plant Physiol. 27,* 834 (1952).
55. P. Boger and G. Sandmann (eds.), *Target Sites of Herbicide Action,* CRC Press, Boca Raton, Florida (1989).
56. J. C. Hall and M. Soni, *Pestic. Biochem. Physiol. 33,* 1 (1989).
57. K. R. Libbenga, A. C. Maan, P. L. G. van der Linde, and A. M. Mennes, in: *Hormones, Receptors, and Cellular Interactions in Plants* (C. M. Chadwick and D. R. Garrod, eds.), p. 1, Cambridge University Press, London (1986).
58. B. J. Mazur and S. C. Falco, *Annu. Rev. Plant Physiol. Plant Mol. Biol. 40,* 441 (1989).
59. B. Rubin and J. E. Casida, *Weed Sci. 33,* 462 (1985).
60. N. D. Polge, A. D. Dodge, and J. C. Caseley, *Proc. Br. Crop Prot. Conf.—Weeds,* p. 1113 (1987).
61. M. Barrett, in: *Crop Safeners for Herbicides: Development, Uses and Mechanisms of Action* (K. K. Hatzios and R. E. Hoagland, eds.), p. 195, Academic Press, San Diego (1988).
62. J. L. Harwood, *Trends Biochem. Sci. 13,* 330 (1988).
63. C. Hall, L. V. Edgington, and C. M. Switzer, *Weed Sci. 30,* 672 (1982).
64. W. A. Olson and J. D. Nalewaja, *Weed Sci. 29,* 566 (1981).
65. P. A. O'Sullivan, H. A. Friesen, and W. H. VandenBorn, *Can. J. Plant Sci. 57,* 117 (1977).
66. F. A. Qureshi and W. H. VandenBorn, *Weed Sci. 27,* 202 (1979).
67. H. F. Taylor and M. P. C. Loader, *Outlook Agric. 13,* 58 (1984).
68. J. R. Campbell and D. Penner, *Weed Sci. 30,* 458 (1982).
69. H. Coble, *Proc. South. Weed Sci. Soc. 37,* 89 (1984).
70. K. A. Croon and M. G. Merkle, *Weed Technol. 2,* 36 (1988).
71. T. C. Mueller, W. W. Witt, and M. Barrett, *Weed Technol. 3,* 86 (1989).
72. G. Retzlaff, R. Berghaus, K. Kobek, M. Focke, and H. K. Lichtenthaler, *Proc. EWRS Symp. Factors Affecting Herbicidal Activity and Selectivity,* p. 139 (1988).
73. A. R. Rendina, J. M. Felts, J. D. Beaudoin, A. C. Craig-Kenard, L. L. Look, S. L. Paraskos, and J. A. Hagenah, *Arch. Biochem. Biophys. 265,* 219 (1988).
74. S. P. Yenne and K. K. Hatzios, *Pestic. Biochem. Physiol. 35,* 146 (1989).
75. K. K. Hatzios and P. A. Moon, *Proc. South. Weed Sci. Soc. 38,* 42 (1985).

76. G. N. Rhodes and H. D. Coble, *Weed Sci. 32*, 595 (1984).
77. A. Weber and U. Luttge, *Z. Naturforsch., C 43*, 257 (1988).
78. A. Weber, E. Fischer, H. S. von Branitz, and U. Luttge, *Z. Naturforsch., C 43*, 249 (1988).
79. M. A. Shimabukuro, R. H. Shimabukuro, W. S. Nord, and R. A. Hoerauf, *Pestic. Biochem. Physiol. 8*, 199 (1978).
80. R. H. Shimabukuro, W. C. Walsh, and R. A. Hoerauf, *Plant Physiol. 80*, 612 (1986).
81. W. Olson and J. D. Nalewaja, *Weed Sci. 30*, 59 (1982).
82. B. G. Todd and E. H. Stobbe, *Weed Sci. 28*, 371 (1980).
83. B. Kafiz, J. P. Caussanel, R. Scalla, and P. Gaillardon, *Weed Res. 29*, 299 (1989).
84. R. H. Shimabukuro, W. C. Walsh, and J. P. Wright, *Physiol. Plant. 77*, 107 (1989).
85. P. J. Fitzsimons, P. Barnwell, and A. H. Cobb, *Proc. EWRS Symp. Factors Affecting Herbicidal Activity and Selectivity*, p. 63 (1988).
86. J. D. Nalewaja, G. A. Skrzypezak, and G. R. Gillespie, *Weed Sci. 34*, 564 (1986).
87. H. Kocher, B. Butimger, E. Schmidt, K. Lotzsch, and A. Schulz, *Proc. Br. Crop Prot. Conf.—Weeds*, p. 495 (1989).
88. W. J. Vermaas, C. J. Arntzen, L.-Q. Gu, and C.-A. Yu, *Biochim. Biophys. Acta 723*, 266 (1983).
89. W. J. Vermaas and C. J. Arntzen, *Biochim. Biophys. Acta 725*, 483 (1983).
90. W. Oettmeier and H.-J. Soll, *Biochim. Biophys. Acta 724*, 287 (1983).
91. A. W. Rutherford and P. Heathcote, *Photosynth. Res. 6*, 295 (1985).
92. B. L. Upham, K. K. Hatzios, and L. S. Jahnke, *Photosynth. Res. 12*, 63 (1987).
93. B. L. Upham and K. K. Hatzios, *Z. Naturforsch., C 42*, 824 (1987).
94. E. Lewinsohn and J. Gressel, *Plant Physiol. 76*, 125 (1984).
95. R. J. Youngman, A. D. Dodge, E. Lengfelder, and E. F. Elstner, *Experientia 35*, 1295 (1979).
96. H. D. Rabinowitch and I. Fridovich, *Planta 164*, 524 (1985).
97. D. D. Woods, *J. Gen. Microbiol. 29*, 687 (1962).
98. M. A. DeWard, in: *Modern Selective Fungicides—Properties, Applications, Mechanisms of Action* (H. Lyr, ed.), p. 355, Longman, London (1987).
99. P. Veerasekaran, R. C. Kirkwood, and E. W. Parnell, *Pestic. Sci. 12*, 330 (1981).
100. P. Veerasekaran, R. C. Kirkwood, and E. W. Parnell, *Pestic. Sci. 12*, 325 (1981).
101. J. L. Killmer, J. M. Widholm, and F. W. Slife, *Plant Sci. Lett. 19*, 203 (1980).
102. B. R. Kidd, N. H. Stephen, and H. J. Duncan, *Plant Sci. Lett. 26*, 211 (1982).
103. A. M. Hinshalwood and R. C. Kirkwood, *Can. J. Plant Sci. 68*, 1025 (1988).
104. L. H. Zhang and K. H. Lin, *Pestic. Biochem. Physiol. 32*, 11 (1988).
105. K. H. Lin and Y. B. Chen, *Pestic. Biochem. Physiol. 23*, 205 (1985).
106. G. M. Kishore and D. Shah, *Annu. Rev. Biochem. 57*, 627 (1988).
107. P. J. Lea, K. W. Joy, J. L. Ramos, and M. G. Gurrero, *Phytochemistry 23*, 1 (1984).
108. K. Tachibana, T. Watanabe, S. Sekizawa, and T. Takematsu, *J. Pestic. Sci. 11*, 33 (1986).
109. E. G. Jaworski, *J. Agric. Food Chem. 20*, 1195 (1972).
110. P. M. Gresshoff, *Aust. J. Plant Physiol. 6*, 177 (1979).
111. W. E. Cooley and C. L. Foy, *Pestic. Biochem. Phyysiol. 26*, 365 (1986).
112. P. C. Anderson and K. A. Hibberd, *Weed Sci. 33*, 479 (1985).
113. D. L. Shaner and M. L. Reider, *Pestic. Biochem. Physiol. 25*, 248 (1986).
114. T. Ray, *Plant Physiol. 75*, 827 (1984).
115. J. L. Killmer, J. M. Widholm, and F. W. Slife, *Plant Physiol. 68*, 1299 (1981).
116. K. Tachibana, T. Watanabe, Y. Sekizawa, and T. Takematsu, *J. Pestic. Sci. 11*, 27 (1986).
117. W. A. Gentner and J. L. Hilton, *Weeds 8*, 413 (1960).
118. D. E. Moreland, W. A. Gentner, J. L. Hilton, and K. L. Hill, *Plant Physiol. 34*, 432 (1959).
119. K. A. Sund, E. C. Putala, and H. N. Little, *J. Agric. Food Chem. 8*, 210 (1960).
120. J. L. Hilton, J. S. Ard, L. L. Jansen, and W. A. Gentner, *Weeds 7*, 381 (1959).
121. S. J. Parka and O. F. Soper, *Weed Sci. 25*, 79 (1977).
122. J. B. Huffman and N. D. Camper, *Weed Sci. 26*, 527 (1978).

123. J. L. Hilton and N. M. Christiansen, *Weed Sci. 20,* 290 (1972).
124. N. M. Christiansen and J. L. Hilton, *Crop Sci. 14,* 489 (1974).
125. E. F. Elstner, E. Lengfelder, and G. Kwaitkowski, *Z. Naturforsch., C 35,* 303 (1980).
126. A. Schmidt and K. J. Kunert, in: *Molecular Strategies for Crop Protection* (C. J. Arntzen and C. Ryan, eds.), Vol. 48, UCLA Symp. Mol. Cell Biol., Allan R. Liss, New York (1987).
127. B. B. Finckh and K. J. Kunert, *J. Agric. Food Chem. 33,* 574 (1985).

7

Fate of Soil-Applied Herbicides: Factors Influencing Delivery of Active Ingredients to Target Sites

CHRISTOPHER R. LEAKE

7.1. INTRODUCTION

An understanding of the fate of soil-applied herbicides is of value in the search for new molecules which could be more specifically orientated to reach the target site. This offers the advantages of minimizing the quantity of chemical applied to the soil necessary to perform effective biological control and reducing the possibility of any adverse environmental effects. During the last 40 years, a substantial quantity of in-depth research has been carried out on the fate of xenobiotics or "foreign" compounds applied to the soil. As a result of this research, there is perhaps a better understanding of what happens to these molecules in soil than there is of the many unknown and "natural" molecules reaching the soil from the degradation of plant material. Soil is often described as the final sink, the place where herbicide

CHRISTOPHER R. LEAKE • Environmental Sciences Department, Schering Agrochemicals Ltd., Chesterford Park Research Station, Saffron Walden, Essex CB10 1XL, United Kingdom. The author's present address is: NOR-AM Research Center, Pikeville, North Carolina 27863.

molecules which fail to reach their target site are found. The soil is equipped with a wide range of mechanisms for coping with this role, but it does have finite capabilities. In this chapter, the mechanisms which determine the fate of soil-applied herbicides and how they contribute toward the delivery of the active moiety to the target site will be reviewed.

The fate of a herbicide may be described in terms of the dissipation mechanisms which result in its redistribution in the environment and the degradation or transformation mechanisms by which the chemical structure is altered. Although this provides a convenient way of dividing the subject, the two types of mechanisms operate simultaneously within a constantly changing environment.

It is the molecular structure of the herbicide which determines its physical and chemical properties. These properties, in turn, influence the effect that the soil dissipation mechanisms will have on the spatial distribution of the molecule. Concurrent with these mechanisms are transformation pathways through which the molecule and its metabolites may pass. The same mechanisms also determine the delivery of the herbicide, and dissipation and delivery will therefore be considered together. Many other reviews of these topics have been made over the last 40 years as knowledge has increased, and, in the course of this chapter, the reader is directed to the more specialist reviews cited.

7.2. DISSIPATION MECHANISMS

7.2.1. Volatilization

The mechanism of volatilization, in which herbicide molecules leave the soil or plant surface in the vapor phase and move into atmospheric air, is a significant cause of dissipation. Since the volume of atmospheric air is substantial compared with the number of herbicide molecules, the dilution factor is very large and therefore subsequent quantification is difficult. The volatility of a herbicide depends on its inherent vapor pressure, which of course is determined by its chemical structure. As shown by Guenzi and Beard,[1] quite small changes in chemical structure can result in substantial changes in vapor pressure. The actual quantity of herbicide lost to the atmosphere also depends upon the concentration in the soil, the soil water content, air flow over the soil surface, humidity, temperature, diffusion rates within the soil air and at the surface, adsorption to soil particles, and water solubility.

The rate of volatilization from a solid surface to which a herbicide is not adsorbed, as can occur on a plant leaf, at constant temperature depends only on the vapor pressure of the herbicide and the rate of movement away from the surface. Diffusion is the major mechanism governing movement in the air layer immediately above a solid surface.[2] Thus, it is the wind speed and turbulence which mainly control the volatilization from leaf surfaces.

The volatilization from soil is more complicated because of the numerous additional mechanisms that are present in the soil, as shown in Figure 7.1. To be volatilized from soil, the herbicide may be lost directly from its solid form to its gaseous form, or it may be released from being adsorbed to soil particles into the soil solution, diffuse into the soil air, and move to the surface to be dispersed into the turbulent air flow above the soil.[2] How close this turbulent region is to the surface depends on the condition of the soil surface. Cultivation practice and soil texture determine the number and size of the clods on the surface, and thus the relative smoothness will determine the evenness of the application of the herbicide and influence the air flow and microclimate at the air-soil interface. The air within the soil is virtually static; above this, there is an area of laminar

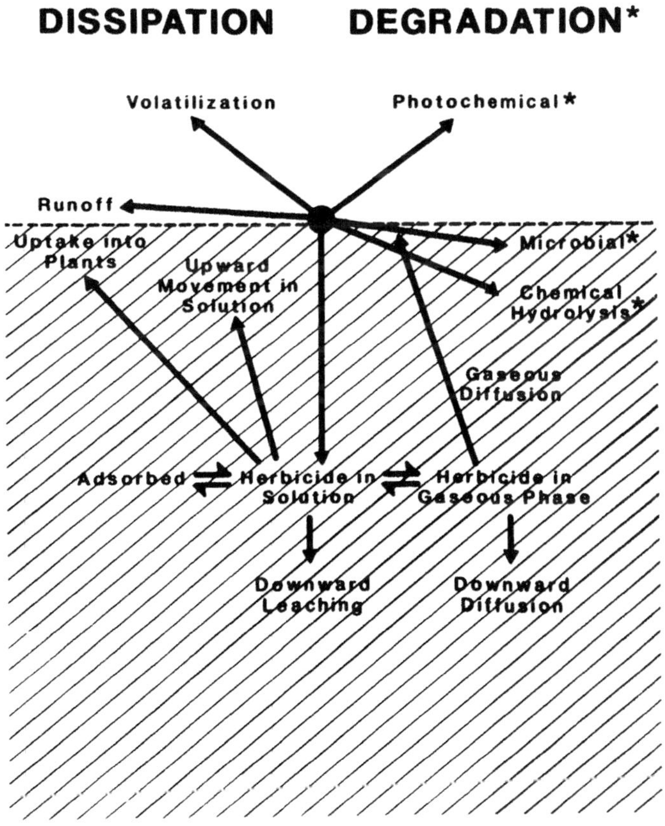

Figure 7.1. Dissipation and degradation mechanisms.

flow at the surface where movement is controlled by viscosity[2]; above this is the turbulent air flow. Thus, if a herbicide is incorporated into the soil, then the potential for volatilization is substantially reduced because of the low diffusion rates in the soil air. As the chemical moves closer to the surface, the rate of loss is substantially increased.[3]

Although losses by volatilization are potentially possible for virtually any surface-applied herbicide given certain environmental conditions, herbicides which have particularly high vapor pressures and have been shown to be volatilized from soils include EPTC,[4] triallate,[5] and trifluralin.[6]

7.2.1.1. Vapor Pressure

The molecular property of vapor pressure is one of the key parameters controlling the volatilization rate of herbicides, and methods of measurement have been reviewed by Spencer and Cliath[7] and Plimmer.[8] A lack of consistency is apparent in numerical values reported in the literature, and results are often quoted at a wide variety of temperatures. Both Plimmer[8] and Taylor and Glotfelty[9] have pointed out the problems in obtaining reliable measurements. Some authors refer to vapor density rather than vapor pressure, and Eq. (7.1) relates the two terms:

$$\mathrm{VP} = \mathrm{VD}\frac{RT}{M} \tag{7.1}$$

where VP is the vapor pressure, VD is the vapor density, R is the molar gas constant, T is the absolute temperature, and M is the molecular weight of the herbicide. The concentration of the herbicide directly determines its vapor pressure, and therefore the amount of penetration and dilution in the soil is a determining factor. If all the herbicide remains in the top 2–5 mm of soil, the surface concentration will be high. However, if it dissolves in the soil water and, following rain, is moved to lower depths within the profile, the vapor pressure at the surface will be substantially reduced. Movement of a herbicide by diffusion within the soil is considerably greater in the gaseous soil atmosphere than in the soil water. A figure of 10,000 has been suggested[10] for the increased movement in air compared with water. Movement is then dependent on the pore size and distribution of pore spaces within the soil together with the fraction of the pore space occupied by water.

7.2.1.2. Temperature

Temperature influences volatilization rates mainly through its effect on vapor pressure. In general, for every 10°C increase in temperature, there is a three- to fourfold increase in volatility.[11] The situation is complicated, however, due to the effect of temperature on soil moisture content. After the heat energy has been used as the latent heat of evaporation of the soil water, the soil surface temperature rises rapidly, causing increased adsorption of the herbicide to the soil.[9]

7.2.1.3. Soil Water Content

The effect of soil water on volatilization rates is substantial and has been investigated over a number of years.[6,12,13] Results have shown that reduced volatilization occurs in dry soil conditions due to increased adsorption to soil surfaces. When the soil water content increases, desorption from the soil surfaces occurs, resulting in increased concentration of herbicide in the soil air. It was suggested by Acree *et al.*,[14] among others, that as the soil water evaporates, the herbicide does so in the same ratio; this concept is known as "codistillation." Spencer and Cliath,[6] using the herbicide trifluralin, have shown that this concept does not apply, as greater volatilization occurred at 100% relative humidity, in which case there is no evaporation of water. More detailed discussion of the reasons why this concept does not apply has been given by Taylor and Glotfelty[9] and Spencer *et al.*[3]

7.2.1.4. Methods of Measurement

Methods of measurement of herbicide volatilization in the laboratory and the field have recently been published by Glotfelty and Schomburg[15] and Taylor and Glotfelty.[9] Laboratory methods enable control of the factors, listed above, which determine herbicide volatilization. This has resulted in a greater understanding of how vapor pressure, temperature, and soil water content influence volatilization since each factor can be adjusted in turn. Two laboratory methods have been described by Spencer and Cliath. The first involves the use of a volatilization cell,[7] in which air is passed over the surface and volatile products are collected in a series of traps. The water content of the soil is controlled by suction, and glass or other inert surfaces may be inserted into the cell to enable the effect due to the soil to be quantified. The second system[16] is a gas saturation column, in which nitrogen is passed through a column of treated sand or soil. The variables such as temperature, soil water content, soil type, and herbicide concentration may be altered to suit requirements. Volatile products are again collected in solvent traps or in polyurethane bungs. Analysis of the contents of these traps not only determines the quantity evolved but also confirms the presence of the original herbicide together with any volatile transformation products. These two laboratory systems enable investigation of the many parameters that influence volatilization; the actual quantity of herbicide volatilized under field conditions is so dependent on field environmental factors, however, that it can only be determined from field measurements. In general, because of the complexity of field measurements, there are relatively few comprehensive investigations. Clearly the environmental conditions are critical and therefore accurate determination of temperature (air and soil), wind speed, soil moisture, and residue concentrations together with any changes during the experiment are necessary. The positioning and method of obtaining air samples is another important factor. Taylor and Glotfelty[9] have shown how variable the quantities volatilized can be. For example, 90% of the applied trifluralin was volatilized in 7 days following a surface application to moist silt loam soil compared with only 3.4% in 90 days when the herbicide was incorporated to 7.5 cm into a loam soil.

7.2.1.5. Air/Water Distribution Coefficient

Volatilized herbicides therefore pass from being adsorbed to the soil into solution and from solution into the gaseous phase for transfer to atmospheric air. The distribution between the soil water and soil air is described by Henry's law,

$$\mathrm{VP} = K_H C \tag{7.2}$$

where VP is the vapor pressure, K_H is Henry's constant, and C is the concentration in solution at equilibrium; this equation is only applicable at low concentrations. Herbicides with a low water solubility, therefore, have a greater tendency to be volatilized and have a higher Henry's constant.[17] The application of this equation to modeling volatilization from soils has been discussed by Thomas.[18] It is evident that compounds with relatively low vapor pressures can be lost from soil by volatilization, particularly if they have a low water solubility.

7.2.1.6. Agronomic Factors

For the maximum concentration of a herbicide to reach its target site with the minimum of losses by volatilization, application should be made to dry rather than wet soil. Increased losses may occur from soils where the surface is smooth and compacted such as can occur in direct drilling or no-ploughing cultivation regimes. If rainfall occurs soon after application, some herbicide will be dissolved or physically washed into the soil; increased losses by volatilization, however, is likely at the surface. Greater efficacy for soil-acting herbicides with relatively high vapor pressures such as trifluralin and EPTC is achieved by incorporation into the soil. Changes in formulation type such as microencapsulation[4] may also reduce losses.

7.2.2. Adsorption

One of the major dissipation mechanisms by which herbicides fail to reach their target site and are "deactivated" is by adsorption to soil particles. It is this mechanism which causes soil to be regarded as the ultimate place of rest ("sink") for many xenobiotics. Several reviews[19-24] have covered the subject of the adsorption of xenobiotics to soils with specific references to herbicides, so only a brief overview is presented here.

Herbicides are adsorbed to soil particles which are composed of sand, silt, and clay together with the rather amorphous binding material, organic matter. Adsorption generally increases with increase in surface area of the soil constituents and with soil organic matter content.[25] The available surface area is primarily determined by the clay content of the soil because it is the clay fraction (i.e., particles with diameters of $<2\ \mu m$) that has the largest surface area. Some clays such as vermiculite and montmorillonite, the expandable clays, can have surface areas as high as 800 m^2/g.[23] This is due to their property of being able to expand in water, thus making available for adsorption surfaces that were previously internal. Other soil factors that influence adsorption include the pH and the cation exchange capacity. For nonionic herbicides, the organic matter content is the predominant soil property determining adsorption.

Adsorption of herbicides from the vapor phase can be strong as already discussed; most soils, however, have at least a monolayer of water on their surface and therefore adsorption is often via the soil solution.

7.2.2.1. Determination of Adsorption

Herbicide adsorption to soils through the solution phase is determined by obtaining an adsorption isotherm. This is done by measuring the amount of solute adsorbed at several equilibrium solute concentrations at constant temperature. In practice, most measurements determine the quantity of herbicide in solution and obtain the quantity adsorbed by difference compared with the initial amount of herbicide. The most widely used technique is the batch equilibrium method as recommended by the U.S. Environmental Protection Agency (EPA)[26] and the Organization for Economic Cooperation and Development (OECD).[27] After it has been established that there is no adsorption of the herbicide in solution to the walls of the test vessel, the time to reach equilibrium is determined with a portion of soil. In general, a plot of solution concentration against time levels out within 24 h,[28] only gradually declining with time thereafter (Figure 7.2).

It is important that equilibrium be established or else the amount of adsorption will be underestimated. Other factors such as solution/soil ratio and centrifugation techniques which determine the accuracy of the result

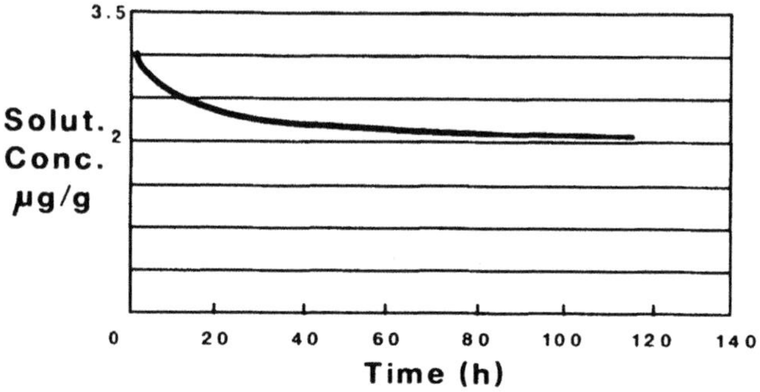

Figure 7.2. Determination of the time of adsorption equilibrium.

have been discussed by McCall *et al.*[29] For most herbicides, solution/soil ratios between 5:1 and 20:1 are suitable, and only in the case of compounds which are very readily adsorbed is it necessary to increase the ratio to 100:1 in order to determine the quantity of herbicide in solution. If the quantity of herbicide in solution at equilibrium is low, then it may approach the limit of detection of the analytical method. The use of ^{14}C-labeled herbicides substantially improves the sensitivity of the measurement. It also provides the capability of determining the quantity of herbicide adsorbed to the soil as well as in solution at equilibrium. A total account of the herbicide confirms the validity of the experiment. Increased accuracy may be obtained by making all measurements by weight. The handling of the soil prior to the experiment has been shown to change the adsorption capacity.[30] In addition, sterilization of the soil to prevent microbiological decomposition has been shown to cause decreased adsorption.[31] Sterilization does not prevent chemical hydrolysis which may be catalyzed by clay surfaces. It is surprising that many workers fail to perform chromatographic analysis of the supernatant solution to confirm the presence of parent herbicide at equilibrium.

Other methods of determining adsorption involve flowing systems similar to that used to measure mobility; these have been reviewed recently by Hance.[19] Wahid and Sethunathan[32] proposed a method for determination of rapidly adsorbed compounds for which the inclusion of a centrifugation stage precludes measurement of the kinetics of adsorption. Their method involved the use of a hypodermic syringe, with a filter to remove herbicide in solution.

7.2.2.2. Adsorption and Desorption Equations

A plot of the concentration in solution versus concentration adsorbed gives the adsorption isotherm. Adsorption isotherms have been classified

Figure 7.3. Classification of adsorption isotherms according to Giles *et al.*[33]

according to their shape by Giles *et al.*,[33] who used the shape of the adsorption isotherm to characterize the type of mechanism that might describe the adsorption. Four basic types were described: S, L, H, and C (Figure 7.3).

The S-shaped isotherm is explained by increasing adsorption with increasing concentration, the L-shaped by decreasing adsorption with concentration, H is a special case in which herbicide is highly adsorbed and C is where the herbicide penetrates more readily than the soil water. A more detailed explanation of the mechanisms involved in each type of isotherm and which herbicides adopt which isotherm shape has been given by Weber and Miller.[34]

Many attempts have been made to devise an equation which will mathematically fit soil adsorption isotherms, but because of the complexity of soil surfaces, none have been found to be particularly successful. In the simplest case, the quantity adsorbed per unit weight, x/m, increases in direct proportion to the concentration in solution, C, and a constant called the distribution coefficient, K_d, relates the two terms:

$$x/m = K_d C \tag{7.3}$$

This, however, takes no account of the filling up of adsorption sites as the solution concentration increases. The Langmuir equation[35] makes such an adjustment on the basis of there being a uniform surface, with adsorption dependent on the area available for adsorption and the concentration in solution. Thus, the Langmuir equation is

$$x/m = \frac{KCY}{(1 + KC)} \tag{7.4}$$

where Y is the maximum adsorption, assuming monolayer coverage, and K is a constant. For low solution concentrations, the factor KC becomes small compared with 1, and thus the equation reverts to the simple form. Since soil surfaces are far from uniform, the equation has proved of limited use; in addition, more than one-layer adsorption can take place. A form of the simple equation known as the Freundlich isotherm equation has been found by experience to fit many herbicide adsorption isotherms:

$$x/m = K_d C^{1/n} \tag{7.5}$$

where n is the exponential constant and is added to account for nonlinearity at higher concentrations. The equation therefore contains two independent variables, K_d and n, which vary with the soil and the herbicide. This renders the equation limited only to interpolation and not extrapolation.[19] For low solution concentrations, the value of n generally approaches 1; deviations from 1 have been discussed by Hamaker and Thompson.[21]

The adsorption of a herbicide is not necessarily a one-way process. Once adsorbed, a herbicide may also be desorbed or released back into the soil solution, where it will be free to travel toward its target site. In general, desorption of herbicides from soils has received relatively little attention. It is well known under laboratory conditions that many herbicides, once adsorbed, are less readily desorbed, thus exerting a hysteresis effect. Graham-Bryce[36] showed that drying a soil can cause reduced desorption. A typical example of the hysteresis effect seen with adsorption–desorption isotherms is shown by atrazine[37] (Figure 7.4). Each point represents a new

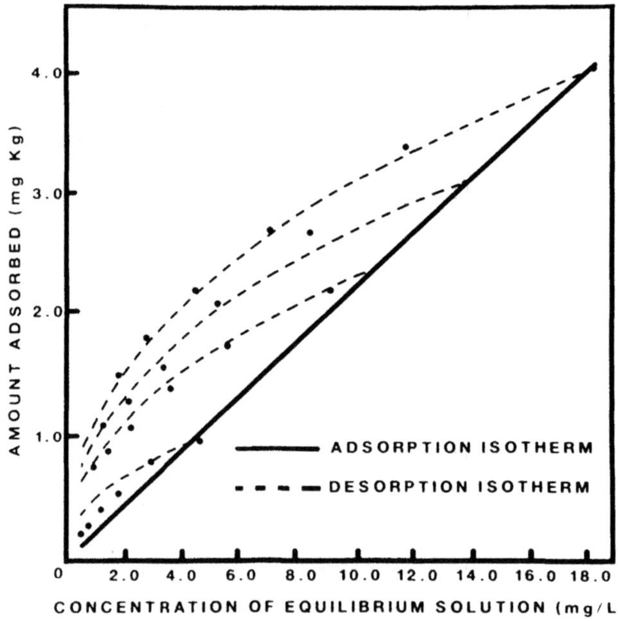

Figure 7.4. Adsorption/desorption isotherms of atrazine. Reproduced from Swanson and Dutt,[37] with permission of the Soil Science Society of America.

equilibrium position reached after atrazine is removed from the solution phase and the solution phase is replenished with fresh solution. With each desorption stage, progressively less atrazine is removed from the soil. It is important that the true new equilibrium position be reached. There have been relatively few attempts to derive equations for the desorption curve. van Genuchten *et al.*[38] proposed the use of the Freundlich equation (Eq. 7.5) with one equation for the adsorption phase and another for each desorption curve at each concentration. In the field, however, it is unlikely that true equilibrium conditions for either adsorption or desorption ever exist, as there is insufficient opportunity for mixing and the system is rarely static. Laboratory measurements, therefore, may overestimate the amount of adsorption. Under field conditions, increasing adsorption with time has been reported,[39] and the extent of desorption under field conditions remains largely unquantified.

7.2.2.3. Mechanisms of Adsorption

Mechanisms of adsorption have been reviewed by Calvert,[20] Hance,[19] and Green.[23] For some herbicides, the mechanism of adsorption can be deduced from the properties of the molecule, but for many others, the mechanisms are unknown. Most herbicide molecules are uncharged; for those that are charged, however, ionic adsorption is the most likely mechanism. The overall charge on soil surfaces is negative, though positive charges are present as are pH-dependent charges. Herbicides which are adsorbed by ionic bonding include the cations paraquat and diquat, weak acids 2,4-D and dicamba, and weak bases such as the triazines. The pH of the soil solution is an important factor determining the amount of dissociation of these herbicides. Other potential mechanisms of adsorption include liquid exchange, hydrogen bonding, charge transfer, dipole, and van der Waals forces.[19] For unionized herbicides, a direct correlation between the amount of adsorption and the soil organic matter content has been found by many workers, and it is now generally accepted that the organic matter is the predominant soil property determining adsorption. In the strict use of the word, adsorption describes the binding to a surface; in the case of organic matter, another possible mechanism is absorption, in which the herbicide is incorporated into its complex structure. Since organic matter is derived from a wide variety of plant debris, including root exudates and a multitude of other sources, it is perhaps surprising that such a consistent correlation between the quantity of organic matter and the amount of herbicide adsorption is found. Though some variations based on the type of organic matter have been observed, Hance and Cotterill[40] found that the presence of organic matter from freshly burnt straw increased the quantity of adsorbed chlortoluron.

7.2.1.4. Adsorption Coefficient

A direct relationship between the organic matter content of the soil and the adsorption coefficient K_d having been established, a new coefficient K_{oc} has been derived which is mainly independent of the properties of the soil. K_{oc} is defined as the ratio of the amount of herbicide adsorbed per unit weight of organic carbon in the soil to the concentration of the herbicide in solution at equilibrium:

$$K_{oc} = \frac{\mu\text{g adsorbed/g organic carbon}}{\mu\text{g/ml solution}} \tag{7.6}$$

This coefficient can be applied to all nonionic organic chemicals. Soil organic matter analysis has shown that approximately 58% is organic carbon,[41] and a multiplication factor of 1.724 is therefore used to convert organic carbon to organic matter. K_{oc} is an adsorption coefficient which describes the adsorption of a herbicide to soil based only on the properties of the molecule and therefore can be used to compare the relative amounts of adsorption of different herbicides. Table 7.1 gives some values reported

Table 7.1. K_{oc} Values for Some Herbicides

Herbicide	K_{oc} value	Reference
Dicamba	2.2	43
Chloramben	12.8	42
2,4-D	20	43
Picloram	26	43
Propham	51	42
Bromacil	71	42
Monuron	83	42
Simazine	140	43
Propazine	152	42
Dichlobenil	164	42
Atrazine	172	42
Fluometuron	175	43
Chlorpropham	245	42
Prometon	300	42
Ametryn	380	42
Diuron	485	42
Prometryn	513	42
Linuron	860	43
Diallate	1,800	44
Neburon	3,100	43
Chloroxuron	4,986	42
Trifluralin	13,700	44
Paraquat	20,000	42

in the literature for the K_{oc} for a series of herbicides. Some of the reported K_{oc} values are highly variable.[21] The K_{oc} value has also been found to correlate with the water solubility (S), the octanol/water partition coefficient K_{ow}, the bioconcentration factor in aquatic life (BCF), and the parachor (P). These correlations have been reviewed by Seiber,[45] Lyman,[46] and Briggs.[47] Lyman[46] has shown that there is a general relationship between K_{oc} and the properties of the molecule:

$$\log K_{oc} = a \log X + b \tag{7.7}$$

where X represents the water solubility, the octanol/water partition coefficient, or the bioconcentration factor, and a and b are constants. The parachor is directly related to the K_{oc} value. Lyman[46] has reviewed a number of equations with the general form of Eq. (7.7) and has shown that the constants a and b vary according to the number of molecules considered and type of compound. If a sufficiently large number of compounds with a wide variety of structural characteristics were considered, then a general equation relating K_{oc} to the other properties would be obtained. Some compounds may not fit this general equation either because of insufficient accuracy in the determination of the parameter used to calculate the K_{oc} value or because of particular structural characteristics. K_{oc} can be calculated by a variety of methods, each having its own potential error. Lyman[46] considered each of the potential sources of error, in turn, and concluded that the uncertainty is not unreasonable, considering that the values of K_{oc} can range over seven orders of magnitude and that a typical error in estimation is less than one order of magnitude. Hance[48] referred to the fourfold range in K_{oc} values over a range of soils in the paper by Briggs[47] and pointed out the errors associated with the accurate determination of the organic carbon content of the soil.

The adsorption coefficient K_{oc} provides a useful indication of the adsorption of a herbicide to soil and therefore enables an assessment of the proportion of herbicide remaining in the soil solution that is potentially mobile.

7.2.2.5. Agronomic Factors

Soil-acting herbicides which remain bound to the soil are unable to reach the target weed species. Temporary binding may increase the period of control. For soils with higher adsorptive capacity, such as those high in organic matter, increased herbicide treatment rates are recommended by manufacturers to enable a sufficient quantity of herbicide to reach the target site. For herbicides which are gradually desorbed from the soil, subsequent crop tolerance is important.

7.2.3. Transport in Solution

For herbicides with low vapor pressures, the major mechanism of transport within the soil is movement in the soil water. The ability of a soil-acting herbicide to provide effective weed control depends upon its ability to move to its target site within the soil profile. Some selectivity between weeds and crops can be achieved by preparation of molecules with appropriate penetration depth within the soil. For the shallow-rooted weeds, the herbicide only needs to move to approximately the top 5 cm of the soil profile; further movement takes the chemical beyond its target site and increases the potential for environmental contamination. An understanding of water movement in soils is therefore fundamental in determining the transport of herbicides to their target sites.

7.2.3.1. Mass Flow

Water movement within soil is complex and has been extensively investigated, initially to provide a greater understanding of nutrient supply and subsequently due to environmental interest. Recent reviews include those by Rao *et al.*[49] and Bouchard *et al.*[50] The basic principles, however, have been known for some time.[51] When it rains, water moves through the soil in the spaces between the solid particles under a series of forces including gravity. For a herbicide dissolved in the soil, movement in water is essentially determined by the rate of flow of the water. This is termed mass flow and is the predominant factor governing herbicide and nutrient movement in soils.[52] As the water passes through the soil, it is held by soil particles because of the energy of attraction called matric potential. For this reason, a rain shower may only cause penetration of water into the surface layer of the soil. The infiltration rate is determined by the permeability, which is dependent on the texture and structure of the soil. In soils which are initially near saturation, water flow will be greater because of the lower quantity of entrapped air pockets and the increased continuity of the soil water.

Soils containing a large clay or organic matter content will swell with wetting, and this will reduce the number and continuity of the pore spaces. Darcy's equation (Eq. 7.8) has been used to express the movement of water in soils:

$$F = -K_v \frac{dh}{dz} \tag{7.8}$$

where F is the volumetric water flux, that is, the amount of water moving per unit area in a given direction per unit time; K_v is the hydraulic conductivity constant, which is related to v, the volumetric water content of the soil; h is the hydraulic head; and z is the soil depth. Water movement in the field is complicated by the fact that the water moves in three dimensions. Due to the effect of matric potential, water movement from wet soil to dry causes upward movement in moist soils to the dry surface. In addition, water can move quickly along very narrow or capillary pores due to capillary attraction. Therefore, in the field, upward movement of herbicide in solution can be substantial, depending upon climatic factors. The effect of climatic factors such as rainfall, temperature, and evaporation rates on the movement of herbicides has been shown by Nicholls[53] and others.

7.2.3.2. Diffusion

As water moves through the soil by mass flow, diffusion of herbicide molecules takes place within the moving soil water. Movement by diffusion is much less extensive than by mass flow; however, it enables herbicide molecules to penetrate into the smaller pores within aggregates and other areas of the soil that are more isolated from the main stream of mass flow. The process of diffusion is described by Fick's law:

$$\mathrm{DF} = -D\frac{dc}{dx} \tag{7.9}$$

where DF is the diffusive flux, that is, the amount of herbicide diffusing per unit area in a given direction at any point per unit time; dc/dx is the concentration gradient, that is, the rate of change of concentration with distance per unit time; and D is the diffusion coefficient, which may be concentration dependent. Herbicides, therefore, diffuse in the direction of decreasing concentration. In general, diffusion rates of herbicides in soil are low[51]; however, over a period of time, particularly in moist soils with relatively static water, movement by diffusion can be significant.

7.2.3.3. Hydrodynamic Dispersion

As movement of herbicide in solution takes place through the soil, there is increasing dispersion caused by the variation in mass flow rates through the different pores and channels and diffusion within the soil water. This results in dispersion through the soil and is an important mechanism of distribution of the herbicide within the soil helping to achieve a uniform weed control. Hydrodynamic dispersion is therefore of importance for the efficacy of soil-acting herbicides. Rapid water flow through macropores has been increasingly investigated as a cause of potential loss of herbicide from the root zone with the resulting possibility of groundwater contamination.[54]

7.2.3.4. Measurements of Mobility

Under laboratory conditions, herbicide mobility has been determined using soil columns[55] and soil thin-layer[56] and thick-layer[55] chromatography. The leaching of herbicides through soil columns more closely resembles field conditions and allows for variation in the amounts and periods of water addition. Columns are normally hand-packed with contrasting soils to a constant bulk density, although intact cores collected from the field have beeen used.[57] Soil thin-layer chromatography is based on traditional thin-layer chromatographic movement on silica plates with soil used as the adsorbent. The soil is screened through a fine mesh (generally less than 500 μm) and spread as a slurry onto 20 × 20 cm glass plates. An advantage of this technique compared with the use of columns is that reference compounds with known mobility can be eluted directly alongside compounds of unknown mobility, without the need for a separate analytical method. Soil thick-layer chromatography[58] is a variation in which the soil does not need to be screened below 2 mm and is applied to a plate (often aluminum) 30 cm long, 5 cm wide, and 0.5 cm deep. A wick is inserted at one end, and the test chemical is applied as a narrow band to the soil. If radiolabeled herbicide molecules are available, then detection by radiochemical analysis is preferable; otherwise, chemical analysis or bioassay may be used. Wu and Santelmann[58] compared all three techniques (columns, soil thin-layer and thick-layer chromatography) and found similar results. In addition, in general, a good agreement has been found between laboratory measurements of mobility and the adsorption coefficient K_d (see Section 7.2.2.2). Briggs[47] has shown (Eq. 7.10) that for nonionized compounds, excellent agreement is achieved between measured R_f or K_d values in laboratory studies and estimates based on the octanol/water partition coefficient K_{ow}, which may be obtained from Hansch constants or the parachor:

$$\log\left(\frac{1}{R_f} - 1\right) = 0.52 \log K_{ow} + \log \text{OM} - 1.33 \qquad (7.10)$$

where R_f is the distance moved by the herbicide divided by the distance moved by the water, and OM is the soil organic matter. Helling[56] devised an index from the R_f value based on five classifications: 1 (immobile), 2 (low mobility), 3 (intermediate mobility), 4 (mobile), 5 (highly mobile). This index has proved a useful basis to describe herbicide mobility.

Comparisons with mobility measurements in the field, however, are more variable. This is not surprising because of the greater variability of the factors influencing mobility under field conditions. As already discussed, water movement in the field, unlike that in laboratory studies, can be both upward as well as downward. In addition, other factors such as volatilization, decomposition, plant uptake, the soil structure, and climatic factors increase variability.[59] Outdoor lysimeters[60] with intact soil cores offer an intermediate alternative method, in that total collection of the leachate together with destructive analysis of the soil enables an account of the distribution of the herbicide and its degradation products to be achieved under more natural conditions than those in the laboratory. Outdoor and field studies are crucial to establishing the true extent of mobility. Laboratory studies generally only provide an estimate of the potential mobility of the parent herbicide. Even when potentially mobile degradation products are identified and investigated under laboratory conditions; degradation may be more rapid than leaching under outdoor conditions.[61]

In general, soil-acting herbicides which are applied preemergence are in Helling classes 1 to 3, that is, they are immobile or have low to intermediate mobility. Herbicides such as trifluralin, triallate, and chloroxuron are all virtually immobile (class 1) in soil and provide preemergence control.[56] Diuron, dichlobenil, linuron, and EPTC are in class 2 and are all soil-acting compounds. Compounds in class 3 include atrazine, monuron, chlortoluron, and simazine and generally require some movement within the soil to provide weed control. Herbicides in classes 4 and 5 are generally acids, for example, MCPA, dicamba, TCA, and 2,4-D. Some, such as dicamba have soil as well as foliar activity. Others such as MCPA are foliar acting and therefore reach the soil by spray drift and runoff from leaf surfaces. In general, preemergence systemic herbicides applied in the autumn and subject to subsequent winter leaching conditions are found in classes 1 and 2. For application in the spring with summer conditions to follow, herbicides in classes 1 to 3 may be used. Exceptions to these generalizations are likely either to have short half-lives or, like chlorsulfuron, are found at depths below that necessary to provide weed control.[62]

7.2.3.5. Uptake by Plants

Herbicides which are taken up by the shoots rely on diffusion, as there is no mass flow of soil water to the shoot. As already discussed, movement in the soil air by diffusion is much greater than that in the soil water, so herbicides with high vapor pressures are more likely to be active via the shoot. These compounds are likely to have low mobility in soil, remaining near the surface to provide weed control. Under cool and wet conditions, some crop damage might be expected.

Herbicides which are taken up by the root depend upon mass flow of soil water to the root and uptake of the chemical in the water removed from the soil by the plant. The concentration within the root has been shown by Briggs *et al.*[63] to be related to the octanol/water partition coefficient. For compounds with a K_{ow} of less than zero, there is virtually no accumulation in the plant root; as the K_{ow} rises, so the concentration in the root rises compared to the soil solution concentration. Briggs[64] reported that the optimum K_{ow} for translocation to the leaves is 2. Herbicides with K_{ow} values higher than 4 are accumulated in the roots but not translocated. Soil-acting compounds with systemic activity are, therefore, unlikely to have a K_{ow} greater than 3.5.[64] Further details are discussed in Chapter 9.

7.3. DEGRADATION

The term degradation is commonly used to describe a change in the chemical structure of a herbicide; if used only in the strictly correct sense, it would apply only to destructive changes. Hill[65] and Smith[66] in reviews have preferred the term transformation so as to encompass all the biological and chemical changes which can take place in soils. In general, most herbicides are degraded to small, less biologically active molecules, often resulting ultimately in the formation of molecules such as water and carbon dioxide. It is perhaps for this reason that the term degradation is still frequently used, as it conveys the general direction of change, as do the similar words decomposition and breakdown. Dissipation was used earlier in this chapter to convey a transfer from one location to another without a change in chemical structure. Figure 7.1 shows the three major mechanisms of degradation: photochemical, microbial, and chemical. It is often difficult to ascertain the contribution made by each mechanism to the overall degradation rate. Attempts have been made, however, by sterilizing the soil to exclude microbial activity or incubating a dark control in photochemical studies.

7.3.1. Photochemical Degradation

Photochemical degradation is probably the least well investigated degradation mechanism, particularly under field conditions, where microbiological and chemical degradation are concurrent. As recently pointed out by Hance,[67] however, it is likely that many photochemical reactions occur under the natural sunlight of field conditions. Several laboratory techniques have been devised to simulate natural sunlight, and these have been reviewed by Guth[68] and Cavell.[69] Guth[68] pointed out that some of the techniques have used wavelengths below 290 nm, which do not reach the soil surface, and are therefore of limited value. For soil-applied herbicides, it is likely that once the compound reaches the soil, exposure to sunlight is considerably reduced. Hance[67] referred to the possibility of the presence of photosensitizers in solution which give rise to free radicals capable of initiating degradation reactions. Crosby and Li[70] have reported that a number of herbicides are subject to photochemical degradation. Most laboratory methods of determining photochemical degradation involve a light source located over a thin layer of the soil which is maintained at constant temperature by a cooling jacket. In the case of studies involving radiolabeled molecules, collection of volatile products enables an account of applied radiactivity to be obtained.

7.3.2. Microbiological and Chemical Degradation

Microbiological and chemical degradation will be considered together because sometimes they are difficult to distinguish from each other and many of the techniques used in their investigation are similar. Microbiological degradation in soil is caused by active organisms such as bacteria, actinomycetes, and fungi utilizing the herbicide for growth and/or energy gain. Chemical degradation often takes place in the soil solution and may be catalyzed by clay surfaces or other soil constituents.

7.3.2.1. Experimental Systems

A wide variety of experimental techniques have been devised for studies under laboratory conditions, and these have been reviewed by Guth,[68] Hill and Arnold,[71] and Laskowski *et al.*[72] Generally, open systems such as polyethylene bags, glass jars, and soil pots are no longer used for in-depth investigations due to the difficulty of controling the water content and of collecting evolved volatile products. Therefore, closed systems such as perfusion apparatuses, biometer flasks, and flow-through systems are more commonly utilized. Perfusion apparatuses are based upon laboratory soil fermenters in which oxygenated water is circulated through soil using continuous flow, and the resulting experimental conditions are unlikely to be found in the field. In general, the organisms present are bacteria,[68] and this enables the use of perfusion apparatuses for the generation of bacterially produced metabolites *in vitro*. Biometer flasks such as those designed by Bartha and Pramer[73] involve a virtually sealed system comprising a conical-shaped flask with a side arm in which a liquid adsorber (KOH or NaOH) is placed with a needle installed below the surface and vented to the atmosphere. Moist soil treated with herbicide is added to the main body of the flask, which is capped with a filter open to the atmosphere to remove incoming carbon dioxide. Volatile products other than carbon dioxide evolved as a result of herbicide degradation may not be collected in KOH or NaOH solution. Oxygen supply within the flask may be limited as 1.0 g of soil exhausts 1 ml of air in 24 h.[74] The geometric shape of the flasks together with the soil-to-void volume ratio has been shown to influence degradation rates.[75] Flow-through systems are perhaps the closest simulation of field conditions, where in the laboratory it is possible to control the temperature, moisture content, and air flow rate/oxygen supply. Several systems have been reported; examples are those used by Parr and Smith,[76] Goswami and Koch,[77] and Kearney and Konston.[78] All systems operate using the same basic principles. The treated soil is held in a sealed glass container such as a 250-ml conical flask, which may be wrapped in aluminum foil to preclude photochemical degradation. A current of moist air, normally with the carbon dioxide content removed or substantially reduced to avoid saturating the effluent trap, is gently blown or sucked over the soil surface, and volatile products are collected in a series of traps. The system may be used for determining volatility by fitting a polyurethane trap immediately after the flask.[78] A series of liquid trapping solutions follow for collection of organic, basic, or acidic volatiles, these solutions comprising ethanediol, sulfuric acid, and ethanolamine, respectively. The system used by Schering Agrochemicals Ltd., and shown in Figure 7.5 is based on that of Guth,[68] the flow rate being controlled at around 20 ml/min by a capillary flow cell similar to that used by Greaves *et al.*[79] This system has been used successfully for aerobic, anaerobic (flooded), and sterile investigations and has been modified for volatilization experiments. The water moisturizer generally keeps water losses below 0.2 g per month.

Virtually all the herbicide degradation studies conducted in these systems involve the use of ^{14}C- or ^{3}H-molecules. Sometimes, several radiolabeled positions are required within the molecule in order to fully

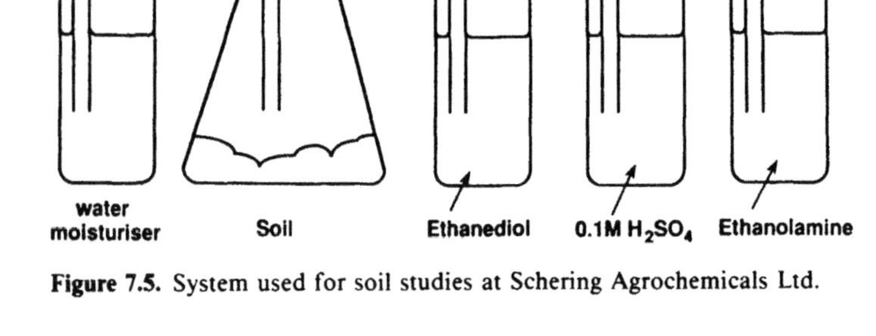

Figure 7.5. System used for soil studies at Schering Agrochemicals Ltd.

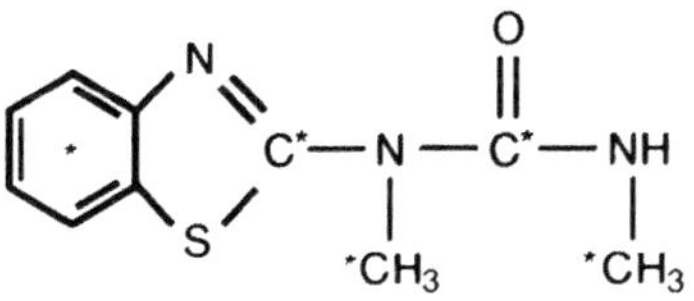

Figure 7.6. Positions labeled with ^{14}C (marked with asterisks) in methabenzthiazuron used by Cheng *et al.*[80]

elucidate the degradation pathway. Cheng *et al.*[80] used five separate preparations of the herbicide methabenzthiazuron together with radiolabeled metabolites to elucidate the route and rate of degradation of each component part of the molecule. Normally, separate positions within a molecule are labeled in order to determine the fate of larger moieties than studied by Cheng *et al.*[80] (Figure 7.6).

7.3.2.2. Microbial Biomass

The microbial biomass has been defined as the microbially active portion of the soil microflora[81] and is of vital importance in the degradation of organic matter in soil. The effects of herbicides on the microbial biomass are outside the scope of this chapter and have been reviewed by Greaves.[82] The subject of soil biomass has been reviewed by Sparling.[83] Addition of organic substrate to the soil generally increases the microbial biomass. In general, soils with the highest organic matter content have the greatest microbial biomass and would therefore be expected to have the highest degradation rate of herbicides. Since the report by Anderson[84] of the influence of initial biomass activity on the degradation of diallate and triallate, there have been relatively few studies relating herbicide degradation rate to microbial biomass. Clearly, increased biomass does not necessarily mean an increased number of organisms capable of degrading a particular herbicide. In addition, methods of determination of microbial biomass are subject to controversy.[83] The most commonly used method is probably the chloroform fumigation technique of Jenkinson and Poulson.[85] The respiration response method of Anderson and Domsch,[86] which utilizes substrate amendment however, is more rapid. The respiration response method does however, require calibration using the fumigation technique. The basis of the chloroform fumigation method is measurement of the additional CO_2 evolved following reinoculation after kill by fumigation. The following assumptions are made: (1) the carbon in the dead organisms is mineralized to CO_2 more rapidly than that in living organisms, (2) the kill is substantially complete, (3) the biomass dying in the unfumigated control is negligible compared with that killed by fumigation, (4) the fraction of the carbon from the killed biomass that is mineralized is the same in different soils, and (5) the fumigation has no effect other than the killing of the biomass.[87] Further work is required both on obtaining reliable and reproducible methods of determination of microbial biomass and on assessing the contributions of microbial biomass to the degradation of herbicides.

7.3.2.3. Degradation Pathways

Since there are numerous variations in the structures of organic herbicides, there are, correspondingly, many potential degradation products. Some compounds, such as 2,4-D, have been on the market for many years and have been extensively investigated.[88] Degradation products of other older molecules still remain unreported.

Comprehensive reviews by Smith,[66] Cripps and Roberts,[89] and Hassall[90] have covered many of the main classes of herbicides such as the ureas (e.g., diuron, linuron, chlorbromuron, chloroxuron, and chlortoluron), triazines (e.g., atrazine, simazine, trietazine, and prometryn), phenoxyalkanoates (e.g., 2,4-D, MCPA, and mecoprop), phenylcarbamates (e.g., propham and barban), benzoic acids (e.g., dichlobenil, dicamba, and chloramben), and dinitroanilines (e.g., trifluralin, nitralin, and isopropalin). Smith[66] classified the major types of reactions into 12 general classes, 9 of which are biologically mediated and 3 of which can occur as a result of either biochemical or chemical mechanisms. Hill[65] has also devised a scheme of 11 general classes.

It is apparent that soil microorganisms are capable of causing a wide range of reactions; other types of reactions may be discovered when new classes of herbicides are found or invented. Often, the ability of soil microorganisms to undertake these reactions depends on their ability to obtain the necessary nutrients, water, and oxygen supply. The microbial population can change rapidly if, for example, anaerobic conditions are created by the filling of the soil pores with water following heavy rain. Anaerobiosis occurs in soil when the oxygen consumption rate exceeds the supply rate. The oxygen supply rate is reduced in flooded soils due to the lower diffusion rate of oxygen in water than in the soil air. Soils with large aggregates may contain anaerobic centers which expand and contract according to the aeration status of the soil.[91] Obligate anaerobic organisms will be present in the centers of these aggregates, and facultative anaerobic organisms in the regions of more temporary anaerobic conditions. These organisms have been shown to degrade herbicides, for example, trifluralin,[92] by quite different metabolic routes. Willis *et al.*[93] reported that a certain E_h value between +50 and +150 mV was required for more rapid trifluralin degradation (Figure 7.7), suggesting that the microorganisms responsible required a certain level of anaerobic conditions before they were able to function. The changes in redox potential and the resulting chemical transformations

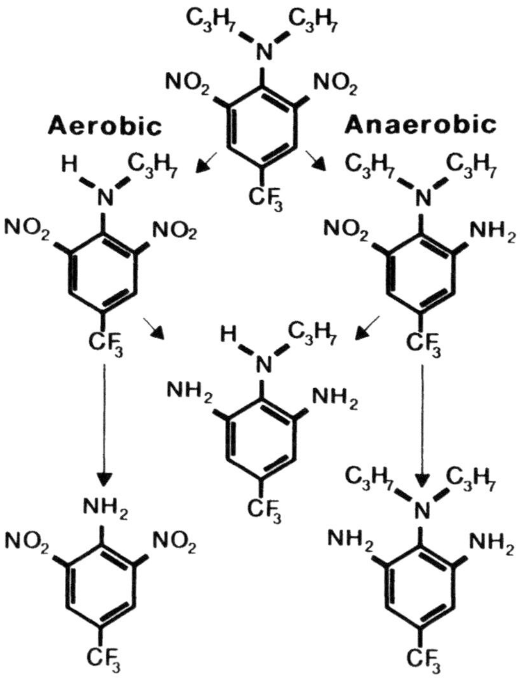

Figure 7.7. Postulated degradation pathway of trifluralin in soil. Reprinted from Probst *et al.*,[92] with permission of the American Chemical Society.

have been well documented by Ponnamperuma.[94] In general, there has been much less emphasis on anaerobic pathways of herbicides in soil. In agronomic terms, anaerobic conditions can extend the control period of the herbicide by preventing degradation or can cause more rapid degradation, resulting in reduced effectiveness.

In terms of understanding the delivery of the active moiety to the target site, it is apparent that some compounds are applied in their active form whereas others depend upon the soil to degrade or transform the applied chemical to its active form. The importance of the route and rate of degradation and the interaction of the herbicide molecule with the many soil factors can be illustrated by the following examples.

a. Aryloxyphenoxypropionates. The aryloxyphenoxypropionates (Figure 7.8) are a group of herbicides effective against a broad range of annual and perennial grass weeds.[95] All the compounds have a chiral center and, therefore, can exist as a mixture of two optical isomers called enantiomers (Figure 7.9). Herbicidal activity has been reported under both pre- and postemergence applications. When fluazifop-butyl was applied postemergence in the *R*, *S*, and *RS* enantiomeric forms, the *R* form was observed

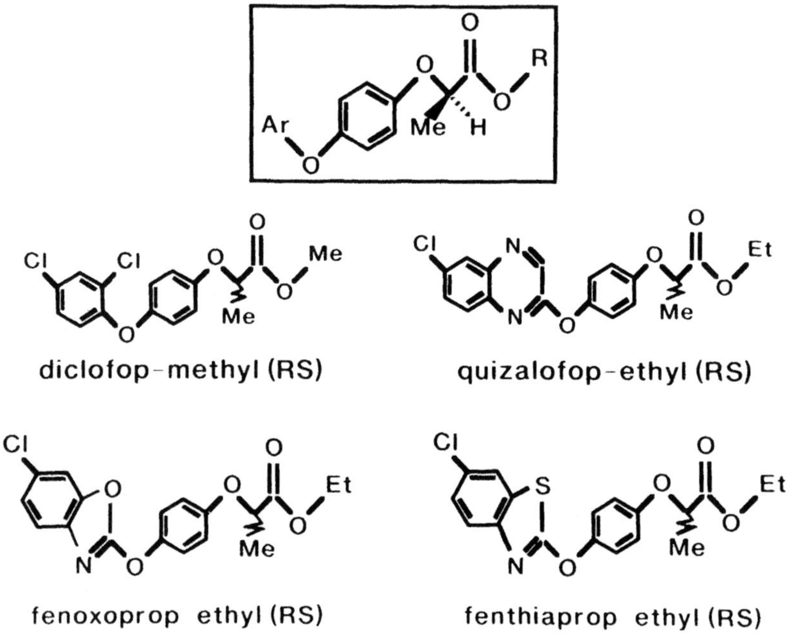

Figure 7.8. General structure and examples of aryloxyphenoxypropionates.

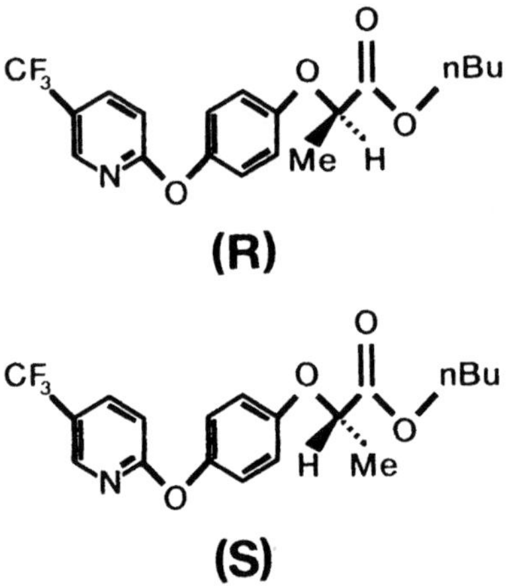

Figure 7.9. Enantiomers of fluazifop-butyl.

to be twice as active as the *S* and *RS* forms.[95] When applied preemergence, all three forms had a similar activity. In a moist sandy loam soil under aerobic conditions, fluazifop-butyl is rapidly degraded by hydrolysis to the corresponding acid, fluazifop, with a half-life of 2 h.[96] The *R* and *S* forms of fluazifop-butyl were found to hydrolyze to the corresponding *R* and *S* forms of fluazifop (acid). The fluazifop *S*-acid was then converted to fluazifop *R*-acid with a "half-life" of 1 to 2 days. This conversion of the *S* enantiomer to the *R* enantiomer did not occur in sterile soils, suggesting the involvement of microorganisms. The herbicidally active moiety of the aryloxyphenoxypropionates in plants has been shown to be the acid, ester hydrolysis occurring rapidly in plants.[97-99] The soil is, therefore, causing activation of the *S*-enantiomeric form by conversion to the herbicidally active *R*-enantiomeric form of fluazifop (acid). Similar results have been shown for the other aryloxyphenoxypropionates—diclofop-methyl, fenoxaprop-ethyl, and the esters of haloxyfop.[100,101] Further degradation in soils has been reported by Smith.[102]

b. Sulfonylureas. The sulfonylurea herbicides are highly potent herbicides with activity at rates as low as 2 g/ha following pre- and postemergence application.[103] The general structure of sulfonylurea herbicides is shown in Figure 7.10. In soil, sulfonylureas can exist in the neutral or the ionized form depending on the pK_a of the molecule and the soil pH (Figure 7.11).

The degradation of sulfonylureas in soil depends mainly on the soil pH, temperature, moisture, and organic matter content. The effect of pH is particularly important in that the pH directly influences the rate of chemical hydrolysis. Under acidic conditions, sulfonylureas are in the neutral form, and hydrolysis takes place to the corresponding sulfonamide and heterocyclic amine[104] (Figure 7.11), whereas very little degradation occurs under alkaline conditions. In the case of sulfometuron-methyl, saccharin was also identified as a degradation product. In the soil under sterile conditions, the same pH effect is observed.[105] More rapid degradation occurs in microbially active soils with corresponding pH values. Thus, chemical

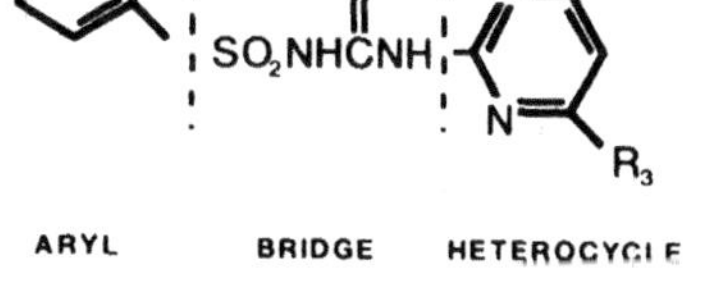

Figure 7.10. General structure of sulfonylurea herbicides. Reproduced from Savers and Levitt,[103] with permission of the American Chemical Society.

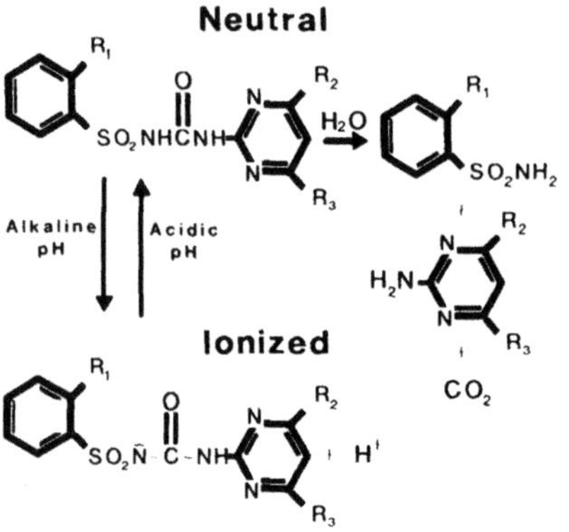

	R_1	R_2	R_3
Chlorsulfuron	Cl	CH_3	OCH_3
Sulfometuron methyl	CO_2CH_3	CH_3	CH_3
DPX-M6316	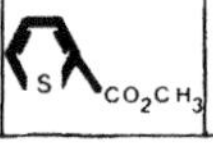	CH_3	OCH_3

Figure 7.11. Degradation in soil of sulfonylureas. *Chlorsulfuron and DPX-M6316 contain a *triazine* ring.

hydrolysis and soil microbial activity contribute to the degradation in acid soils, whereas under alkaline conditions microbial degradation is the predominant mechanism, chemical hydrolysis being virtually absent. The effect of temperature and soil moisture on the degradation of chlorsulfuron has been demonstrated by Walker and Brown.[62] An increase in temperature of 10°C increased the rate of loss of the herbicide by a factor of between 2 and 3 and an increase in soil moisture content from 6 to 12% increased the rate of loss by a factor of approximately 2.

Application of sulfonylureas in some U.K. soils has been reported to cause damage to subsequent crops (e.g., sugar beet), and this has been attributed to high biological potency and low degradation rate in cool alkaline soils.[62] It is worth pointing out the influence of mobility. The neutral form of the sulfonylurea is more lipophilic than the ionized form. In consequence, as the soil pH rises, so the molecule distributes itself toward the soil water and moves upward and downward in the soil solution, resulting in the presence of sulfometuron-methyl at depths of 32 cm in loam soil.[106] More recently introduced sulfonylureas (e.g., DPX-M6316) have shorter soil half-lives.

7.4. CONCLUSIONS

The fate of a soil-applied herbicide has been shown to depend not only on the physicochemical properties of the molecule but also on soil-related factors. In general, knowledge of the physicochemical properties of the herbicide enables some predictions to be made about likely movement in the soil water and, to a certain extent, potential losses by volatilization. These predictions, however, are limited by the influence of herbicide degradation, elucidation of which continues to rely on experimental data to determine both the products formed and the rate of their formation. Plant uptake of an active moiety from soil may first require degradation or transformation of the applied compound in soil. There are still many unknown factors in soil-herbicide behavior, mainly because there are so many competing mechanisms, any one of which may predominate for a given compound in a particular soil under certain climatic conditions. It is unlikely that any one experimental design will enable an understanding of the contribution of each mechanism. Greater knowledge of herbicide-soil relations should enable manufacturers to produce more effective and safer herbicides.

REFERENCES

1. W. D. Guenzi and W. E. Beard, in: *Pesticides in Soil and Water* (W. D. Guenzi, ed.), p. 107, Soil Science Society of America, Madison, Wisconsin (1974).
2. J. L. Montieth, *Principles of Environmental Physics*, Chapter 2, Elsevier, New York (1973).
3. W. F. Spencer, W. J. Farmer, and M. M. Cliath, *Residue Rev. 49*, 1 (1973).
4. B. E. Groenwold, F. Pereiro, T. J. Purnell, and H. B. Scher, *Proceedings of the British Crop Protection Conference—Weeds*, p. 185 (1980).
5. R. J. Hance, J. Holroyd, and C. E. McKone, *Pestic. Sci. 4*, 13 (1973).
6. W. F. Spencer and M. M. Cliath, *J. Agric. Food Chem. 22*, 987 (1974).
7. W. F. Spencer and M. M. Cliath, in: *Test Protocols for Environmental Fate and Movement of Toxicants*, p. 110, Association of Official Analytical Chemists, Arlington, Virginia (1981).
8. J. R. Plimmer, in: *Herbicides: Chemistry, Degradation and Mode of Action* (P. C. Kearney and D. Kaufman, eds.), Vol. 2, p. 891, Marcel Dekker, New York (1976).
9. A. W. Taylor and D. E. Glotfelty, in: *Environmental Chemistry of Herbicides* (R. Grover, ed.), Vol. 1, p. 89 CRC Press, Boca Raton, Florida (1988).
10. C. G. L. Furmidge, in: *Soils and Crop Protection Chemicals* (R. J. Hance, ed.), British Crop Protection Council Monograph No. 27, p. 49 (1984).
11. W. F. Spencer, in: *Fate of Pesticides in the Environment* (J. W. Biggar and J. N. Seiber, eds.), p. 61, University of California, Berkeley, California (1987).
12. S. C. Fang, P. Theisen, and V. H. Freed, *Weeds 9*, 569 (1961).
13. W. F. Spencer, M. M. Cliath, and W. J. Farmer, *Soil Sci. Soc. Am. Proc. 33*, 509 (1969).
14. F. Acree, M. Beroza, and M. C. Bowman, *J. Agric. Food Chem. 11*, 278 (1963).
15. D. E. Glotfelty and C. J. Schomburg, in: *Reactions and Movement of Organic Chemicals in Soils* (B. L. Sawhney and K. Brown, eds.), Soil Science Society of America Special Publication No. 22, p. 181 (1989).
16. W. F. Spencer and M. M. Cliath, *Environ. Sci. Technol. 3*, 670 (1969).
17. I. J. Graham-Bryce, in: *The Chemistry of Soil Processes* (D. J. Greenland and M. H. B. Hayes, eds.), p. 621, J. Wiley & Sons, Chichester, Pennsylvania (1981).

18. R. G. Thomas, in: *Handbook of Chemical Property Estimation Methods* (W. J. Lyman, W. F. Reehl, and D. H. Rosenblatt, eds.), Chapter 16, McGraw-Hill, New York (1982).
19. R. J. Hance, in: *Environmental Chemistry of Herbicides* (R. Grover, ed.), Vol. 1, p. 1, CRC Press, Boca Raton, Florida (1988).
20. R. Calvert, in: *Interactions between Herbicides and the Soil* (R. J. Hance, ed.), p. 1. Academic Press, London (1980).
21. J. W. Hamaker and J. M. Thompson, in: *Organic Chemicals in the Soil Environment* (C. A. I. Goring and J. W. Hamaker, eds.), Vol. 1, p. 49, Marcel Dekker, New York (1972).
22. G. W. Bailey and J. L. White, *Residue Rev. 32,* 29 (1970).
23. R. E. Green, in: *Pesticides in Soil and Water* (W. D. Guenzi, ed.), p. 3, Soil Science Society of America, Madison, Wisconsin (1974).
24. S. B. Weed and J. B. Weber, in: *Pesticides in Soil and Water* (W. D. Guenzi, ed.), p. 39, Soil Science Society of America, Madison, Wisconsin (1974).
25. D. S. Murray, P. W. Santelmann, and J. M. Davidson, *J. Agric. Food Chem. 23,* 578 (1975).
26. U.S. Environmental Protection Agency, Guidelines for Registering Pesticides in the United States Federal Register 40, 123 26881, U.S. Environmental Protection Agency, Washington, D.C. (1982).
27. Organization for Economic Cooperation and Development (OECD), Test Guidelines No. 106, 12 May 1981, OECD, Paris.
28. R. J. Hance, *Weed Res. 7,* 29 (1967).
29. P. J. McCall, D. A. Laskowski, R. L. Swann, and H. J. Dishburger, in: *Test Protocols for Environmental Fate and Movement of Toxicants,* p. 89, Proceedings of the Association of Official Analytical Chemists 94th Meeting, Washington, D.C. (1980).
30. B. Von Oepen, W. Kördel, and W. Klein, *Chemosphere 18,* 1495 (1989).
31. T. H. Dao, D. B. Marx, T. L. Lavy, and J. Dragon, *Soil Sci. Soc. Am. J. 46,* 963 (1982).
32. P. A. Wahid and N. Sethunathan, *Soil Sci. 126,* 56 (1978).
33. C. H. Giles, T. H. MacEwan, S. N. Nakhwa, and D. Smith, *J. Chem. Soc.* 3973 (1960).
34. J. B. Weber and C. T. Miller, in: *Reactions and Movement of Organic Chemicals in Soils* (B. L. Sawhney and K. Brown, eds.), Soil Science Society of America Special Publication No. 22, p. 305 (1989).
35. I. Langmuir, *J. Am. Chem. Soc. 38,* 2221 (1916).
36. I. J. Graham-Bryce, *J. Sci. Food Agric. 18,* 72 (1967).
37. R. A. Swanson and G. R. Dutt, *Soil Sci. Soc. Am. Proc. 37,* 872 (1973).
38. M. Th. van Genuchten, J. M. Davidson, and P. J. Wierenga, *Soil Sci. Soc. Am. Proc. 38,* 29 (1974).
39. A. Walker, *Weed Res. 27,* 143 (1987).
40. R. J. Hance and E. G. Cotterill, in: *Soils and Crop Protection Chemicals* (R. J. Hance, ed.), British Crop Protection Council Monograph No. 27, p. 65 (1984).
41. J. E. Gieseking, in: *Soil Components,* Vol. 1, Springer-Verlag, New York (1975).
42. J. W. Hamaker, in: *Environmental Dynamics of Pesticides* (R. Haque, ed.), p. 115, Plenum Press, New York (1975).
43. P. S. C. Rao and J. M. Davidson, in: *Retention and Transformation of Pesticides and Phosphorus in Soil Water Systems,* U.S. Environmental Protection Agency, Washington, D.C. (1980).
44. E. E. Kenaga and C. A. I. Goring, ASTM Special Technical Publication 707 (1980).
45. J. N. Seiber, in: *Fate of Pesticides in the Environment* (J. W. Biggar and J. N. Seiber, eds.), University of California Publication No. 3320, p. 53 (1987).
46. W. J. Lyman, in: *Handbook of Chemical Property Estimation Methods* (W. J. Lyman, W. F. Reehl, and D. H. Rosenblatt, eds.), Chapter 1, McGraw-Hill, New York (1982).
47. G. G. Briggs, *J. Agric. Food Chem. 29,* 1050 (1981).
48. R. J. Hance, *Ann. Appl. Biol. 110,* 195 (1987).

49. P. S. C. Rao, R. E. Jessup, and J. M. Davidson, in: *Environmental Chemistry of Herbicides* (R. Grover, ed.), Vol. 1, p. 21, CRC Press, Boca Raton, Florida (1988).
50. D. C. Bouchard, C. G. Enfield, and M. D. Piwoni, in: *Reactions and Movement of Organic Chemicals in Soils* (B. L. Sawhney and K. Brown, eds.), Soil Science Society of America Special Publication No. 22, p. 349 (1989).
51. J. Letey and W. J. Farmer, in: *Pesticides in Soil and Water* (W. D. Guenzi, ed.), p. 67, Soil Science Society of America, Madison, Wisconsin (1974).
52. A. Wild, in: *The Chemistry of Soil Processes* (D. J. Greenland and M. H. B. Hayes, eds.), p. 37, John Wiley and Sons, New York (1981).
53. P. H. Nicholls, *Aspects Appl. Biol. 21,* 173 (1989).
54. R. E. White, *Adv. Soil Sci. 3,* 95 (1985).
55. C. S. Helling and J. Dragun, in: *Test Protocols for Environmental Fate and Movement,* p. 43, Association of Official Analytical Chemists, Arlington, Virginia (1981).
56. C. S. Helling, *Soil Sci. Soc. Am. Proc. 35,* 732 (1971).
57. J. B. Weber and T. F. Peeper, in: *Research Methods in Weed Science* (B. Truelove, ed.), p. 78, Society of Weed Sciences, Auburn, Alabama.
58. C. H. Wu and P. W. Santelmann, *Weed Sci. 23,* 508 (1975).
59. P. S. C. Rao, K. S. V. Edvardsson, L. T. Ou, R. E. Jessup, P. Nkedi-Kizza, and A. G. Hornsby, in: *Evaluation of Pesticides in Groundwater* (W. Y. Garner, R. C. Honeycutt, and H. N. Nigg, eds.), p. 100, ACS Symposium Series, No. 315, American Chemical Society, Washington, D.C. (1986).
60. C. R. Leake, *Pestic. Sci. 31,* 363-373 (1991).
61. C. R. Leake, D. J. Arnold, S. E. Newby, and L. Somerville, *Proceedings of the British Crop Protection Conference—Weeds,* p. 577 (1987).
62. A. Walker and P. Brown, *Bull. Environ. Contam. Toxicol. 30,* 365 (1983).
63. G. G. Briggs, R. H. Bromilow, and A. A. Evans, *Pestic. Sci. 13,* 495 (1982).
64. G. G. Briggs, in: *Soil and Crop Protection Chemicals* (R. J. Hance, ed.), British Crop Protection Council Monograph No. 27, p. 35 (1984).
65. I. R. Hill, in: *Pesticide Microbiology* (I. R. Hill and S. J. L. Wright, eds.), p. 137, Academic Press, New York (1978).
66. A. E. Smith, in: *Environmental Chemistry of Herbicides* (R. Grover, ed.), p. 171, CRC Press, Boca Raton, Florida (1988).
67. R. J. Hance, *Aspects Appl. Biol. 21,* 147 (1989).
68. J. A. Guth, in: *Progress in Pesticide Biochemistry,* Vol. 1 (D. H. Hutson and T. R. Roberts, eds.), p. 85, John Wiley and Sons, New York (1981).
69. B. D. Cavell, *Pestic. Sci. 10,* 177 (1979).
70. D. G. Crosby and M. Y. Li, in: *Degradation of Herbicides* (P. C. Kearney and D. D. Kaufman, eds.), p. 321, Marcel Dekker, New York (1969).
71. I. R. Hill and D. J. Arnold, in: *Pesticide Microbiology* (I. R. Hill and J. T. L. Wright, eds.), p. 203, Academic Press, New York (1978).
72. D. A. Laskowski, R. L. Swann, P. J. McCall, H. J. Dishburger, and H. D. Bidlack, in: *Test Protocols for Environmental Fate and Movement of Toxicants,* p. 122, Association of Official Analytical Chemists, Arlington, Virginia (1981).
73. R. Bartha and D. Pramer, *Soil Sci. 100,* 68 (1965).
74. D. Pramer and R. Bartha, *Environ. Lett. 2,* 214 (1972).
75. J. E. Blum, C. A. Stevenson, and D. M. Stainken, *Environ. Toxicol. Chem. 2,* 141 (1983).
76. J. F. Parr and S. Smith, *Soil Sci. 107,* 271 (1969).
77. K. P. Goswami and B. L. Koch, *Soil Biol. Biochem. 8,* 527 (1976).
78. P. C. Kearney and A. Konston, *J. Agric. Food Chem. 24,* 424 (1976).
79. M. P. Greaves, S. L. Cooper, H. A. Davies, J. A. P. Marsh, and G. I. Wingfield, in: Weed Research Organisation Technical Report 45, Agricultural Research Council, U.K. (1978).

80. H. H. Cheng, F. Fuhr, H. J. Jarczyk, and W. Mittelstaedt, *J. Agric. Food Chem. 26,* 595 (1978).
81. J. P. E. Anderson, in: *Pesticide Effects on Soil Microflora* (L. Somerville and M. P. Greaves, eds.), p. 45, Taylor and Francis, London (1987).
82. M. P. Greaves, *Proceedings of the British Crop Protection Conference—Weeds,* p. 501 (1987).
83. G. D. Sparling, in: *Soil Organic Matter and Biological Activity* (D. Vaughan and R. E. Malcolm, eds.), p. 223 Martinus Nijhoff/Dr. W. Junk, The Hague (1985).
84. J. P. Anderson, *Soil Biol. Biochem. 16,* 483 (1984).
85. D. S. Jenkinson and D. S. Poulson, *Soil Biol. Biochem. 8,* 209 (1976).
86. J. P. E. Anderson and K. H. Domsch, *Soil Biol. Biochem. 10,* 215 (1978).
87. D. S. Jenkinson and J. N. Ladd, in: *Soil Biochemistry 5* (E. A. Paul and J. N. Ladd, eds.), p. 415, Marcel Dekker, New York (1981).
88. W. R. Mullinson, in: *Fate of Pesticides in the Environment* (J. W. Biggar and J. N. Seiber, eds.), University of California Publication No. 3320, p. 121 (1987).
89. R. E. Cripps and T. R. Roberts, in: *Pesticide Microbiology* (I. R. Hill and S. J. L. Wright, eds.), p. 669 Academic Press, New York (1978).
90. K. A. Hassall, in: *The Chemistry of Pesticides* (K. H. Buchell, ed.), p. 297, Macmillan, New York (1982).
91. J. M. Tiedje, *Plant Soil 76,* 197 (1984).
92. G. W. Probst, T. Golab, R. J. Herberg, F. J. Holzer, S. J. Parka, C. Van der Schaus, and J. B. Tepe, *J. Agric. Food Chem. 15,* 592 (1967).
93. G. H. Willis, R. C. Wander, and L. M. Southwick, *J. Environ. Quality 3,* 262 (1974).
94. F. N. Ponnamperuma, *Adv. Agron. 24,* 29 (1972).
95. D. Cartwright, *Proceedings of the British Crop Protection Conference—Weeds,* p. 707 (1989).
96. D. W. Bewick, *Pestic. Sci. 17,* 349 (1986).
97. P. F. Boldt and A. R. Putman, *Weed Sci. 29,* 237 (1981).
98. R. H. Shimabukuro, W. C. Walsh, and R. A. Hoerauf, *J. Agric. Food Chem. 27,* 615 (1979).
99. P. Hendley, J. W. Dicks, T. J. Monaco, S. M. Slyfield, O. J. Tunnon, and J. C. Barrett, *Weed Sci. 33,* 11 (1985).
100. O. Wink and U. Luley, *Pestic. Sci. 22,* 31 (1988).
101. B. C. Gerwick, L. A. Jackson, J. Handly, N. R. Grey, and J. W. Russell, *Weed Sci. 36,* 453 (1988).
102. A. E. Smith, *J. Agric. Food Chem. 33,* 483 (1985).
103. R. F. Savers and G. Levitt, in: *Pesticide Syntheses through Rational Approaches* (P. S. Magee, G. K. Kohn, and J. J. Mean, eds.), p. 21, American Chemical Society, Washington, D.C. (1984).
104. J. Harvey, J. J. Dulka, and J. J. Anderson, *J. Agric. Food Chem. 33,* 590 (1985).
105. M. M. Joshi, M. M. Brown, and J. A. Romesser, *Weed Sci. 33,* 888 (1985).
106. J. J. Anderson and J. J. Dulka, *J. Agric. Food Chem. 33,* 596 (1985).

8

Pathways and Mechanisms of Uptake of Foliage-Applied Herbicides with Particular Reference to the Role of Surfactants

RALPH C. KIRKWOOD

The activity of a foliage-applied herbicide depends upon its effect on the target enzyme(s), and, in turn, this must be influenced by the efficiency of target site delivery. It is well known that delivery is influenced by a series of interrelated factors including the efficiency of cuticle retention and penetration, absorption into the leaf tissues, and possibly translocation from the absorption site. Immobilization or metabolism of the active ingredient (a.i.) at any stage may reduce the concentration reaching the target site. Interspecies differences in activity (selectivity) of a herbicidal compound may be due to variation in the rate or efficiency of any of these processes. In this chapter, the mechanism and factors influencing cuticle wetting, retention, and penetration prior to absorption into the leaf tissues will be

RALPH C. KIRKWOOD • Department of Bioscience and Biotechnology, University of Strathclyde, Glasgow G4 0NR, United Kingdom.

considered, with particular reference to the role of surfactants. The processes and factors determining the transport of the a.i. away from the absorption site are reviewed in Chapter 9.

8.1. THE CUTICLE AS A BARRIER TO HERBICIDE UPTAKE

8.1.1. The Structure of the Cuticle

The outer surfaces of leaves are covered by a waxy cuticle which minimizes the loss of water from the plant and may act as a barrier to the uptake of agrochemicals, including herbicides. The cuticular membrane formed on the outer wall of the epidermal cells is composed of a framework of polymeric cutins in which are embedded cuticular waxes[1]; on the surface, soluble waxes are deposited to form epicuticular waxes.[2] Carbohydrate fibers may extend into the cuticular membrane from the epidermal wall and middle lamella.[2,3] In consequence, while the outer surface is essentially nonpolar, the inner region is relatively polar (Figure 8.1).

The cutins are insoluble high-molecular-weight polyesters, composed of long-chain substituted aliphatic acids combined with hydroxyl, aldehyde, ketone, epoxide, and unsaturated groups.[4] In plants of the Angiospermae, the cutins are formed of monomers having chain lengths of C_{16}, C_{18}, or a mixture of both; the composition of the monomers depends upon the plant species and location on the plant.[4]

The physicochemical properties of the leaf surface largely depend on the morphological and chemical nature of the epicuticular waxes. The configuration of these crystalline waxes varies according to plant species and the surface and age of particular leaves and may take the form of plates,

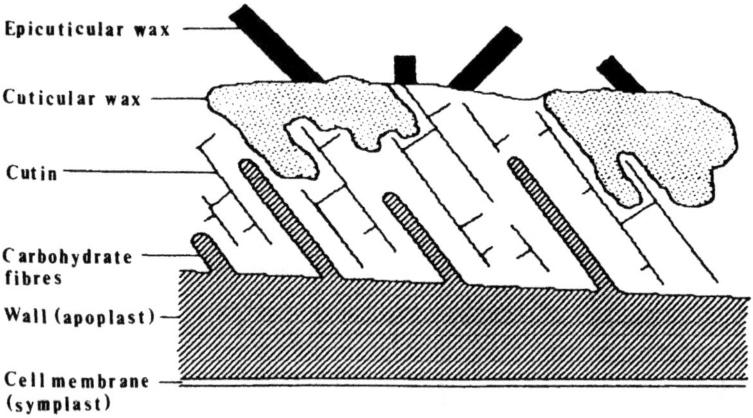

Figure 8.1. Generalized scheme for the structure of the plant cuticle membrane.[14]

tubes, ribbons, rods, filaments, or dendrites.[2] The morphology of the epicuticular waxes is determined largely by the composition of the wax exuding from the cuticular membrane,[5,6] crystallization rates and structure being influenced by environmental effects. Chemically, the epicuticular waxes are composed of a number of long-chain aliphatics, pentacyclic triterpenoids, sterols, and flavonoids; in contrast, the embedded waxes are composed of short-chain fatty acids (e.g., C_{16}, C_{18}).

8.1.2. Cuticle Wetting, Spreading, and Retention

Foliar application of herbicides may involve volumes ranging from high (e.g., 1000 liter/ha) to ultralow (e.g., 1 liter/ha), and maximum capture by the leaf surface depends upon such factors as droplet size, impaction, wetting, spreading, and retention. The interactions at various "interfaces" involved in pesticide application and their effects on transfer and performance have been reviewed by Tadros.[7]

The first interface occurs between the spray solution and the atmosphere (air) and determines the droplet spectrum, rate of evaporation, drift, etc. The second interface occurs between the liquid droplets and the leaf surface. The droplets impinging on the surface are subject to a number of processes which determine their adhesion, retention, and further spread. The nature of the deposit formed is governed by the rate of droplet evaporation and the concentration gradient of the surfactant across the droplet.

When a droplet impinges on a leaf surface it may bounce, or fragment into smaller droplets which bounce, and return to the surface at reduced kinetic energy; alternatively, it may adhere to the surface, forming a hemispherical cap.[7] The factors determining droplet fate may involve: (1) the mass (volume) of the droplet, its velocity in flight, and the distance between the spray nozzle and the target surface; (2) the difference between the surface energy of the droplet in flight and its surface energy after impact; and (3) displacement of air between the droplet and the leaf.

Droplet size is a very important factor influencing adhesion/reflection.[7] Droplets with a diameter of 20–50 μm normally do not undergo reflection, whereas large droplets of a few thousand microns undergo fragmentation. Droplets in the intermediate range (100–400 μm) may be reflected or retained, depending on leaf surface roughness, droplet elasticity, and surface tension. For any given spray solution and surface tension, a critical droplet diameter exists below which adhesion is high and above which adhesion is low; critical droplet size increases as the surface tension decreases. The percentage of adhering droplets decreases as the angle of incidence on the target surface increases.

Droplet adhesion is determined by the "in flight" kinetic energy of the droplet and its surface energy on impaction with the leaf surface. Adherence will occur only if the difference in surface energy between the free and the attached drop is greater than the kinetic energy of the droplet. Tadros[7] calculated the critical contact angle required for adhesion to smooth surfaces of water droplets of different sizes and velocities. For any given droplet size, the critical contact angle necessary for adhesion declined with increase in droplet velocity. As droplet size increased, so the contact angle necessary for adhesion declined. Surfactants (surface-active agents) normally would be required for enhanced adhesion, and this would certainly be so in the case of larger droplets (e.g., 300–400 μm), which require much lower contact angles.[7]

Leaf surfaces, which normally have hairs or crystalline epicuticular waxes, cause hysteresis in the contact angle, producing different values for the advancing and receding contact angles. Depending on the topography of the leaf surfaces, adhesion of droplets can occur at contact angle values which are above or below the critical values calculated for a smooth surface.

The problem of droplet sliding in relation to spray retention has also been considered by Tadros.[7] Spray retention increases as the volume applied increases, up to the point of "incipient runoff." Above this point, the volume of spray retained begins to decline, and retention is determined by the movement of drops on the leaf surface. Factors influencing sliding include surface roughness, presence of hairs, leaf age, position on the leaf surface, environmental conditions, surfactant incorporation and concentration, droplet size and velocity, and wind speed.

The final distribution of herbicide over the treatment area is determined by surface wetting, which can be assessed from the value of the contact angle,[7] though its measurement is complicated by hysteresis. Surface wetting is influenced also by (1) the "spread factor," which depends on the ratio between the diameter of the wetted area and the diameter of the drop, and (2) the spreading coefficient(s), which reflects the tension when a solid-liquid and a liquid-air interface are replaced by a solid-air interface.

In order to achieve enhanced wetting and spreading of liquids on leaf surfaces, the contact angle has to be reduced, and normally this is achieved by addition of an appropriate surfactant. These compounds increase biological efficiency by enhancing wetting and spreading, though their effects may be highly complex and determined by hydrophilic-lipophilic balance (HLB). (See Section 8.2.3.3.)

8.1.3. Spray Deposits

The amount of herbicide deposited on a crop/weed canopy is determined by plant, environmental, spray, or formulation factors.[8] The plant factors include growth habit and leaf size, shape, orientation, and age; surface factors include corrugation, hairs, and nature of epicuticular waxes.

Environmental factors such as wind speed, temperature, light intensity, and rainfall influence drift, rate of evaporation, photochemical degradation, and wash off, respectively. Spray application factors affecting deposition include spray volume, droplet size, and boom height. Chemical and formulation factors include the molecular and solubility characteristics, vapor pressure, partition coefficients, and incorporation of surfactants. The factors affecting the dissipation of a pesticide following application have been reviewed by Ripley and Edgington.[9] They used a modeling approach to explain the decline in surface residues by volatilization, hydrolysis, or photodecomposition or attributed it to physical properties such as vapor pressure, water solubility, or partition coefficient.

Dried spray deposits on leaf surfaces have been studied using scanning electron microscopy (SEM),[10,11] while secondary electron emission, cathodoluminescence, and X-ray microanalysis have been used to examine the phytotoxic effects of pesticides or surfactants on the leaf epidermis. Using these techniques, it has been demonstrated that uptake continued after spray deposits have visibly dried.[10,12]

Fluorescence microscopy, autoradiography, and energy-dispersive X-ray analysis have been used to examine the retention and redistribution of pesticides of widely differing physicochemical properties applied as solutions of commercial formulations to leaves of differing surface characteristics.[13] The fine structure of herbicide deposits has been studied using SEM with a secondary electron detector, and the information has been used to interpret uptake data. For example, the relatively low rate of absorption of maleic hydrazide (MH) by maize, recorded over a 72-h period, was related to the annular distribution of the a.i. in groups of fine crystals.[13]

The form of spray deposits depends on a number of factors including the rate of evaporation of the carrier liquid.[7] Evaporation from a spray drop generally occurs most rapidly from the periphery, causing a higher concentration of surfactant and preferential precipitation of a.i. near the edge. Thus, the nature of the spray deposit depends largely on the nature of the formulation and the type of dispersant or emulsifying agent used in suspensions or emulsions.

8.1.4. Cuticle Penetration by Herbicides

The properties of the cuticle which determine herbicide absorption have been reviewed by Price[14] and Kirkwood.[8] Permeability of the cuticular membrane is determined largely by the wax component of the cuticle and is influenced by such factors as cuticle thickness, wax viscosity, molecular radius of the penetrating molecule, partition coefficient, effective area, formulation, and temperature. Pores and canals, which have been demonstrated to exist in the cuticular membrane of many species, appear to act as avenues for the extrusion of waxes or their precursors[15]; their role, if any, in transcuticular movement of herbicidal compounds is unknown.

The techniques for measuring cuticle penetration of a herbicide following foliage application can be divided into those involving isolated cuticles and those using intact plants or leaves. Stomatous or astomatous cuticles may be isolated using chemical enzyme methods.[16] To be most meaningful, the results should state the permeability coefficient, which depends on membrane thickness and on the properties of the compound. The use of intact leaves for the study of cuticle penetration requires analysis of the various factors involved in the loss of compound from the treatment region.[17] These factors may include evaporation from the cuticle surface and translocation away from the underside of the cuticle, particularly in whole plants.[18]

The barrier to penetration of water and solutes across the cuticular membrane occurs at the air–cuticle interface. This solid-state barrier is composed of cutin and waxes, and the latter are presumed to be relatively inaccessible to polar compounds. The determining factor in cuticle penetration by herbicides appears to involve the HLB of the compound and its ability to partition from air into the nonpolar wax components and thence into the more polar regions within the basal regions of the cuticle.

8.1.4.1. Nonpolar Compounds

The characteristics of cuticle penetration by nonpolar (lipophilic) chemicals have been investigated in depth by Schönherr and his co-workers, with particular reference to accumulation in plant cuticles,[19] permeation across plant cuticles,[20] and desorption from plant cuticles.[21]

Plant cuticles can be regarded as potential sorption compartments for lipophilic compounds,[22] and sorption of moderately lipophilic phenolic compounds can be predicted from the octanol/water partition coefficients (K_{ow}). Partition coefficients in the cuticle/water system are important parameters for the transport of compounds across plant cuticles and for the uptake of such compounds.[23]

Kerler and Schönherr[19] have studied the accumulation of lipophilic chemicals in plant cuticles. Using cuticles isolated enzymatically from mature leaves of *Citrus aurantium* L., they examined the sorption of a range of ^{14}C-labeled compounds of widely differing lipid solubilities. They found that logarithms of cuticle/water ($K_{cm/w}$) and octanol/water (K_{ow}) partition coefficients between 10^2 and 10^8, the logarithms were correlated by linear regression equations with coefficients of correlation from 0.980 to 0.988.

Using the same range of ^{14}C-labeled compounds, Kerler and Schönherr[20] determined the permeability of astromatous cuticular membranes from the leaves of *C. aurantium* by measuring permeability coefficients (flow per unit area, time, and driving force) and extrapolated "holdup" times. A linear correlation ($r = 0.91$) between the logarithms of the permeability coefficients ($\log P$) and the $\log K_{ow}$ values was obtained using the relationship

$$\log P \text{ (cm)} = 0.704 \log K_{ow} - 11.2 \tag{8.1}$$

Using this simple equation, permeability coefficients for compounds having $\log K_{ow}$ values ranging from 2.0 to 7.9 can be estimated to within an order of magnitude of values determined experimentally.

The precision of prediction using Eq. (8.1) was improved ($r = 0.95$) when $\log P$ was related to the logarithm of the cuticle/water partition coefficient ($\log K_{cm/w}$) using the following equation:

$$\log P \text{ (cm)} = 0.734 \log K_{cm/w} - 11.26 \tag{8.2}$$

The best agreement, however, was evident ($r = 0.98$) when the molar volumes (MV) of the test compounds were taken into account by relating $\log P$ to $(\log L_{cm/w})/\text{MV}$:

$$\log P \text{ (cm)} = 238 \frac{\log K_{cm/w}}{\text{MV}} - 12.48 \tag{8.3}$$

Using cuticles isolated from the leaves or fruits of four plant species and preloaded with [^{14}C]-2,4-D, Schönherr and Riederer[21] examined the simultaneous bilateral desportion of [^{14}C]-2,4-D. In the case of cuticular membranes, only 2–3% of the 2,4-D initially contained in the cuticles could be desorbed from the outer surface, while 86–92% was desorbed from the inner surfaces within 6 h. Schönherr and Riederer[21] attributed this asymmetrical desorption mainly to the presence of apolar and crystalline waxes on the surface and in the outer layers of the cuticle.

Schönherr and Riederer[21] concluded that the fate of chemicals sorbed into cuticles depends on the presence or absence of a sink. In the presence of a sink, all cuticles will exhibit a pronounced inward permeability and accumulation in the cuticles is unlikely to occur; accumulation will occur only in the absence of a sink.

On the basis of these results, Schönherr and Riederer[21] proposed a simple model of cuticles which explicitly includes heterogeneity (Figure 8.2). Cuticular membranes are composed of an outer and an inner volume element (V_o and V_i), each being characterized by its thickness (l), its partition coefficient (K), and its diffusion coefficient (D). The inner surface area (A_i) is considerably greater than the outer (A_o) due to the cutinization

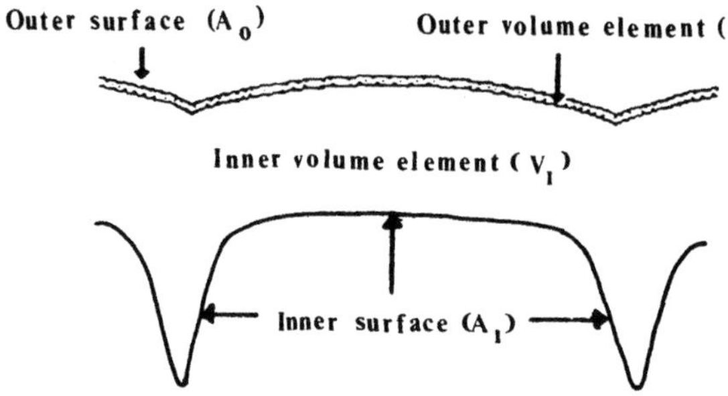

Figure 8.2. Schematic diagram showing the cross section of a leaf cuticular membrane.[21]

of the anticlinal and periclinal walls. The presence of soluble lipids in the outer volume elements imparts a high transport resistance and a low partition coefficient,[22,23] resulting in the desorption of 2,4-D molecules through the inner surfaces.

In consequence, if a sink exists in a leaf (or fruit), 2,4-D molecules will move out of the inner volume element much more rapidly than the outer skin can be penetrated. Under these conditions, 2,4-D will not accumulate in cuticles as long as sink activity continues and permeability of the outer volume elements is lower than the rates of removal into sinks. In this context, the "sink" may involve sorption into membrane and storage lipids, metabolization, conjugation, or other mechanisms of removal.

The range of permeances measured for cuticular membranes of various plant species is 1×10^{-10} to 270×10^{-10} m/s.[23] The time required for 5% of a solute to penetrate from a donor droplet through the outer volume element into a semi-infinite receiver (the inner volume element of the cuticle and adjacent cells) can be calculated from Eq. (8.4)[17]:

$$-PA_t/V_{\text{donor}} = \ln C_{\text{donor}}/C_o \tag{8.4}$$

Given a hemispherical droplet with a volume of 10 μl (V_{donor}) having a contact area (A) with the cuticle, the time required for a 5% decrease in the donor concentration ranges from 160 h to 35.5 min for the range of permeances observed. Schönherr and Riederer[21] concluded that penetration through plant cuticles is a slow process when compared to transport across cellular membranes and that, with most compounds, penetration across the outer layer of cuticles will likely be the rate-limiting step, rather than uptake into cells and organelles.[21]

Hamburg and McCall[24] have examined the factors affecting the rate of cuticle penetration of yellow foxtail [*Setaria glauca* (L.) Beauv.] by a series of aryloxyphenoxypropionate herbicides. They selected this family of herbicides because of (1) the diversity of structures (three classes) and (2) the importance of the group in the postemergence control of grass weeds. The cuticular penetration was found to be dependent primarily on the structural class (pyridinyloxy, quinoxalinyloxy, and phenoxy) and secondarily on formulation. Correlation of the penetration rate constant (K_{pen}) with the physical properties melting point, log P, vapor pressure, and water solubility was found to be unreliable; separation of the penetration behavior of the six test compounds into structural classes reduced the number of compounds available for such correlations.

The herbicides used were haloxyfop, haloxyfop-methyl, haloxyfop-ethyl, fluazifop-butyl, quizalofop-methyl, and diclofop-methyl; each was ^{14}C-phenyl labeled. The base formulation was 50% acetone in water with 0.1% (v/v) Ortho X-77 (−COC); for comparison, the base formulation was also formulated with 0.3% (v/v) Agri-dex crop oil concentrate (+COC), which contained 17% nonionic surfactant in a crop oil base.

In these experiments, the amount of herbicide distributed on the plant surface (SRF), beneath the plant cuticle (PEN), and in the air (AIR) was monitored at 7–9 time points over a 24–48-h period. The values of k_v and k_{pen}, the rate constants for volatilization and penetration of the cuticle, respectively, were obtained from the compound distribution versus time data using a kinetic model and mathematical modeling. The simplest model for foliar absorption of a volatile compound is given in Eq. (8.5):

$$\text{AIR} \xleftarrow{k_v} \text{SRF} \xrightarrow{k_{pen}} \text{PEN} \tag{8.5}$$

This model was termed simple penetration (SP) since only one rate constant was required to describe penetration behavior. The rate constant k_{pen} is given by:

$$k_{pen} = \frac{A}{V(\Delta x)} \times DK \tag{8.6}$$

where A is the area of the applied herbicide spot on the leaf surface, V is the volume of the applied spot, D is the diffusion constant for the compound penetrating the cuticle of effective thickness Δx, and K is the partition coefficient for the transfer of the solute from the applied spot to the cuticle.[25] According to this analysis, the penetration rate constant should be first order and directly related to the structural or physical properties of the penetrating cuticle.

Under some conditions, Hamburg and McCall[24] found that the SP model failed to fit the experimental data. A biphasic penetration (BP) model that included a fast-penetrating compartment (SRF, f) and a slow-penetrating compartment (SRF, s) with corresponding $k_{pen,f}$ and $k_{pen,s}$ was necessary. The BP model is shown in Eq. (8.7)

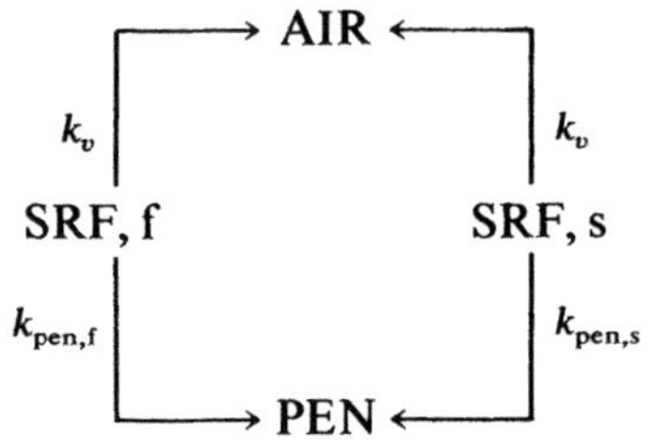

(8.7)

Their results demonstrated the importance of structure (and adjuvant) in controlling the penetration mechanism. Cuticle penetration by haloxyfop, haloxyfop-methyl, haloxyfop-ethyl, and fluazifop-butyl without COC followed BP kinetics, being impeded by an initial fast penetration stage. With the exception of haloxyfop, the addition of COC to the formulation simplified the penetration kinetics to SP. The BP behavior in the absence of COC may be due to the relatively rapid penetration of Ortho X-77, leaving the polar pyridinyloxy ester on the leaf surface. These polar esters may require the assistance of the surfactant and/or COC to facilitate cuticle penetration.[24] In contrast, the behavior of quizalofop-methyl and diclofop-methyl was SP in the presence or absence of COC.

8.1.4.2. Polar Compounds

The possible existence and significance of sites of preferential entry for polar compounds is a controversial matter. There do appear to be regions of the leaf surface which may be less heavily waxed, including the stomatal guard cells, the basal cells of trichomes (glandular hairs), the cuticle associated with veins, especially toward the leaf base, and that above the anticlinal walls of epidermal cells. These areas appear to be more permeable to polar compounds. Certainly, localized areas of retention and/or possibly penetration of water-soluble dye, [^{14}C]-2,4-D, or [^{14}C]-MCPA have been reported for cuticles (+epidermal cells) isolated from the abaxial surface of bean (*Vicia faba* L.).[26] Again, studies of the foliage uptake (and translocation) of asulam in bracken have demonstrated considerably higher levels of cuticle penetration under high humidity conditions.[27] Many years ago, van Overbeek[28] commented that under such conditions the enhanced water continuum in the leaf tissues may cause expansion of the cutin, pushing apart the wax platelets. While there appears to be no evidence to support this idea, it cannot be discounted entirely.

8.1.4.3. The Effect of Environmental Factors

The environmental conditions which prevail during plant growth can affect the permeability of cuticle membranes since wax production varies according to ambient conditions.[29] Hunt and Baker[30] have studied the effect of environmental conditions on cuticle wax developments on leaves of *Brassica oleracea* L. Wax yields increased with increase in radiant energy rate and with reduced temperature and relative humidity; similarly, the wax deposits on the surface of pea (*Pisum sativum* L.) increased with decrease in soil moisture content. Penetration of 1-naphthaleneacetic acid (NAA) into the leaves of pea was inversely correlated with changes in wax deposit, being most rapid through the wax deposits formed under conditions of low radiant energy flux, high soil moisture, and high humidity conditions.

More recently, Wills and McWhorter[31] have reviewed the effect of environmental factors on absorption (and translocation) of herbicides. Increased absorption of a number of herbicides was obtained with increased temperature (<35°C), increased relative humidity, and increased soil moisture content. Generally, factors which individually increased herbicide absorption (and translocation) caused even further increase in these functions when combined. Thus, high temperature combined with high relative humidity resulted in greatest herbicide absorption; similar results have also been reported for uptake of asulam into bracken.[27] Exceptions have been reported, however, in which certain herbicides were more effective at lower temperatures associated with high relative humidity.[32–34]

In general, it would appear that herbicide uptake and activity is adversely affected by pretreatment conditions which are unfavorable to plant growth, namely, low temperature, low light intensity, and moisture stress.

8.2. SURFACTANTS IN RELATION TO CUTICLE RETENTION AND PENETRATION

8.2.1. The Role of Surfactants

Activator adjuvants are often added to spray mixtures of foliage-applied herbicides to enhance the level of weed control.[31] Such adjuvants increase herbicidal activity by increasing cuticle retention, penetration, absorption, and possibly translocation. Activator adjuvants include surfactants, wetting agents, penetrants, and oils. Oils used as additives have been found to influence wetting properties, evaporation, leaching, and foliar uptake and translocation of herbicides while surfactants increase their solubility, spreading, and penetration.[35–37]

The range of formulation additives have been reviewed with particular reference to surfactants by Foy.[38] Four classes of surface-active agents have been defined:

- *Anionic,* in which the surface-active properties are provided by a negatively charged ion: a hydrophobic group (e.g., paraffinic chains) is balanced by a negatively charged hydrophilic group (e.g., carboxyl).
- *Cationic,* in which the surfactant properties are provided by a positively charged ion: the hydrophobic group is balanced by a positively charged hydrophilic group (e.g., quaternary ammonium).
- *Nonionic,* in which there are no electrical charges: the hydrophobic group consists of alkylphenols, fatty alcohols, or fatty acids and is balanced by nonionized hydrophilic groups (e.g., polymerized ethylene oxide).
- *Amphoteric,* having a molecular arrangement of hydrophilic groups which have the potential to become cationic in an acid medium or anionic in alkaline conditions.

Surfactants thus contain hydrophilic and lipophilic groups, and the balance between the opposing effects is called the hydrophilic–lipophilic balance (HLB). Compounds of low HLB are relatively lipophilic while those of high HLB are relatively water soluble; thus, it is possible to select surfactants of certain HLB values for specific purposes.

Surfactants can be divided into two categories according to their application and mode of action.[39] "Spray modifiers" reduce the surface tension of the spray droplets and are added to improve the wetting and spreading properties of formulations; they are particularly useful in the case of nonsystemic compounds, where uniform coverage of the canopy is important. "Activators" are added specifically to improve foliage absorption and thus activity of systemic a.i. Their activity depends upon specific physicochemical interactions between the surfactant, the a.i., and the plant. While the performance of spray modifiers generally can be predicted from physical measurements prior to formulation, there are no guidelines for the selection of suitable activator surfactants for commercial usage. In practice, a range of polyoxyethylene surfactants with HLB values ranging from 10 to 13 are often used as activator adjuvants.

Surfactants have an important role to play in regard to formulation of pesticides, including herbicides.[40] They may act as emulsifiers for emulsifiable concentrates and microemulsions or as wetting and dispersing agents for wettable powders, water-dispersible granules, and flowables. Alternatively, they can be used as spray tank additives to enhance droplet adhesion, wetting of the foliage, and uptake into the plant.

The effect of surfactants on spray retention efficiency is influenced by droplet size and velocity, leaf angle, and surfactant type and concentration.[41] Examination of spray retention and surface spreading on a range of outdoor-grown crop plants was examined by Anderson *et al.*[42] Retention was determined by dynamic surface tension, whereas droplet spreading was related to equilibrium values. They concluded that the basis of selection could be determined by cost, lack of phytotoxicity, and ability to solubilize or aid penetration of the a.i.; this would not prejudice the retention properties of the formulation.

The properties which influence the retention (and penetration) of the herbicide have been reviewed by Scher.[40] The physical properties of the herbicide, particularly the oil/water partition coefficient, are of primary importance, together with the presence of organic solvents and surfactant type and concentration. The surfactant characteristics have a major effect on the surface tension of the spray droplets at the air–water interface and the contact angle at the water–plant interface. Surface tension influences the range of droplet sizes and, in turn, the efficiency of transport to the leaf surface and losses due to drift; effects on surface retention are determined by the reflection losses. The product of the surface tension and the cosine of the contact angle influences adhesion, spreading, wetting, coverage, and runoff from the plant. Surfactant type and concentration also may have a major effect on the rate of efficiency of cuticle penetration and absorption (and indirectly translocation) of the herbicide.

The influence of surfactants and plant species on the retention of spray solution has been examined by de Ruiter and Uffing.[43] They reported a linear relationship between retention and the logarithm of surfactant concentrations. It would appear that there is a need to add a surfactant concentration far in excess of the critical micelle concentration (CMC) and that the diffusion of surfactant from the bulk to the surface of the flattening drop is a retention-determining factor. The dynamic surface tension is regarded as a useful parameter determining surfactant type and concentration with respect to efficient surface retention.

8.2.2. Cuticle Penetration of Surfactants

The mechanism by which surfactants facilitate cuticle penetration is uncertain. A number of studies involving the use of radiolabeled surfactants have helped to clarify their possible role. Penetration and distribution studies with [^{14}C]-Tween 20,[44] and [^{14}C]-Tween 80,[45,46] showed very limited foliar penetration, the acid and polyoxyethylene regions of the molecule being cleaved, probably on the leaf surface. Conversely, Stolzenberg *et al.*[47] reported a substantial level of uptake from dilute solutions of Triton X-100 [average ethoxylate (E) content 9.5%]. They found that a high level of metabolism, mainly to polar residues, occurred within the treated leaf. Anderson and Girling[48] investigated the uptake of four linear alcohol (C_{13}–C_{18}) polyethoxylates (E12–17) into wheat leaves and found up to 80% uptake within 48 h.

At Long Ashton, Silcox and Holloway[49–52] carried out foliar uptake experiments with dilute aqueous solutions of ^{14}C-labeled linear alcohol and alkylphenol polyoxyethylene surfactants (0.1% w/v) having HLB values of 10 to 14. Considerable absorption was recorded, the rate and total amount of uptake varying according to species and chemical nature of the surfactant. Maximum uptake occurred with compounds having a C_{12}/C_{13} alkyl chain as the hydrophobic moiety; there was little movement of the surfactants within the treated tissues, but subsequently they were metabolized.

The uptake of ^{14}C-labeled surfactants (0.1% w/v) in the presence of difenzoquat indicated an intimate association between surfactant and chemical.[52] In the presence of C_{12} E8, uptake of difenzoquat was increased fivefold, particularly during the first 48 h after treatment. There was evidence that surfactant and herbicide penetrated at similar rates, confirming the findings of Stevens and Bukovac.[53] The interaction between OPE 9.5 and difenzoquat, however, was different. The overall rate of uptake was lower than that for the herbicide alone, though a greater proportion of the absorbed herbicide was translocated.

8.2.3. The Mechanism of Surfactant Action

The sites and mechanisms of surfactant action have been considered by Holloway and Stock.[39] Surfactants may act at four possible sites in treated leaves, namely, on the surface, cuticle, and epidermis or within the internal tissues. As mentioned previously, the addition of surfactants to formulations of agrochemicals invariably improves their wetting and spreading characteristics as measured by contact angle, deposit area, spread, and surface tension. Generally, however, the effect of surfactants on coverage can seldom be related to herbicide uptake or activity.

There is little evidence from SEM studies that leaf surfaces are dissolved or disrupted following droplet application of surfactant solutions.[39] The presence of surfactant, however may result in preferential precipitation near the edge of the deposit,[54–56] although the significance with respect to activating herbicide uptake is uncertain. There is evidence that the surfactant may facilitate uptake of the agrochemical by aiding its dissolution in the deposit.[39] This could be achieved mainly by solubilization involving incorporation of the compound into surfactant micelles.[7] While such micelles are too large to penetrate leaf cuticles, their lifetime is short (<1 m/s), and they may function as "transport packages" releasing the dissolved molecules near the absorption sites.

It is possible that surfactants can aid the uptake of herbicides by increasing their water solubility and thus facilitating their partitioning from lipid to aqueous phases in the cuticle. While increases in water solubilities of certain pesticides by surfactants have been demonstrated *in vitro*, only small differences were evident in the solubilizing capacities of nonionic surfactants of different classes and E contents. It seems unlikely that differences between surfactants in their relative solubilizing efficiencies account for any differences in their activation efficiency.[39]

The moisture-absorbing capacity of certain surfactants may enhance the activation of water-soluble compounds; the capacity to do so appears to depend upon surfactant chemical class, E content, and ambient relative humidity.[39] Hydrophilic surfactants of high E content are the most hygroscopic, and under field conditions they may act as condensation nuclei just below water vapor saturation. Under these circumstances, a water droplet may re-form over the original droplet, thus enhancing uptake of the a.i.

In penetration of the cuticle and epidermis, the mechanisms of movement involve a partition–diffusion process which involves three compartments:

1. an outer mainly aqueous external phase at the surface
2. a central heterogeneous cuticle phase composed of hydrophobic and hydrophilic components
3. an inner aqueous phase composed of plant apoplast or free space

As previously noted, the penetration flux will depend on the concentration gradient between the outer and inner compartments and on the ability of the herbicide molecule to partition between compartments.

Studies with radiolabeled surfactants have demonstrated that a penetrating surfactant may be present in all three compartments and thus may influence any rate-limiting steps associated with herbicide uptake.[39] Indeed, parallel studies with radiolabeled formulations containing either ^{14}C-surfactants or ^{14}C-agrochemicals indicate that nonionic surfactants which exhibit penetration characteristics may activate by copenetration or prepenetration. In the former case, activation is associated with closely related rates of uptake of surfactant and herbicide, whereas, in the latter case, activation only occurs after the surfactant has penetrated the leaf. The efficiency of these mechanisms appears to depend on the concentration of both surfactant and a.i. and on the plant species and agrochemical involved. The prepenetration requirement suggests that the need for the activating surfactant is located mainly in the inner cell wall compartment,[39] or possibly is related to cell membrane permeability.

In animal tissues, a relationship between membrane activity and lipophilicity has been reported by Attwood and Florence.[57] The optimum chain length for the hydrophobic moiety was C_{12}, and the optimum E content for the hydrophilic moiety was around E10. At low concentrations, these surfactants penetrate the cell membrane, increasing permeability; at higher concentrations, the membrane may be disrupted due to solubilization and extraction of the lipid and protein components. In plants, a similar pattern has been reported for nonionic surfactants,[58] their action being exerted in the tissues immediately beneath the zone of application; transport of surfactants has not been reported.

Holloway and Stock[39] have proposed a predictive model which offers considerable potential for optimizing the performance of formulations (Figure 8.3). The model is based on the observations that optimum promoting properties for compounds with low log *P* occur with surfactants of high E content; compounds with high log *P* require surfactants of low E content. Other factors influencing uptake may also be superimposed on the model, including physicochemical properties of the agrochemical (e.g., melting point and surfactant concentration) and its phytotoxicity.

Schönherr and his co-workers have examined the effects of surfactants on cuticle permeability and their mechanism of action. Geyer and Schönherr[59] reported that polyoxyethylene *p*-t-octylphenol and polyoxyethylene nonylphenol (HLB 4–16) increased water permeability by a factor of 2 while polyoxyethylene alkylether (HLB 5–13) increased it 7- to 8-fold. The cationic surfactant dodecyltrimethylammonium chloride (HLB 18.5) also

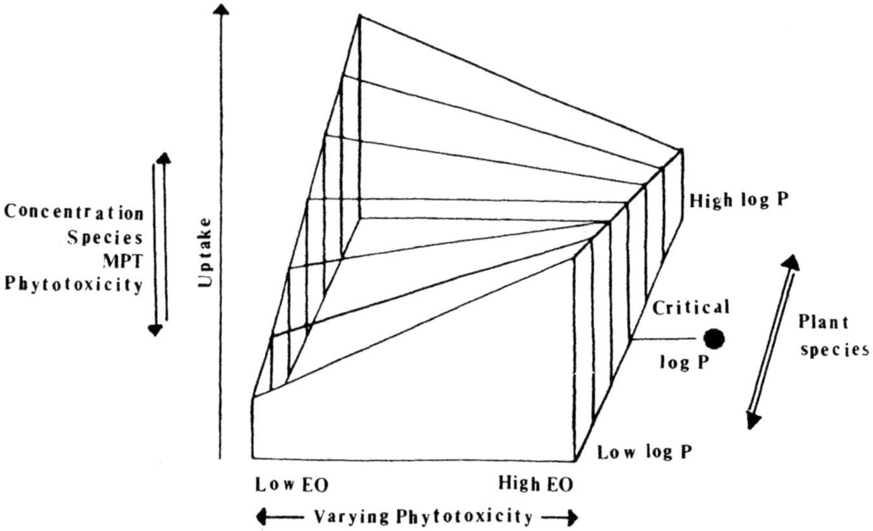

Figure 8.3. Model relating the foliar uptake of neutral compounds to their log *P* and the interaction of this variable with the E content of polyethoxylate surfactants applied in combination with the compounds.[74]

increased water permeability (by a factor of 22). These authors speculated on the mechanisms by which surfactants affect the water permeability of cuticles. They attributed the rate-limiting barrier to transport of water and solutes to soluble cuticular lipids[21] and believed that solute permeability was most likely to be increased by surfactants which increased membrane permeability. In the case of lipophilic solutes, any partitioning of the solute into the waxes would reduce permeability. Geyer and Schönherr[59] used an *in vitro* test in which partitioning was absent so that the effects of surfactants on mobility of water or solutes could be measured in isolation of the partitioning effect.

8.2.3.1. Nonpolar Herbicides

Insoluble a.i. requires solubilization into surfactant micelles.[60] This involves the formation of a thermodynamically stable, isotropic solution, by the introduction of an amphiphilic component(s). Solubilization is affected by the structure of the surfactant and the solubilisate, temperature, and addition of electrolytes and nonelectrolytes.[57] The structural characteristics of the surfactant influence its solubilizing capacity, particularly with regard to the existence of different solubilization sites within the micelles.[7] In cases in which insoluble a.i. are located either in the micelle core or deeply within the micelle, solubilization should increase with increase in the alkyl chain length of the surfactant. If solubilization occurs in the hydrophilic portion of the surfactant, it will increase in accordance with the size of that group; solubilization is influenced also by polarity, chain branching, and molecular size, shape, and structure.[7]

The biological efficiency of an insoluble pesticide will be significantly affected by the presence of micelles, due in part to improved solution of the a.i.[7] In addition, surfactant micelles influence the rate of dissolution of a solute as well as its membrane permeability. The driving force for transportation across a membrane may involve dissolution, diffusion, or convection in bulk liquid and crossing of the membrane. Diffusion in bulk liquid obeys Fick's law:

$$J_D = D\frac{dc}{dx} \tag{8.8}$$

where J_D is the amount of solute crossing a unit cross section in unit time, D is the diffusion coefficient, and dc/dx is the concentration gradient. While the location of the chemical in a micelle will reduce D, the presence of the solubilizing agent will increase the concentration gradient in direct proportion to the increase in solubility. Thus, solubilization will normally increase transport by diffusion, since the saturation value can be increased by many orders of magnitude.[7] the reduction in resistance of steps involving diffusion and convection indirectly accelerates transport though the membrane, despite the possibility that the resistance of this transmembrane step remains unaltered.

The importance of surfactant micelles in the transport of pesticides has been summarized by Tadros.[7] Since the droplets reaching the leaf surface undergo rapid evaporation, the surfactant concentration in the spray deposit may reach very high values, enabling extensive solubilization of the a.i. to occur; this will enhance transport of the chemical. Further, since the lifetime of the surfactant micelles is less than 1 m/s, these units are constantly releasing their contents near the sites of action, thus increasing the concentration gradient and producing a large flux.

8.2.3.2. Polar Herbicides

The foliar penetration of water-soluble herbicides is relatively inefficient in comparison with that of lipophilic compounds, no doubt reflecting the incompatibility of the polar droplets and the nonpolar epicuticular waxes, and also the relative inefficiency of transcuticular polar transport. The choice of a suitable surfactant certainly reduces the degree of incompatibility and may also enhance passage through the membrane.

The effects of three surfactants, including Tween 20, on foliar penetration of NAA and on NAA-induced ethylene production of cowpea [*Vigna unguiculata* (L.) Walp.] have been examined by Lownds *et al.*[61] They observed that the surfactants decreased surface tension of NAA solutions, causing increased wetting and droplet–leaf interface area. The beneficial effect of NAA penetration was most evident during the droplet drying phase, and it continued after drying.

Baker and Hunt[55] examined the foliar uptake of 12 radiolabeled agrochemicals with and without a nonionic surfactant and found that the lowest rates of uptake occurred in the case of water-soluble compounds. In contrast, the highest rates of uptake were recorded for highly lipophilic compounds with values of log P greater than 3, though translocation efficiency was low. Addition of nonionic NP 8 surfactant increased uptake (but not translocation) of all compounds, possibly reflecting the increased area of contact. In the case of water-soluble compounds such as glyphosate or paraquat, the a.i. may be resolubilized under high humidity or dew conditions. Baker and Hunt concluded that in the presence of a hygroscopic surfactant these herbicides spread extensively, entering the leaf tissues via preferential sites of uptake such as the venation or leaf axils.[55]

The possibility that relatively polar compounds such as MCPA and 2,4-D penetrate via sites of preferential entry cannot be discounted,[26] particularly in the presence of surfactants which reduce the surface tension to less than 20 mN/m. Under such conditions, the mass flow of solute molecules into the leaves from spray droplets has been reported.[62,63] Surface tensions of this level can be achieved using nonionic organosilicone surfactants such as Silwet L-77.

The effect of organosilicone surfactants on the uptake of ^{14}C-labeled deoxyglucose applied to the adaxial surface of selected dicotyledon species has been investigated by Zabkiewicz *et al.*[64] Low concentrations of the Silwet organosilicone surfactants conveyed good wetting and spreading properties to the solutions, and droplet drying was enhanced. The aqueous surface tension was reduced to the extent that stomatal penetration (10–15%) could occur.

The effect of certain surfactants, including Silwet L-77, on the foliar uptake of deoxyglucose, glyphosate, and picloram by bean (*V. faba*) and *Eucalyptus botryoides* has been examined by Stevens and Zabkiewicz.[65] Silwet L-77 provided significantly greater uptake of the three chemicals into bean than the other surfactants, and this could not be attributed to differential wetting. Stevens and Zabkiewicz suggested that the uniformity of uptake of these chemicals into bean in the presence of Silwet L-77 may be related to the very low aqueous surface tension (20 mN/m) of this surfactant at above the critical micelle concentration (CMC). They found that 21–42% of the volume of Silwet L-77 solutions applied to bean infiltrated the stomata and concluded that this surfactant must also have enhanced cuticular penetration after the droplets dried. The beneficial effects of Silwet L-77 may also be attributed to its low phytotoxicity.[66]

Low phytotoxicity of herbicide and surfactants will be of greatest importance in regard to the efficient target site delivery of foliage-applied systemic herbicides such as asulam or glyphosate. These polar herbicides are useful in the control of perennial rhizomatous weeds such as bracken [*Pteridium aquilinum* (L.) Khun.] in which the target site exists in the buds associated with the underground rhizome system.[67]

Studies using pinna explants of frond tissues,[68] as well as whole plants,[69] have investigated the criteria influencing absorption and translocation of asulam to the target sites, which ultimately involve the folic acid biosynthetic pathway.[70] The use of selected nonionic surfactants as a means of enhancing target site delivery in explant systems has been investigated by Gaskin and Kirkwood.[71] The nonionic surfactants Silwet L-77, Triton X-45, and Agral 90 were used at concentrations above the CMC to enhance the uptake (and translocation) of [^{14}C]-asulam and [^{14}C]-glyphosate. Silwet L-77 (HLB 10.4) and Triton X-45 (HLB 10.4) were most effective, and Agral 90 (HLB 7.0) was less so. Surfactant incorporation had no effect on the levels of glyphosate found in soluble wax extracts from the cuticle, but all three surfactants increased (by a factor of 2) the level of asulam detected, suggesting that partitioning of asulam into the waxes was enhanced in the presence of these surfactants. Leaf membrane permeability to both herbicides was enhanced by Silwet L-77, presumably by stimulating movement and increasing the concentration gradient across the cuticle.

Foliage uptake and translocation of [^{14}C]-glyphosate in gorse (*Ulex europaeus* L.) was similarly enhanced by the addition of nonionic surfactants, Silwet L-77 being particularly effective.[72]

8.2.3.3. Surfactant Hydrophilic–Lipophilic Balance

The HLB of the penetrating compound is a most important factor influencing surfactant choice. Stevens and Bukovac examined the effect of surfactants of differing E content on the uptake of 2-deoxy-D-glucose, atrazine, and DDT (water solubilities of ca. 50 g/liter, 40 mg/liter, and 17 g/liter, respectively).[54] Uptake of hydrophilic glucose was directly related to the water content of the surfactants at 80% relative humidity, and this content increased with both relative humidity and the logarithm of the E content of the surfactant. Stevens and Bukovac suggested that the surfactant-enhanced uptake of 2-deoxy-D-glucose was due to maintenance of the chemical in solution on the leaf surface.

Uptake of the lipophilic compounds atrazine and DDT, however, was increased by surfactants in inverse relationship to their HLB, suggesting copenetration with the surfactant. The water solubilities of these relatively lipophilic compounds were markedly increased by surfactants, particularly at intermediate HLBs. Stevens and Bukovac concluded that surfactants with short E chains would be surfactants of choice to maximize cuticle penetration of nonpolar a.i., while those with long E chains would be preferable for more hydrophilic a.i.[54]

Similar findings have been reported by Kirkwood *et al.* for the effect of surfactant E number on the uptake of asulam (hydrophilic, log K_{ow} 0.3) and diflufenican (lipophilic, log K_{ow} 3.9) into frond tissue of bracken.[73] Uptake of asulam was increased by Ethylan surfactants (0.1%) of E4–10 (optimum E8), whereas absorption of diflufenican was enhanced only by surfactant of E4; translocation was increased indirectly in proportion to absorption.

The mechanism of surfactant action is uncertain. In the case of asulam, incorporation of surfactant of E6.5–8 may increase the rate of transcuticular passage by facilitating the partitioning of a.i. from droplet/deposit to cuticle waxes and from waxes to the inner, relatively aqueous phase. Alternatively, if a polar route across the cuticle does indeed exist, transport may be enhanced by prolonged retention of the a.i. in the aqueous phase of the droplet or by surfactant-induced transport via this polar route. The relatively low levels of [^{14}C] asulam detected in the cuticle waxes tend to support either view. The beneficial effect of E4 surfactant on uptake of diflufenican combined with reduced cuticle wax levels suggests that this surfactant enhances partition of this herbicide out of the cuticular waxes. Increased retention of [^{14}C] diflufenican in the cuticle waxes was evident when it was combined with Ethylan surfactant of higher E number. Further, the enhancement of the water solubility of diflufenican by surfactants of higher E number appeared not to be beneficial in terms of uptake.

Similar findings have been reported by Stock *et al.*, who investigated the effect of surfactant ethoxylation ("E effect") on uptake activation of [^{14}C]-3-*O*-methylglucose.[74] When assessed one day after treatment, the efficiency of uptake was found to increase progressively with increasing E content (E6–20). In the case of cynazine (lipophilic), however, uptake was inversely related to E number, the optimum occurring at E6, with uptake declining at higher E levels.

Bukovac and his co-workers have carried out a comprehensive series of studies on the effect of octylphenoxy (OP) surfactants on aspects of foliar uptake.[75,54,76,77] Reduction of the oxyethylene chain length (and thus HLB) resulted in decreasing surface tension, contact angle, and hygroscopicity but increased surface wetting (maize leaf); increased rate of evaporation with decreasing HLB was attributed to the increased wetting.[75] Uptake by maize leaves of the hydrophilic compound 2-deoxy-D-glucose was related to its dissolution in the water retained within the deposits on the leaf surface by surfactants of high E number.[54]

Shafer and Bukovac examined sorption of Triton X-100 by isolated tomato cuticles and concluded that orientation of absorbed surfactant molecules on the surface of the cuticle membrane (CM) may be important.[76] If the OP surfactant molecules formed a monolayer on the CM, the hydrophilic moiety would be oriented farthest away from the CM, thus imparting a relatively more polar chemical environment. If multiple layers of surfactant molecules were adsorbed by the CM, then the overall surface chemistry would depend on complex solution–sorbed molecule–CM interactions. This would influence the sorption and subsequent penetration of an a.i. through the cuticle.

Examination of the toxicity of a range of OP surfactants applied to the adaxial surfaces of cowpea revealed that phytotoxicity increased with increase in ethoxy chain length, concentration, droplet volume, and temperature; toxicity decreased, however, with increasing humidity.[77]

8.3. CONCLUSIONS

The importance of the cuticle as a barrier to penetration of herbicides, especially polar compounds, is well documented. Among a range of factors, the physicochemical characteristics of the epicuticular waxes may be of particular significance since they can affect the retention and distribution of the active ingredient. The heterogeneous nature of the cuticular membrane appears to necessitate a series of partitioning steps, and there is evidence to suggest that sorption into the outer region of the membrane takes place less readily than desorption from the inner region. The HLB of the herbicide appears to determine the fate of the compound in its transcuticular passage. Relatively polar compounds of low log K_{ow} appear to require formulation with an appropriate surfactant to facilitate efficient spreading, retention, and penetration. Compounds of high log K_{ow} appear to partition into the waxes without difficulty, though slow partitioning from the waxes into the inner aqueous phase may lead to accumulation of such compounds in the waxes.

The existence of relatively polar routes across the cuticle remains controversial, though there may be localized regions of greater permeability. If this is the case, these may facilitate preferential penetration of properly formulated polar herbicides, especially under high-humidity conditions. The use of certain organosilicone surfactants appears to result in the reduction of surface tension to a level which permits mass flow through the stomatal aperture.

The incorporation of surfactants may be required to achieve spreading or activation of the active ingredient. There are few guidelines determining choice of surfactant, reflecting the highly specific nature of plant–herbicide–surfactant interactions. Recent information, however, does suggest that in the case of lipophilic herbicides, nonionic surfactants of low E number may be preferable, whereas more hydrophilic compounds would require a surfactant of intermediate to high E number. In the case of octylphenol and linear alcohol surfactants, and doubtless other surfactant groups, high E number is consistent with lower phytotoxicity.

Prepenetration or copenetration of surfactant appears to facilitate transcuticular or transmembrane penetration of the a.i. but to have no direct effect on the efficiency of translocation. In the case of systemic herbicides of low phytotoxicity, increased absorption can result in a proportional increase in the efficiency of translocation to the target sites. The relationships between uptake and translocation are discussed in depth in Chapter 9.

ACKNOWLEDGMENTS

Thanks are due to Dr. P. J. Holloway and Dr. D. Stock of Long Ashton, U.K., and to Dr. J. Schönherr of Technische Universität, Munich, for kindly sending prepublication papers. Grateful thanks are expressed to Mrs. Sheila McMillan for preparation of this manuscript.

REFERENCES

1. P. J. Holloway, in: *The Plant Cuticle* (D. F. Cutler, K. L. Alvin, and C. E. Price, eds.), Linnean Society Symposium Series, No. 10, p. 1, Academic Press, London (1982).
2. E. A. Baker, in: *The Plant Cuticle* (D. F. Cutler, K. L. Alvin, and C. E. Price, eds.), Linnean Society Symposium Series, No. 10, p. 139, Academic Press, London (1982).
3. N. D. Hallam, in: *The Plant Cuticle* (D. F. Cutler, K. L. Alvin, and C. E. Price, eds.), Linnean Society Symposium Series, No. 10, p. 197, Academic Press, London (1982).
4. P. J. Holloway, in: *The Plant Cuticle* (D. F. Cutler, K. L. Alvin, and C. E. Price, eds.), Linnean Society Symposium Series, No. 10, p. 45, Academic Press, London (1982).
5. C. F. Jeffree, E. A. Baker, and P. J. Holloway, *New Phytol. 75*, 539 (1975).
6. C. F. Jeffree, E. A. Baker, and P. J. Holloway, in: *Microbiology of Aerial Plant Surfaces* (C. H. Dickinson and T. F. Preece, eds.), p. 119, Academic Press, London (1976).
7. T. F. Tadros, *Aspects Appl. Biol. 14*, 1 (1987).
8. R. C. Kirkwood, in: *Pesticides on Plant Surfaces* (H. J. Cottrell, ed.), Critical Reports on Applied Chemistry, Vol. 18, p. 1, John Wiley and Sons, Chichester (1987).
9. B. D. Ripley and L. V. Edgington, Proceedings of the 10th International Congress of Plant Protection, Brighton, Vol. 2, p. 545 (1983).
10. C. A. Hart, *Pestic. Sci. 10*, 341 (1979).
11. E. A. Baker and G. M. Hunt, *Ann. Appl. Biol. 106*, 579 (1985).
12. E. A. Baker, G. M. Hunt, and P. J. G. Stevens, *Pestic. Sci. 14*, 645 (1983).
13. G. M. Hunt and E. A. Baker, *Aspects Appl. Biol. 14*, 113 (1987).
14. C. E. Price, in: *The Plant Cuticle* (D. F. Cutler, K. L. Alvin, and C. E. Price, eds.), Linnean Society Symposium Series, No. 10, p. 237, Academic Press, London (1982).
15. R. H. Miller, *Ann. Bot. 55*, 459 (1985).
16. D. G. Davies, J. S. Mullins, G. E. Stolzenberg, and G. D. Booth, *Pestic. Sci. 10*, 19 (1979).
17. G. S. Hartley and I. J. Graham-Bryce, *Physical Principles of Pesticide Behaviour*, Vol 1, Academic Press, London (1980).
18. L. M. L. Thompson, G. E. Sanders, and K. E. Pallett, *Aspects Appl. Biol. 11*, 45 (1986).
19. F. Kerler and J. Schönherr, *Arch. Environ. Contam. Toxicol. 17*, 1 (1988).
20. F. Kerler and J. Schönherr, *Arch. Environ. Contam. Toxicol. 17*, 7 (1988).
21. J. Schönherr and M. Riederer, *Arch. Environ. Contam. Toxicol. 17*, 13 (1988).
22. M. Riederer and J. Schönherr, *Ecotoxicol. Environ. Safety 8*, 236 (1984).
23. M. Riederer and J. Schönherr, *Ecotoxicol. Environ. Safety 9*, 196 (1985).
24. A. Hamburg and P. J. McCall, in: *Pesticide Formulations, Innovations and Developments* (B. Cross and H. B. Scher, eds.), ACS Symposium Series, No. 371, p. 57, American Chemical Society, Washington, D.C. (1988).
25. C. R. Merritt, *Ann. App. Biol. 101*, 517 (1982).
26. R. C. Kirkwood, I. McKay, and R. Livingstone, in: *The Plant Cuticle* (D. F. Cutler, K. L. Alvin, and C. E. Price, eds.), Linnean Society Symposium Series, No. 10, p. 253, Academic Press, London (1982).
27. P. Veerasekaran, The Mode of Action of Asulam in Bracken, Ph.D. thesis, University of Strathclyde, Glasgow (1975).
28. J. van Overbeek, *Annu. Rev. Plant Physiol. 7*, 355 (1956).
29. E. A. Baker, *New Phytol. 73*, 953 (1974).
30. G. M. Hunt and E. A. Baker, in: *The Plant Cuticle* (D. F. Cutler, K. L. Alvin, and C. E. Price, eds.), Linnean Society Symposium Series, No. 10, p. 279, Academic Press, London (1982).
31. G. D. Wills and C. G. McWhorter, in: *Pesticide Formulations, Innovations and Developments* (B. Cross and H. B. Scher, eds.), ACS Symposium Series, No. 371, p. 90, American Chemical Society, Washington, D.C. (1988).
32. C. G. McWhorter, T. N. Jordan, and G. D. Wills, *Weed Sci. 28*, 113 (1980).

33. C. G. McWhorter and G. D. Wills, *Weed Sci. 26*, 382 (1978).
34. C. G. McWhorter and T. N. Jordan, *Weed Sci. 24*, 257 (1976).
35. D. J. Turner, *J. Plant Prot. Tropics 1*, 77 (1985).
36. J. D. Nalewaja, *Meded Fac. Landbouwwetensch., Rijksuniv. Gent 51*, 301 (1986).
37. R. Zandvoort, *Proc. EWRS Symp. Factors Affecting Herbicidal Activity and Selectivity*, p. 157 (1988).
38. C. L. Foy, in: *Adjuvants and Agrochemicals* (P. N. P. Chow, C. A. Grant, A. M. Hinshalwood, and E. Simundsson, eds.), Vol. 1, p. 1, CRC Press, Boca Raton, Florida (1989).
39. P. J. Holloway and D. Stock, in: *Industrial Applications of Surfactants II*, p. 303, Royal Society of Chemistry (1990).
40. H. B. Scher, in: *Pesticide Formulations, Innovations and Developments* (B. Cross and H. B. Scher, eds.), ACS Symposium Series, No. 371, p. 1, American Chemical Society, Washington, D.C. (1988).
41. B. W. Young, C. A. Hart, and F. R. Hall, *Aspects Appl. Biol. 14*, 23 (1987).
42. N. H. Anderson, D. J. Hall, and D. Seaman, *Aspects Appl. Biol. 14*, 233 (1987).
43. H. de Ruiter and A. J. M. Uffing, *Proc. EWRS Symp. Factors Affecting Herbicidal Activity and Selectivity*, p. 163 (1988).
44. L. W. Smith and C. L. Foy, *J. Agric. Food Chem. 14*, 117 (1966).
45. C. G. McWhorter, in: *Weed Physiology, Vol II: Herbicide Physiology* (S. O. Duke, ed.) p. 141, CRC Press, Boca Raton, Florida (1985).
46. Y. Sugimura and T. Takeno, *Pestic. Sci. 10*, 233 (1985).
47. G. E. Stolzenberg, P. A. Olson, R. G. Zaylskie, and E. R. Mansager, *J. Agric. Food Chem. 30*, 637 (1982).
48. N. H. Anderson and J. Girling, *Pestic. Sci. 14*, 399 (1983).
49. P. J. Holloway and D. Silcox, *Proceedings of the British Crop Protection Conference—Weeds*, p. 297 (1985).
50. D. Silcox and P. J. Holloway, *Aspects Appl. Biol. 11*, 1 (1986).
51. D. Silcox and P. J. Holloway, *Aspects Appl. Biol. 11*, 13 (1986).
52. D. Silcox and P. J. Holloway, in: *Adjuvants and Agrochemicals* (P. N. P. Chow, C. A. Grant, A. M. Hinshalwood, and E. Simundsson, eds.), Vol. 1, p. 115, CRC Press, Boca Raton, Florida (1989).
53. P. J. G. Stevens and M. J. Bukovac, *Proceedings of the British Crop Protection Conference—Weeds*, p. 309 (1985).
54. P. J. G. Stevens and M. J. Bukovac, *Pestic. Sci. 20*, 37 (1987).
55. E. A. Baker and G. M. Hunt, in: *Pesticide Formulations, Innovations and Developments* (B Cross and H. B. Scher, eds.), ACS Symposium Series, No. 371, p. 6, American Chemical Society, Washington, D.C. (1988).
56. C. A. Hart and B. W. Young, *Aspects Appl. Biol. 14*, 127 (1987).
57. D. Attwood and A. T. Florence, in: *Surfactant Systems, Their Chemistry, Pharmacy and Biology*, Chapman and Hall, London (1983).
58. D. Silcox, Ph.D. thesis, University of Bristol (1988).
59. V. Geyer and J. Schönherr, in: *Pesticide Formulations, Innovations and Developments* (B. Cross and H. B. Scher, eds.), ACS Symposium Series, No. 371, p. 22, American Chemical Society, Washington, D.C. (1988).
60. T. F. Tadros, in: *Surfactants* (T. F. Tadros, ed.), p. 323, Academic Press, London (1984).
61. N. K. Lownds, J. M. Leon, and M. J. Bukovac, *J. Am. Soc. Hort. Sci. 112*, 554 (1987).
62. J. Schönherr and M. J. Bukovac, *Plant Physiol. Lancaster 49*, 813 (1972).
63. D. W. Green and M. J. Bukovac, *Plant Cell Physiol. 13*, 321 (1974).
64. J. A. Zabkiewicz, D. Coupland, and F. Ede, in: *Pesticide Formulations, Innovations and Developments* (B. Cross and H. B. Scher, eds.), ACS Symposium Series, No. 371, p. 77, American Chemical Society, Washington, D.C. (1988).

65. P. J. G. Stevens and J. A. Zabkiewicz, *Proc. EWRS Symp. Factors Affecting Herbicidal Activity and Selectivity*, p. 145 (1988).
66. P. M. Neumann and R. Prinz, *J. Sci. Food Agric. 25*, 221 (1974).
67. R. C. Kirkwood and A. M. Hinshalwood, *Proc. Roy. Soc. Edin. 86B*, 179 (1985).
68. R. C. Kirkwood, *Aspects Appl. Biol. 14*, 281 (1987).
69. R. C. Kirkwood, *Proc. EWRS Symp. Factors Affecting Herbicidal Activity and Selectivity*, p. 245 (1988).
70. P. Veerasekaran, R. C. Kirkwood, and E. W. Parnell, *Pestic. Sci. 12*, 330 (1981).
71. R. E. Gaskin and R. C. Kirkwood, in: *Adjuvants and Agrochemicals* (P. N. P. Chow, C. A. Grant, A. M. Hinshalwood, and E. Simundsson, eds.), Vol. 1, p. 129, CRC Press, Boca Raton, Florida (1989).
72. J. A. Zabkiewicz and R. E. Gaskin, in: *Adjuvants and Agrochemicals* (P. N. P. Chow, C. A. Grant, A. M. Hinshalwood, and E. Simundson, eds.), Vol. 1, p. 141, CRC Press, Boca Raton, Florida (1989).
73. R. C. Kirkwood, H. Knight, I. McKay, and J. P. N. R. Chandrasena, in: *Adjuvants and Agrochemicals II*, CRC Press, Boca Raton, Florida (in press).
74. D. Stock, P. J. Holloway, P. Whitehouse, and B. T. Grayson, in: *Adjuvants and Agrochemicals II*, CRC Press, Boca Raton, Florida (in press).
75. P. J. G. Stevens and M. J. Bukovac, *Pestic. Sci. 20*, 19 (1987).
76. W. E. Shafer and M. J. Bukovac, *Plant Physiol. 85*, 965 (1987).
77. N. K. Lownds and M. J. Bukovac, *J. Am. Soc. Hort. Sci. 113*, 205 (1988).

9

Pathways and Mechanisms of Transport of Herbicides in Plants

RICHARD H. BROMILOW AND KEITH CHAMBERLAIN

9.1. INTRODUCTION

There now exists an extensive literature on the uptake and translocation of herbicides in plants. In this brief review, it is intended to elucidate the general principles that govern the transport of pesticides in plants. The transport behavior of the different classes of herbicides will be considered, together with ways of using physicochemical properties to predict the likely behavior of new chemicals. It should be noted that the processes of sorption in soil, of cuticular penetration in the case of foliar-applied compounds, and of metabolism in the plant all influence the availability of herbicides for uptake and redistribution in plants, and these factors are the subjects of separate chapters in this book. Interpretation of the literature on systemic transport usually needs substantial consideration of these additional factors.

RICHARD H. BROMILOW AND KEITH CHAMBERLAIN • Department of Insecticides and Fungicides, AFRC Institute of Arable Crops Research, Rothamsted Experimental Station, Harpenden, Herts AL5 2JQ, United Kingdom.

9.2. GENERAL FACTORS DEFINING AND CONTROLLING SYSTEMIC BEHAVIOR

9.2.1. General Description of Transport Systems in Plants

For a full and detailed description of the vascular systems of plants, the reader is directed to a standard text such as Esau.[1] The movement of solutes within plants takes place largely by two pathways. One route is via the extraprotoplasmic continuum (apoplast) of the plant and includes transport over short distances through the intercellular spaces and over long distances in the xylem vessels. Such transport is normally in an upward direction (acropetal). The second route is via the cytoplasmic continuum (symplast) of the plant and includes short-distance cell-to-cell transport through plasmodesmata and long-distance transport in the phloem sieve cells. Phloem translocation takes place in both upward and downward (basipetal) directions to the sites of new growth.

9.2.1.1. Xylem

The elements of the xylem, which are the functional conduits for the long-distance movement of water and dissolved mineral ions, are composed of dead cells whose central lumina contain no cytoplasm. In trees, the diameter of these vessels lies in the range 20–400 μm and in vines can be as large as 700 μm. This allows the rapid and efficient movement of large volumes of solution, the rate of flow in trees being of the order of 1–10 m/h. The mechanism of upward movement depends upon the pull resulting from the loss of water from the leaves during transpiration, with water in the xylem ducts being drawn up the stem under a tension or negative pressure. Water and solutes cannot enter the xylem vessels in the roots by an entirely apoplastic pathway, however, but must move through the symplast of the endodermal cells because the tangential walls of these cells are thickened by the deposition of suberin and lignin to form the Casparian strip (Figure 9.1).

9.2.1.2. Phloem

The phloem tissue of plants is composed mainly of the sieve elements, companion cells, and the phloem parenchyma. The sieve elements are specialized elongated cells lacking a vacuole and connected at each end to other sieve elements by special pores. However, as in all cells, the cytoplasm is contained within the plasmalemma, with connections to other cells via the plasmodesmata. The phloem carries the products of photosynthesis, the assimilates, from the mature leaves to the meristematic and actively growing parts of the plant. The mechanism of phloem transport has long been the

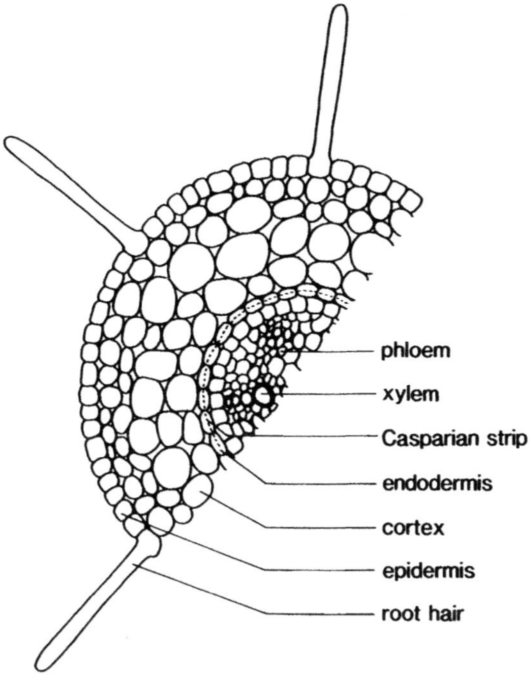

Figure 9.1. Schematic drawing of a root cross section, showing how the endodermis separates the cortex from the vascular tissues.

subject of controversy (see, e.g., Ref. 2), but the theory of mass flow, formulated by Münch,[3] is now widely accepted. During photosynthesis, carbohydrates are produced in the leaves and loaded, usually as sucrose, by a specific mechanism into cells of the phloem complex. Water which has ascended in the xylem is absorbed by these cells as a result of osmotic forces, and this, in turn, brings about an increased hydrostatic pressure in the phloem cells. At the same time, the reduction in sucrose concentration in the phloem at sites of growth and storage causes a lowered hydrostatic pressure. Since the phloem sieve tubes are essentially inelastic, a mass flow of water and solutes takes place from the regions of synthesis (sources) to those of utilization (sinks). In general, the velocities of translocation for assimilates in the phloem are of the order of 0.1–1.0 m/h.

As a result of the close proximity of xylem and phloem tissues, there is a potential for considerable lateral transfer of solutes from one transport system to the other. The volume of flow in the xylem, however, is so much greater (perhaps 50- to 100-fold) that solutes that move freely from one to the other will move mainly in the xylem to the mature leaves.

9.2.2. Sites of Action of Herbicides

The precise primary site of action of many herbicides is not known, but in all cases it is thought that short-distance transport through cellular membranes is essential for herbicides to reach the site of action. Whether long-distance transport, in the phloem or xylem, is also required depends upon the timing of application, that is, pre- or postemergence, and the region of the plant where the lethal action occurs. For example, the dinitroaniline herbicides applied preemergence exert their primary effect by inhibiting cell division in the roots of the plants, and so long-distance transport in the xylem is unnecessary. On the other hand, urea and triazine herbicides, which are also applied preemergence, must be taken up by the roots and translocated in the xylem to their site of action in the chloroplasts of leaf cells. The bipyridinium herbicides also act primarily in the chloroplasts, but, since they are only applied to the green parts of plants, short-distance transport is all that is required.

Some groups of herbicides depend upon phloem translocation to reach their sites of action. For example, the phenoxyalkanoic acid herbicides, which act on cell growth and development, and glyphosate, which inhibits the biosynthesis of aromatic amino acids, are translocated from leaves to all sites of active growth within the plant. Similarly, the sulfonylureas and imidazolinones, which inhibit acetolactate synthase, are, when applied postemergence, translocated in the phloem to all parts of the plant.

9.2.3. Physicochemical Properties and Their Role in Determining Herbicide Transport

9.2.3.1. Vapor Pressure

Movement of herbicides in the vapor phase can be important for transport from soil to plant (Section 9.3.2), and also some compounds transported to leaves from roots via the xylem may be then lost from leaves by volatilization. Worthing[4] and Ashton and Crafts[5] give the vapor pressures for most herbicides.

9.2.3.2. Lipophilicity and Water Solubility

Transport of pesticides in plants is dictated primarily by the rate at which the compounds are able to cross the membranes that encompass the transport vessels. This process appears to be determined mainly by the lipophilicity of the chemicals rather than by any specific property of the molecule such as its shape or the presence of particular functional groups. Usually, lipophilicity is assessed using the octan-1-ol/water partition coefficient (K_{ow}), which for most compounds is best measured directly by shake-flask methods.[6] For log K_{ow} values greater than 4.0, however, such direct measurement is difficult, and indirect methods such as reverse-phase high-pressure liquid chromatography are commonly used.[7] Predictions can be made using fragment constants,[8] though reliable estimates usually depend on a baseline value being measured for a compound of the same structural type.

Partitioning of more lipophilic chemicals onto plant solids is an additional factor that limits their long-distance transport, and this is also a simple function of lipophilicity.

Water solubility is generally rather a poor guide to systemic behavior. Lipophilicity, however, can be estimated from water solubility using relationships such as those given by Briggs[9]:

- for liquids,

$$\log \text{WS} = 0.84 - 1.18 \log K_{ow} \tag{9.1}$$

- for solids,

$$\log \text{WS} = 0.01 - \log K_{ow} - (0.01\, T_m - 0.25) \tag{9.2}$$

where WS is the molar water solubility and T is the melting point in °C.

9.2.3.3. Acid Strength (pK_a)

The picture is more complicated for compounds that are appreciably ionized at physiological pH, because the ionized and nonionized forms will have very different physicochemical properties and hence different permeation rates across membranes. For example, many herbicides are monobasic acids with pK_a in the range of 3 to 6, and the anions are typically 3 to 4 log K_{ow} units more polar than the corresponding undissociated acids. Care needs to be taken that log K_{ow} values for such compounds are measured at pH values where effectively only one form is present; if this is not possible, then appropriate corrections must be made. The pK_a can most simply be regarded as the pH at which a particular acid or base group is 50% ionized, and pK_a can be measured or predicted by standard procedures.[10,11]

If the pH differs in two plant compartments separated by a membrane (e.g., vacuole, pH $\approx$ 5.5; cytoplasm, pH $\approx$ 7.5), then there will be different proportions of the nonionized and ionized forms of an acid in the two compartments. The nonionized form, for example, will cross the membrane more rapidly than does the anion, and so nonionized molecules moving from the low-pH compartment will be substantially ionized in the higher-pH compartment and the anions so produced can only escape slowly. This process is the basis of the "ion trap" effect, whereby weak acids can be substantially accumulated in plant cells of high pH.

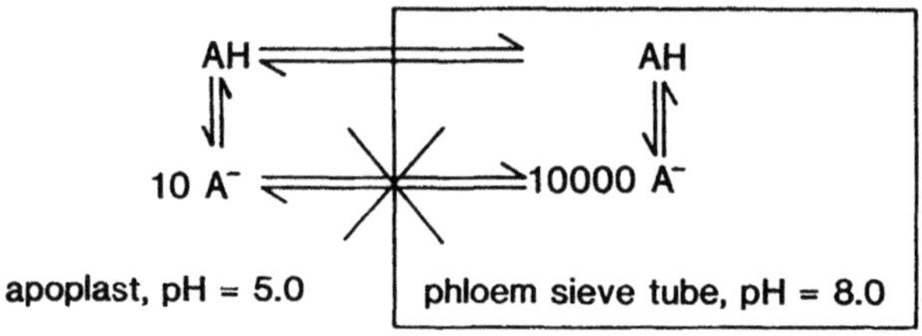

Figure 9.2. Accumulation of weak acids within cells by the ion-trap effect. (Reproduced from Ref. 12.)

The example in Figure 9.2, taken from Ref. 12, shows how a carboxylic acid of pK_a 4.0 might be accumulated in a sieve tube relative to the free space. The degree of accumulation is dependent on the relative permeabilities of the membrane to AH and A^-, with accumulation increasing as the permeability ratio (P_{AH}/P_{A^-}) increases. In the example, the maximum possible accumulation in the phloem would be

$$[(A^- + AH) \text{ inside}/(A^- + AH) \text{ outside}],$$

that is, $(10{,}000 + 1)/(10 + 1) = 909$; in practice, however, leakage of the anion across the plasmalemma would reduce this considerably. This simple calculation neglects the small effect of the membrane charge on accumulation, but Briggs *et al.*[13] have utilized the following equation[14] for accumulation which takes this charge into account:

$$\frac{[AH]_i + [A^-]_i}{[AH]_o + [A^-]_o} = \frac{(1+10^{pH_i - pK_a})\{P_{AH}/P_{A^-} + [(FE/RT)/(1-e^{-FE/RT})]10^{pH_o - pK_a}\}}{(1+10^{pH_o - pK_a})\{P_{AH}/P_{A^-} + [(FE/RT)/(1-e^{-FE/RT})]10^{pH_i - pK_a} \cdot e^{-FE/RT}\}} \tag{9.3}$$

where pH_i and pH_o are, respectively, the pH inside and outside the membrane, F is the Faraday constant (96,487 C/mol), E is the charge on the membrane, R is the universal gas constant (8.31 J/K · mol), T is the absolute temperature (K), and P_{AH} and P_{A^-} are the permeation rates through the membrane of the undissociated and dissociated forms of the acid, respectively.

9.2.4. Effects of Phytotoxicity of Herbicides on Transport in Plants

Many herbicides cause membrane disruption, usually as a secondary effect, and this destroys the integrity of the transport systems, particularly the phloem. If such effects are produced close to the point of application of a herbicide, they will interfere with subsequent translocation to other parts of the plant. Difenzoquat, for example, produced a marked local effect on wild oats (*Avena fatua*) which was distinct from the overall effect on the plant.[15] Factors which increased this local chlorosis and scorching, such as increasing drop size or concentration while maintaining constant dose, reduced the overall effect on the plant. It was concluded that this local phytotoxicity reduced the movement of the herbicide in the plant. In a parallel study,[16] such effects were not observed with paraquat, MCPA, or glyphosate. However, in some detailed studies of the effect of glyphosate on carbon assimilation and phloem translocation in sugar beet (*Beta vulgaris*), Geiger and co-workers[17–19] found that this slow-acting herbicide rapidly inhibits the allocation of carbon to starch during photosynthesis, though sucrose levels and export were maintained. As a consequence of this reduction in starch levels during the light period, sucrose export during the subsequent dark period was inhibited, and, since glyphosate is distributed via bulk flow in the phloem, its movement also was reduced. Likewise, chlorsulfuron has been shown to affect its own translocation by inhibiting assimilate translocation from the treated leaves of common pennycress (*Thlaspi arvense*) and Tartary buckwheat (*Fagopyrum tataricum*)[20]; clopyralid showed no such inhibitory effect.

9.3. UPTAKE AND TRANSPORT PROCESSES FOR HERBICIDES

9.3.1. Experimental Techniques for Measuring Transport

The various approaches to measuring transport of pesticides in plants have been reviewed elsewhere,[21] and here only brief pertinent comment is made. Systemicity is usually assessed by applying radioactive compounds, either to the foliage or to the roots via the growing medium, and measuring the distribution of compound after a certain length of time, typically one to a few days. Techniques of assessment such as autoradiography or the counting of total radioactivity (e.g., by combustion to $^{14}CO_2$) should only be used if the compounds are known to be reasonably stable over the period of the experiments; if they are not stable, the nature of the transported radioactivity should be determined. Metabolites can have very different physicochemical properties[21] and hence different transport patterns from those of the parent compounds. Cuticular penetration may limit the uptake and redistribution of herbicides, and the resistance of this barrier should be reduced as much as possible by cuticle stripping, abrasion, or use of surfactants.

More specialized systems of study involve the collection of xylem or phloem sap. Xylem sap can be collected from the stems or leaf bases of many plants, though care must be taken to avoid contamination with cell contents and phloem sap. Hsu *et al.*[22] recently developed a sophisticated technique in which chemicals were applied to nutrient solutions bathing roots of soybean (*Glycine max*), and application of pressure to the root chamber gave realistic rates of xylem sap flow from the stem stump. Phloem sap can only be readily collected from a few plant species,[2] with the techniques most used in pesticide research involving excised leaves of legumes,[23-25] excised flowering stems of yucca.[26] or incisions made in the stem of castor bean (*Ricinus communis*) plants.[27-29]

The use of different experimental systems and the lack of adequate detail in many articles concerning translocation make comparisons difficult; a more consistent approach would be very helpful. While such consistency is less easy to achieve in the case of phloem translocation, results from studies on uptake and translocation following application to roots via nutrient solution can readily be presented in a standard way (see Sections 9.3.2.2 and 9.3.2.3).

9.3.2. Uptake by Plants from Soil

9.3.2.1. Availability of Herbicides in Soil

The availability of herbicides in soil for uptake by germinating seedlings or by plant roots can be considered to depend upon two processes: (i) the extent to which the herbicide distribution in soil is matched to the distribution of the weed target, usually its roots, and (ii) the efficiency of the transport processes from the soil to the target plant. These topics are considered in detail in Chapter 7; for the purpose of the present chapter, it is necessary only to reiterate the importance of sorption to soil in limiting both movement of the herbicide in soil and the availability of the herbicide to the plant. Sorption of nonionized herbicides to soil is primarily due to partitioning into soil organic matter, and so recommended rates for soil-applied herbicides are often higher for soils containing much organic matter in order to compensate for the reduced availability. Sorption is stronger in dry soil, and herbicides may thus be less effective in very dry weather.

9.3.2.2. Uptake by Roots from Soil

Many herbicides are applied to soil, and their effectiveness is dependent upon the germinating weed seedlings acquiring a lethal dose. Successful control of weeds often depends on killing the weeds at this early stage, for older weeds are usually more difficult to kill. Generally, selectivity between crop and weed is achieved either because the crop is much more deeply rooted than the weeds and so acquires little of a surface-applied chemical or because of rapid detoxification of the herbicide in the crop plant.

Uptake of herbicides into roots can take place via the air or water phase. Diffusion coefficients of substances in air are about 10,000 times greater than those in water. Knowing the Henry's constant, that is, the ratio of concentrations in air and in water at saturation, it is possible to predict which phase is the most important for movement by diffusion in soil and uptake into roots. Some examples of pathways of herbicide movement through soil are given in Figure 9.3; only compounds with Henry's constants above 10^{-4} to 10^{-5} are likely to move appreciably as vapor in moist soil. In drier soils, the limit may be slightly lower, perhaps down to 10^{-6}, because the tortuous nature of the fine capillaries containing the soil water limits the movement of solutes via the water phase. In calculating such constants, it should be noted that literature values of vapor pressure are not always reliable due to the difficulties of measuring low vapor pressures. The potential for vapor transport of weak acids in soil is negligible as they are always substantially ionized.

a. Uptake of Vapor. Uptake from the vapor phase is an important pathway only for those herbicides that have both a low affinity for water, that is, are rather lipophilic, and an appreciable vapor pressure. The required

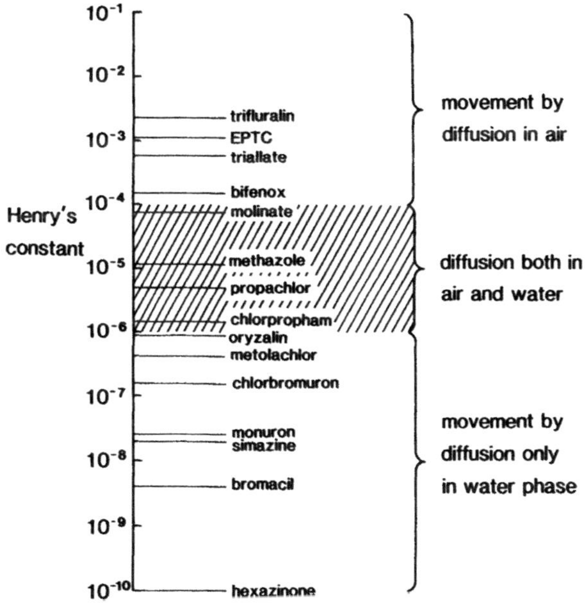

Figure 9.3. Pathways of movement of herbicides through soil as determined by Henry's constants.

property of moderate to high lipophilicity dictates that many such compounds are not translocated after uptake, and they have to be taken up directly at the site of action (e.g., the dinitroanilines such as trifluralin). The volatility of trifluralin is, in fact, so great that applications to the soil surface are rapidly lost by volatilization; usually, the compound is shallowly incorporated into soil in order for it to be retained for a longer time. In very dry soil, trifluralin is ineffective because it is very strongly sorbed by soil under such conditions[30,31]; good weed control requires at least slightly moist soil wherein compounds such as trifluralin can volatilize into the soil air.

b. Uptake via the Aqueous Phase. Herbicides which are active following uptake by weed roots from the soil water include important classes of nonionized herbicides, such as the triazines and phenylureas, and ionized herbicides such as the sulfonylureas. Generally, such compounds active via the water phase are more polar than compounds active via the vapor phase, and so can be leached more readily in soil; this can affect the proportion of roots in contact with the region of treated soil. Other factors affecting herbicide activity include the efficiency of uptake by roots and, for those compounds with an active site in the shoots of plants, the efficiency of transport to this site. Uptake into roots can be considered in terms of the root concentration factor (RCF),[32–34] defined by

$$\mathrm{RCF} = \frac{\text{concentration in roots}}{\text{concentration in the nutrient solution}} \tag{9.4}$$

Nonionized compounds. In order to separate the effects of these processes, researchers have studied uptake by roots and subsequent translocation to shoots of pesticides applied in nutrient solution. Use of nutrient solution rather than soil ensures that all the roots receive an essentially equal and constant supply of the compound over the period of the test. Briggs *et al.*[34] studied the uptake of 20 nonionized compounds, either carbamoyloximes or nonphytotoxic substituted phenylureas, applied in nutrient solution to the roots of small plants of barley (*Hordeum vulgare*). Uptake by roots (Figure 9.4a) consisted of two components: (i) an equilibrium between the aqueous phase in root cells and the external aqueous phase, giving approximately equal concentrations in each, and (ii) a partitioning effect onto root

Figure 9.4. Relationship between the uptake of chemicals by plant roots (expressed as their root concentration factor) and their octan-1-ol/water partition coefficients (as log K_{ow}): (a) in barley roots; ○, *O*-methylcarbamoyloximes; ×, substituted phenylureas; (b) literature values for several herbicides and fungicides in different plant species. (Reproduced from Ref. 34.)

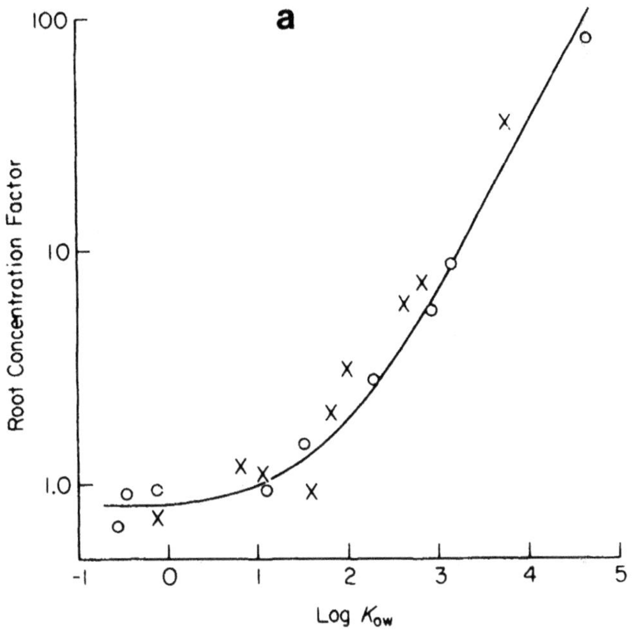
a
Root Concentration Factor
100
10
1.0
-1
0
1
2
3
4
5
Log K_{ow}

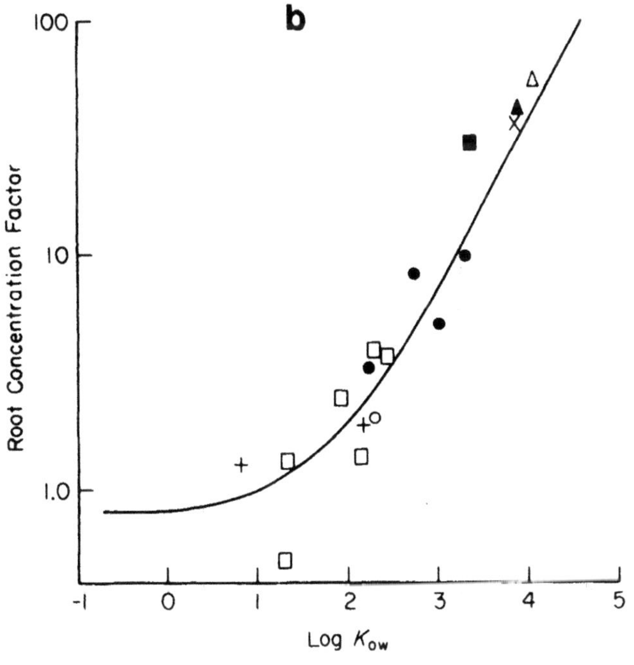
b
Root Concentration Factor
100
10
1.0
-1
0
1
2
3
4
5
Log K_{ow}

solids, which was insignificant for compounds of log $K_{ow} < 1.5$ but then increased rapidly with increasing lipophilicity. Accumulation in roots (expressed as the RCF) could be related to K_{ow} by the following equation:

$$\log(\text{RCF} - 0.82) = 0.77 \log K_{ow} - 1.52 \qquad (9.5)$$

Measurements taken from the literature of the uptake by roots from solution for a wide range of nonionized compounds in different plant species also closely followed this relationship (Figure 9.4b).

In soil, the stronger uptake by roots of lipophilic compounds from the soil water is offset by the reduced availability due to the sorption of lipophilic chemicals onto soil organic matter as discussed above.

Weak acids. Uptake of acidic compounds by roots is very different from that of nonionized compounds in that it is strongly dependent on the pH of the bathing medium. For example, uptake of 2,4-D into barley roots over 24 h from nutrient solution was 36 times greater at pH 4.0 than at pH 7.0.[13] This dependency can be explained by the ion-trap effect discussed in Section 9.2.3.3, whereby weak acids are accumulated in compartments of higher pH by virtue of the greater permeation rates across membranes of the undissociated form compared to the anion.

Within a series of substituted phenoxyacetic acids, Briggs *et al.*[13] noted that uptake by barley roots was optimal for the acids whose undissociated form had log K_{ow} values of 1 to 3 and was less for more polar and more lipophilic compounds. This uptake was due almost entirely to raised concentrations in the cell sap, for only with the most lipophilic compound tested (log K_{ow} 4.51) could any partitioning be measured onto macerated roots from a bathing solution of pH 5.0. The pK_a values of the acids were essentially similar across the series, and so these differences in accumulation were attributed to changes in the permeability ratio P_{AH}/P_{A^-}. This ratio was optimal for 4-chlorophenoxyacetic acid (log K_{ow} 2.25, P_{AH}/P_{A^-} 4×10^5) and declined substantially for more polar and more lipophilic analogues.

Sorption in soil of weak-acid herbicides increases as the pH decreases (see, e.g., Ref. 35), but this effect is much less than the effect of pH on uptake by roots. Accordingly, uptake of weak acids by plant roots is substantially greater from soils of low pH than of high pH, and this may explain why such herbicides are often found to be more effective in more acidic soils.

9.3.2.3. Translocation from Roots to Shoots

a. Nonionized Compounds. On entering roots, nonionized compounds are able to move through the cortex via the free space and thence to equilibrate freely with the cytoplasm and vacuole sap of the cortical cells.

However, at the endodermis, which is the cylindrical sheath of cells that surround the vascular tissues, the apoplastic pathway is blocked by the Casparian strip. If compounds are to enter the stele and reach the xylem vessels, they must pass through the plasmalemma membrane of the endodermal cells and so enter the symplasm. The transpiration stream concentration factor (TSCF) describes the efficiency of this process for compounds relative to that for water[32–34]:

$$\text{TSCF} = \frac{\text{concentration in xylem sap}}{\text{concentration in nutrient solution}} \tag{9.6}$$

As the concentration of compound in xylem sap is not readily measured, it is usually estimated from the amount of compound moved to the shoot for a measured volume of water transpired; a correction needs to be applied if there is appreciable metabolism of the compound in the shoots over the period of the test.

Briggs *et al.*[34] examined the translocation of 20 compounds in barley and observed that transport was optimal for compounds of log $K_{ow} \approx 1.8$; at this point, compounds crossed the membrane barrier at up to about 90% the efficiency of water (i.e., TSCF of 0.9). More polar and more lipophilic compounds were less well transported (Figure 9.5a), despite the fact that the lipophilic compounds were present at high concentrations in the roots. It was concluded that compounds of intermediate lipophilicity are best able to cross plant membranes; this relationship has also been observed with animal membranes.[36] Values of TSCF taken from the literature for other nonionized compounds studied in a range of plant species fitted the same Gaussian relationship reasonably well (Figure 9.5b).

b. Weak Acids. Unlike translocation of nonionized compounds, the translocation of many weak acids depends upon the extent to which the concentration in roots is raised above the concentration in the external solution. Since this accumulation is itself dependent on the pH of the bathing medium (nutrient solution or soil water), translocation to shoots has the same dependency on pH. Briggs *et al.*[13] observed that, in barley, movement of the substituted phenoxyacetic acids across the endodermal barrier was not very efficient. For example, from nutrient solution at pH 5.0 over 24 h, 2,4-D exhibited an RCF of 7.10 and a TSCF of 0.49, and so the concentration of 2,4-D in the xylem sap is only 7% of that in the cell sap (using an average of cytoplasm and vacuole sap concentrations) in the root cortex. Given that such acids of p$K_a \approx 3$ will be very substantially ionized at physiological pH and that anions appear to cross membranes only slowly, such poor transport across the endodermal barrier is not surprising. Several substituted phenoxyacetic acids moved across the endodermis with rather

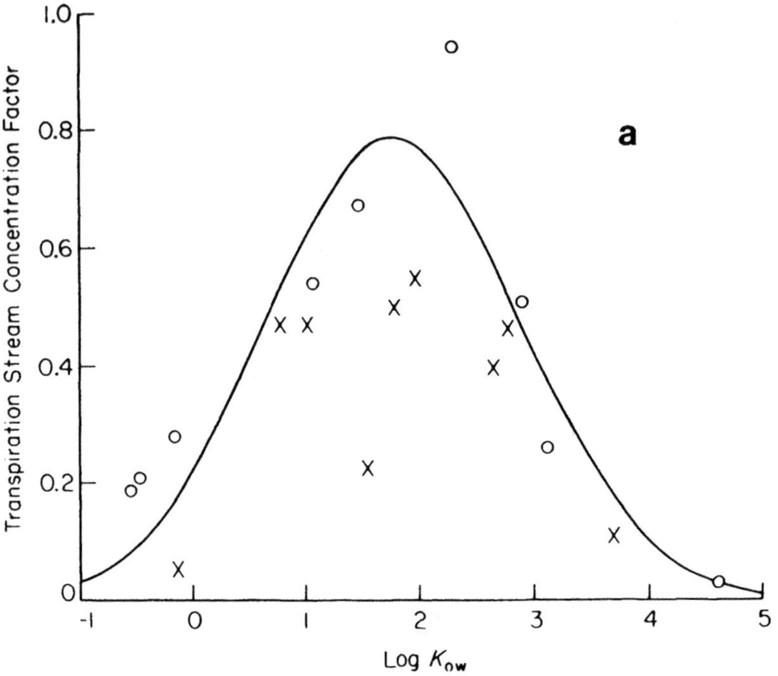
Transcription Stream Concentration Factor
Log K_{ow}
a
1.0
0.8
0.6
0.4
0.2
0
-1
0
1
2
3
4
5

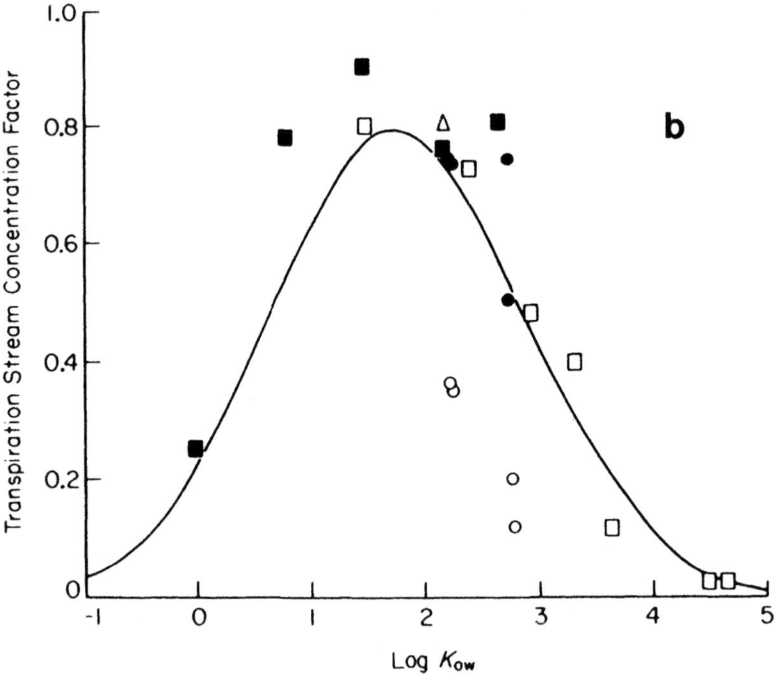
Transcription Stream Concentration Factor
Log K_{ow}
b
1.0
0.8
0.6
0.4
0.2
0
-1
0
1
2
3
4
5

similar efficiencies. The efficiency only seemed to drop with the most lipophilic acid tested (log K_{ow} 4.51), for which partitioning onto root solids accounted for much of the RCF, and this partitioned fraction was not free in the cell sap to cross the membrane barrier.

9.3.3. Uptake and Transport Following Foliar Application

9.3.3.1. Uptake through the Cuticle

The physicochemical properties of compounds affect not only the transport of compounds in plants but also their penetration through the leaf cuticle. This topic is discussed in Chapter 8, and here only brief comment is made on the implications of penetration for subsequent transport. The influence of physicochemical properties on the movement of compounds across cuticles seems to be rather similar to that observed for membranes, in that compounds of intermediate lipophilicity (log K_{ow} values of 1 to 3) appear to penetrate the most rapidly.[37] Uptake of more polar and more lipophilic compounds can, however, be greatly increased by the use of appropriate surfactants. Acidic compounds are taken up relatively slowly, presumably because of poor penetration of the anions; in consequence, they are often applied as esters which enter the plant rapidly and are then metabolized to the active acid. Thus, herbicides of a wide range of structures and physical properties are applied to plant foliage; the use of surfactants and adjuvants increases the uptake of chemical into the leaf and hence its availability for transport via xylem and phloem but does not otherwise influence transport patterns.

9.3.3.2. Transport in Xylem

The same general principles determine xylem translocation of herbicides whether they are foliage applied or soil applied. Provided that the barrier of cuticular penetration can be largely overcome by appropriate choice of formulation, application to foliage is a more reliable way of getting herbicides into the plant. This route avoids the losses caused by sorption to soil, and it bypasses the endodermal barrier surrounding the vascular tissues in the root. Once through the cuticle and into the leaf tissue,

---

Figure 9.5. Relationship between the translocation of chemicals to shoots following uptake by roots (expressed as the transpiration stream concentration factor) and the octan-1-ol/water partition coefficients (as log K_{ow}): (a) in barley; O, *O*-methylcarbamoyloximes; ×, substituted phenylureas; (b) literature values for several herbicides and fungicides in different plant species. (Reproduced from Ref. 34.)

redistribution via the xylem will follow the main water flow to the leaf margins and interveinal spaces.

9.3.3.3. Transport in Phloem

For many years, it was assumed that the failure of many compounds to move via phloem was due to their failure to enter phloem sieve tubes. Tyree *et al.*[38] then postulated that most compounds can, in fact, enter the phloem quite freely; however, only those compounds well retained in phloem are able to move appreciable distances in the phloem sieve tubes. Compounds escaping quickly will be swept upward by the much greater flow of the xylem stream. This elegant concept has become generally accepted and is known as the "intermediate permeability theory."

The efficiency of transport in phloem is now regarded as being determined by two processes. The first of these is the extent to which compounds are accumulated in phloem cells relative to the mesophyll cells, for example, by the ion-trap effect in the case of weak acids (Section 9.2.3.3). The second factor is the extent to which compounds are retained in phloem cells during transport.

a. Nonionized Compounds. In addition to the original proposals by Tyree *et al.*,[38] it was recognized that polar nonionized compounds such as oxamyl[39] would have the required limited permeation rates through membranes that allow movement via phloem.[40] More lipophilic compounds would be expected to cross membranes too quickly to be phloem mobile, whereas extremely polar compounds would be unable to cross membranes at a sufficient rate to ever attain reasonable concentrations in leaf phloem (cf. Figure 9.5). Bromilow *et al.*,[29] were able to provide evidence for this by injecting nonionized chemicals into the petioles of castor bean leaves and examining concentrations in phloem sap taken from the stem below the treated leaves. Relatively polar compounds such as aldicarb sulfone (log K_{ow} – 0.57) and oxamyl (log K_{ow} – 0.47) were 70% retained between the treated leaves and upper stem, whereas slightly more lipophilic compounds were less well retained, and 4-chlorobenzaldehyde *O*-methylcarbamoyloxime (log K_{ow} 2.25) could not be detected in phloem sap at all.

b. Weak Acids. Crisp and Look[41,42] noted that most phloem-mobile compounds are weak acids such as the phenoxyacetic acid herbicides. At first, it was thought that these compounds are able to enter phloem sieve tubes via a specific carrier that recognizes a carboxyl group; however, the current consensus is that these compounds really represent only a special case of the intermediate permeability theory. Due to the relatively high pH of the phloem sap, such acids are very substantially ionized, and the anions escape only slowly through the membranes bounding the sieve tube; furthermore, an additional advantage is obtained with acids in that the ion-trap mechanism permits substantial accumulation in the phloem sap relative to the concentration in the sap of adjacent leaf cells. Rigitano *et al.*[43] observed a 10-fold accumulation of maleic hydrazide in sieve tubes and a smaller accumulation of several substituted phenoxyacetic acids, and the extent of accumulation could be reasonably well predicted from the ion-trap relationship (Eq. 9.3).

9.3.3.4. Distribution Patterns within Plants and Their Relationship to Herbicide Properties

The factors determining uptake by plants and subsequent movement in xylem and phloem have been discussed, and consideration should now be given as to how this information can be used to understand the detailed distribution patterns of compounds in plants. Such patterns are usually obtained by autoradiography of plants following application of radiolabeled herbicides, and it is important that the period of the experiment be sufficiently short that interpretation is not confused by extensive metabolism of the herbicide.

The structural requirements for the transport of pesticides in plants have long been of interest, and the reader is referred to reviews by Crafts and Crisp,[44] Crisp,[41] Christ,[45] Price,[46] Edgington,[47] Jacob and Neumann,[48] Shephard,[49] Lichtner,[25] Giaquinta,[50] Bromilow and Chamberlain,[12] and Bromilow *et al.*[51] Early work can now be reinterpreted in the light of the recent advances in our understanding of the transport of xenobiotics. The only appreciable point of difference is that the important role once postulated for specific carrier mechanisms now seems unlikely, at least for current pesticides.

Nonionized chemicals, which move largely in the xylem, are subject to a kind of inefficient reverse-phase column chromatography by the plant, with the movement of the more lipophilic compounds being retarded by reversible partitioning into the lipid-like materials of the plant solids.[52,53] Barak *et al.*[54] showed that partitioning onto lignin was the most important of these processes. Voss and Geissbühler[55] studied the uptake of three phenylurea herbicides from nutrient solution by roots in French dwarf bean seedlings (Figure 9.6). Chloroxuron, the most lipophilic of the three (log K_{ow} 3.7), was strongly taken up by the roots, but only small amounts entered the aerial parts of the plant, and these were confined to the stem. In contrast, metobromuron (log K_{ow} 2.4) and fluometuron (log K_{ow} 2.2) were less accumulated in roots but were well translocated to leaves; fluometuron was relatively uniformly distributed in the mature leaves, while the slightly more lipophilic metobromuron was more concentrated around

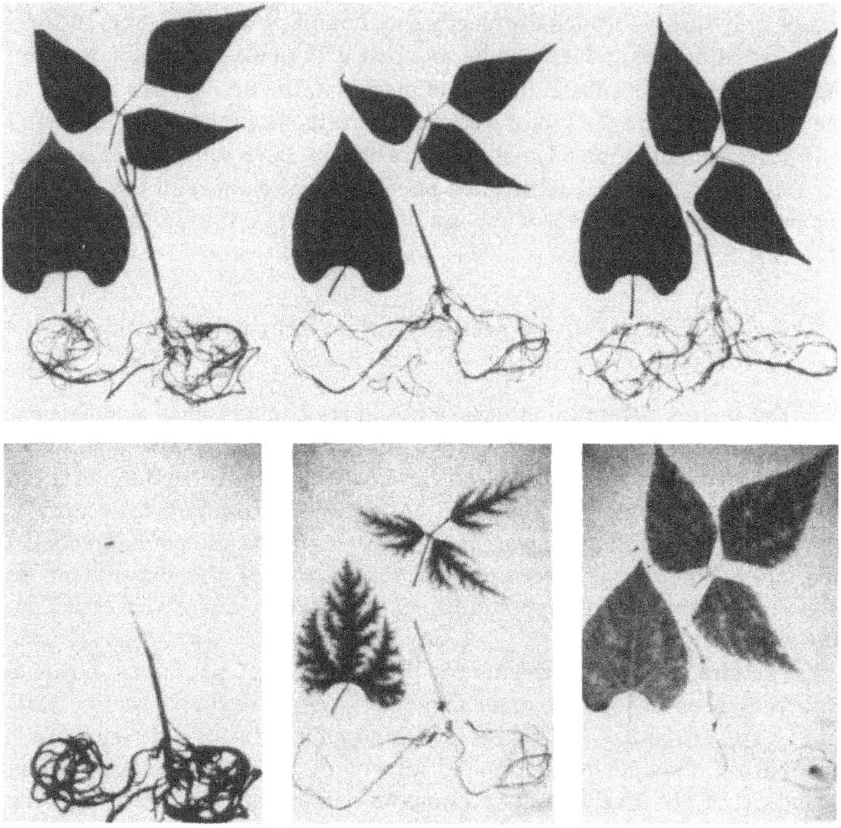

Figure 9.6. Uptake by roots and translocation to shoots of (from left to right) ^{14}C-labeled chloroxuron, metobromuron, and fluometuron in French dwarf bean seedlings, grown in nutrient solution containing 2.5 $\mu g/ml$ (and equal concentrations of radioactivity) of each herbicide. Plants were kept in herbicide solution for 24 h and then transferred to herbicide-free nutrient solution for an additional 24 h. Top: Mounted plants. Bottom: Autoradiographs of the same plants. [Voss and Geissbühler[55]; reprinted from P. C. Kearney and D. D. Kaufman (eds.), *Herbicides; Chemistry, Degradation and Mode of Action*, p. 86, by courtesy of Marcel Dekker, Inc.]

the leaf veins. Thus, these compounds obey the concepts of RCF and TSCF described above (Sections 9.3.2.2.b and 9.3.2.3), and their distribution patterns in the foliage are typical of the behavior of nonionized herbicides.

Over longer periods of time, such xylem-mobile compounds will move increasingly to the leaf margins. If the compound is applied as a point source on the leaf of a dicotyledonous plant, a wedge-shaped distribution pattern (the "apoplastic wedge") is usually observed[44]; in monocotyledonous plants, accumulation occurs at the leaf tips, often to high concentrations.

Turning to compounds mobile in phloem, autoradiography or other analysis indicates the efficiency with which compounds applied, for example, to mature leaves are transported to roots or new foliar growth at the apex. As discussed above (Section 9.3.3.3), this movement of compounds in phloem is limited by their crossing over to the xylem and returning to the transpiring leaves. A second process, however, also influences both long-distance transport in phloem and distribution patterns within leaves. This process is akin to chromatography, but compounds are held by cells against the moving phase (be it phoem or xylem) not by partitioning but apparently by the ion trapping of weak acids (Section 9.2.3.3) into the cells adjacent to the vascular tissues. With very lipophilic acids (log $K_{ow} > 4$), both the processes of partitioning and ion trapping may limit movement.

Bromilow and co-workers[43,51] examined the distribution of ^{14}C-labeled compounds in the leaves of castor bean following application by injection into the petioles of mature leaves (Figure 9.7). Phenylurea (log K_{ow} 0.80) was mobile only in the xylem, moving from the petiole to the leaf blade, where it was concentrated in the interveinal spaces. Aldicarb sulfone, another nonionized but more polar compound (log K_{ow} −0.57), showed a similar distribution pattern in the mature leaf but was also weakly moved in phloem to the developing apical leaf. In contrast, 3,5-dichlorophenoxyacetic acid (pK_a 2.84, log K_{ow} 2.84) was moved well in phloem to the apical leaf but remained in the tissue adjacent to the veins in both apical and mature leaves. The strong accumulation of this compound in this tissue was attributed to ion trapping, as was its lack of transport to roots observed in parallel experiments. Maleic hydrazide (pK_a 5.65, log K_{ow} −0.63) was well translocated to both the apical and mature leaves, and the uniformity of the distribution indicates a lack of ion trapping into tissue cells. While both of these acids are quite strongly accumulated in phloem cells, only 3,5-dichlorophenoxyacetic acid is accumulated by the ordinary cells containing vacuoles; this difference in behavior is explained by the fact that the sensitivity of the ion-trapping mechanism to the pH of the differing cellular compartments varies according to the pK_a of the acid (Eq. 9.3, Section 9.2.3.3). In general, it seems that the more polar and/or strongly acidic compounds are those best able to move over long distances via phloem (e.g., glyphosate, imazapyr, clopyralid).

Chamberlain *et al.*[37] compared the distribution of six substituted phenoxyacetic acids in two plant species. Application to castor bean was by injection into the petioles of mature leaves, and to barley by application (with surfactant) to a central section of the leaf blade. After 24 h, the proportion of each chemical moved in the phloem was greatest for the compounds of intermediate lipophilicity and was very similar for the two plant species. This result indicates that measurements of transport in

Figure 9.7. Autoradiographs showing the distribution of compounds within the leaf blades (apical leaf and one mature leaf presented) of castor bean 24 h after injection of the compounds into the petioles of the mature leaves: (a) phenylurea; (b) aldicarb sulfone; (c) 3,5-dichlorophenoxyacetic acid; (d) maleic hydrazide. (Reproduced from Ref. 51.)

Figure 9.7. (*Continued*)

convenient experimental systems such as castor bean are applicable to crop species.

Autoradiographs have been published illustrating the transport of many herbicides within plants. The compounds discussed above exemplify virtually all the observed patterns of behavior, with no major differences being noted between different plant species except where selective metabolism is occurring.

9.4. MOBILITIES OF HERBICIDES BY CLASS

Phloem transport in plants has been reviewed by Crafts and Crisp,[44] and general aspects of chemistry and systemicity of herbicides have been reviewed by Kearney and Kaufman,[56] Ashton and Crafts,[5] Matolcsy *et al.*,[57] and Hathway.[58] In the following brief discussion of the transport behavior of herbicides, references to the original literature are generally confined to more recent work, and the reader is referred to the above comprehensive reviews for further details of earlier work.

The herbicides considered in this section are listed by chemical class in Table 9.1. Some compounds could qualify for inclusion in more than one class, and these have been classified with compounds sharing a common mode of action (where known).

9.4.1. Nonionized Compounds

9.4.1.1. Amides

a. Chloroacetamides

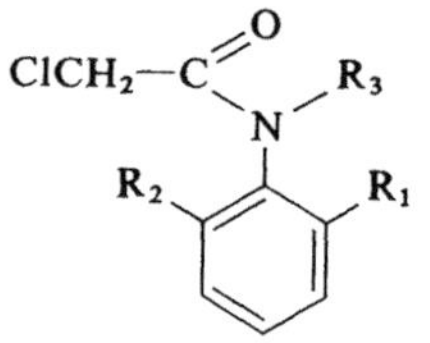

(1)

Most chloroacetamides have the substituent pattern shown in **1**; R_1 and R_2 are hydrogen, methyl or ethyl, and R_3 can be various alkyl or ether groups. They are nonionized compounds of log K_{ow} 1.5 to 4.5 and are usually applied preemergence. Uptake by roots occurs freely, with the compounds being transported through the xylem pathway to the leaves. The mode of action involves inhibition of root growth together with interference with cell division and/or enlargement in the shoots. Some of the compounds seem to exert their major effect directly on the shoot of germinating seedlings, this uptake occurring via both the vapor and aqueous phases in soil.

b. Other Amides. Generally, the effects of other amides on plants are similar to those produced by the chloroacetamides, though the biochemical mode of action is uncertain. These compounds are also of intermediate lipophilicity, having log K_{ow} values of 1.5 to 3.4, and most (e.g., diphenamid and pronamide) are applied to soil, from where they are readily taken up by roots and translocated to shoots. Propanil is applied to foliage, and its main action is believed to be inhibition of photosynthesis.

9.4.1.2. Carbamates

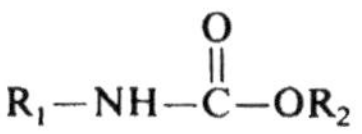

(**2**)

Carbamates (**2**) generally have R_1 as a phenyl or substituted phenyl, and R_2 can be various aliphatic or aryl groups. For these compounds, log K_{ow} ranges from 2.5 to 4.2, except for carbetamide and karbutilate which are more polar (log $K_{ow} \approx 1$). Propham and chlorpropham are soil applied, being taken up by roots and translocated to shoots. The coleoptile may be affected directly, since both compounds are quite volatile and a proportion of uptake may occur via the vapor phase. The other carbamates are less volatile and are applied postemergence to the leaves of weeds except for karbutilate (which contains both carbamate and urea functions), which is applied to soil for residual control of vegetation.

The carbamates appear to act by inhibiting cell division in root or shoot apices. Desmedipham and phenmedipham control annual weeds in sugar beet and act by damaging membranes in leaves under high light intensity. Together with barban and phenisopham, these two compounds are the most lipophilic of the herbicidal carbamates, and their lipophilicity probably permits only limited movement in the xylem.

9.4.1.3. Dinitroanilines

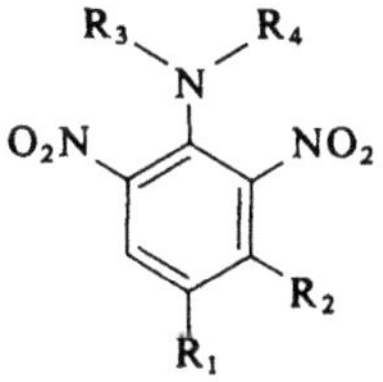

(**3**)

Table 9.1. Herbicides by Classes

NONIONIZED						
AMIDES	CARBAMATES	DIPHENYL ETHERS	THIOCARBAMATES	TRIAZINES	UREAS	UNCLASSIFIED
Chloroacetamides:	Barban	Acifluorfen[a]	Butylate	s-Triazines:	Benzthiazuron	Polar:
Acetochlor	Carbetamide	Aclonifen	Cycloate	Ametryn	Buturon	Amitrole
Alachlor	Chlorbufam	Bifenox	Diallate	Atraton	Chlorbromuron	Intermediate:
Butachlor	Chlorpropham	Chlomethoxyfen	EPTC	Atrazine	Chloroxuron	Benzofluor
CDDA	Desmedipham	(chlormethoxynil)	Molinate	Aziprotryne	Chlortoluron	Buthidazole
(allidochlor)	Karbutilate	Fluorodifen	Orbencarb	Cyanazine	Difenoxuron	Clomazone
Diethatyl-ethyl	Phenisopham	Fluoroglycofen-ethyl	Pebulate	Desmetryn	Dimefuron	Dinoseb
Dimethachlor	Phenmedipham	Fomesafen[a]	Thiobencarb	Dimethametryn	Diuron	Dinoterb
Fluorochloridone	Propham	Lactofen	(benthiocarb)	Dipropetryn	Ethidimuron	Ethofumesate
Metolachlor		Nitrofen	Tiocarbazil	Eglinazine-ethyl	Fenuron	Fluridone
Pentanochlor	DINITROANILINES	Nitrofluorfen	Triallate	Methoprotryne	Fluometuron	Methazole
Pretilachlor	Benefin	Oxyfluorfen	Vernolate	Proglinazine-ethyl	Isoproturon	Norflurazon
Propachlor	(benfluralin)			Prometon	Isouron	Pyrazon
Xylachlor	Butralin	NITRILES	URACILS	Prometryn	Linuron	(chloridazon)
Other amides:	Dinitramine	Bromoxynil[a]	Bromacil	Propazine	Methabenzthiazuron	Lipophilic:
Benzadox	Ethalfluralin	Chlorthiamid[b]	Lenacil	Secbumeton	Metobromuron	Anilofos
Diphenamid	Fluchloralin	Dichlobenil	Terbacil	Simazine	Metoxuron	Bensulide
Isoxaben	Flumetralin	Ioxynil[a]		Simetryn	Monolinuron	Bromofenoxim
Mefenacet	Isopropalin			Terbumeton	Monuron	Butamifos
Monalide	Nitralin			Terbuthylazine	Neburon	Cinmethylin
Napropamide	Oryzalin			Terbutryn	Norea	DCPA
Pronamide	Pendimethalin			Trietazine	(noruron)	(chlorthal dimethyl)
(propyzamide)	Prodiamine			Triazinones:	Siduron	Diflufenican
Propanil	Profluralin			Hexazinone	Tebuthiuron	Oxadiazon
Tebutam	Trifluralin			Metamitron	Thiazafluron	Piperophos
				Metribuzin		Pyridate
						Tridiphane

CATIONS	ACIDS			
Difenzoquat	ACIDIC AMIDES (OTHER THAN SULFONYLUREAS)	ARYLOXYPHENOXYPROPANOIC ACIDS	IMIDAZOLINONES	PYRIDINE CARBOXYLIC ACIDS (OTHER THAN IMIDAZOLINONES)
Diquat	Asulam	Diclofop	Imazapyr	Clopyralid
Morfamquat	Bentazon	Fenoxaprop	Imazaquin	Picloram
Paraquat	Mefluidide	Fluazifop	Imazethapyr	
	MH (maleic hydrazide)	Haloxyfop		SULFONYLUREAS
	Perfluidone	Quizalofop	PHENOXYACETIC ACIDS (INCLUDING ARYLOXYALKANOIC)	Bensulfuron-methyl
			2,4-D	Chlorimuron-methyl
	ALIPHATIC ACIDS	BENZOIC ACIDS	2,4-DB[b]	Chlorsulfuron
	Dalapon	Chloramben (amben)	2,4-DES[b]	Metsulfuron-methyl
	TCA (trichloroacetic)	Dicamba	Dichlorprop	Sulfometuron-methyl
		Naptalam	Fenoprop	Thiameturon-methyl
	ARSENIC ACIDS	2,3,6-TBA	Fluroxypyr	
	Cacodylic (dimethylarsinic)		MCPA	UNCLASSIFIED ACIDS
	MAA (methylarsonic)	HYDROXYCYCLOHEXENONES	Mecoprop	Benazolin
		Alloxydim	2,4,5-T	Benzoylprop
		Clethodim	Triclopyr	Endothall
		Cloproxydim		Fenac (chlorfenac)
		Cycloxydim	PHOSPHORUS ACIDS	Flamprop
		Sethoxydim	Bilanophos[b] (bialaphos)	
			Fosamine	
			Glufosinate	
			Glyphosate	

[a] These weak acids are included here because of similarities in mode of action with the nonionized compounds of these classes.
[b] Precursors of active compounds.

In dinitroanilines (**3**), R_2 is usually hydrogen, while R_3 and R_4 are alkyl or substituted alkyl groups containing at least four carbons in total. Where R_1 is trifluoromethyl or alkyl, the compounds are lipophilic and have a high vapor pressure (e.g., trifluralin; $R_1 = CF_3$, $R_2 = H$, $R_3 = R_4 = CH_2CH_2CH_3$; log K_{ow} 5.3). Where R_1 is a mesyl (nitralin) or sulfonamide (oryzalin), the compounds are less lipophilic (log K_{ow} 4.0–4.7) and the vapor pressure is much less.

The main use of dinitroanilines is to control annual weeds, particularly grasses, following preemergence application to soil. Uptake by roots and germinating shoots may be predominantly via the vapor phase or via the aqueous phase depending upon the compound; the resulting inhibition of mitosis is particularly noticeable from the abnormal features of treated roots. The less lipophilic compounds (dinitramine, nitralin, and oryzalin) are taken up from soil via the water phase and are then translocated to shoots to a small extent. However, such movement does not appear to be necessary for the phytotoxic action of these compounds.

9.4.1.4. Diphenyl Ethers

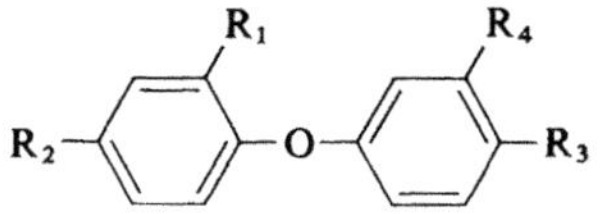

(**4**)

The diphenyl ethers (**4**) are lipophilic compounds (log K_{ow} 4.3–6.0, except for the somewhat less lipophilic fomesafen) of low water solubility and generally modest vapor pressure. R_1 and R_2 are Cl or CF_3 (except in fluorodifen, where $R_1 = NO_2$), R_3 is NO_2 and R_4, if present, is an alkoxy group or carboxy function. Aclonifen has a rather different substitution pattern.

The diphenyl ethers are used as pre- or postemergence sprays to control both grasses and broadleaf weeds in a variety of crops. The mode of action is believed to be by damaging the membrane integrity of plants. Uptake of the compounds from soil appears to be directly into the shoots of germinating weeds, perhaps largely via the vapor phase. When these herbicides are applied to the roots of plants via nutrient solution, the roots strongly accumulate them, but subsequent transport to shoots is either negligible (e.g., nitrofluorfen and oxyfluorfen[59]) or very limited (e.g., fluorodifen[60]); such lipophilic compounds would not be expected to move well in plants.

Acifluorfen and fomesafen are weak acids, having at R_4 a free carboxyl or acidic sulfonamide group respectively, and thus have the potential for phloem transport. When applied to foliage, however, acifluorfen moved very little from the point of application.[61,62] Transport of acifluorfen and also of fomesafen (for which little information is available) may be limited by their high lipophilicity and by the rapid damage which they cause to plant membranes in light.

9.4.1.5. Nitriles

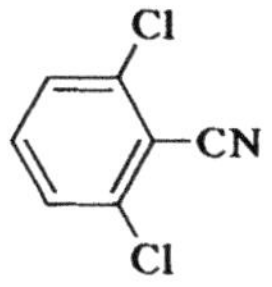

Dichlobenil

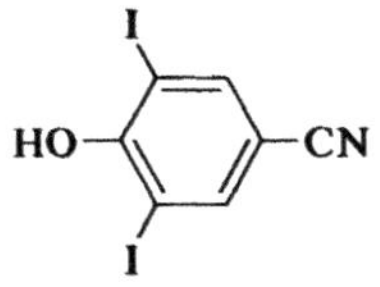

Ioxynil

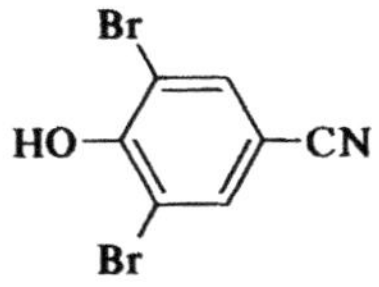

Bromoxynil

Dichlobenil is applied to soil and is incorporated or watered in to reduce losses by evaporation. It is a neutral compound of intermediate lipophilicity (log K_{ow} 2.9) and is translocated slowly in xylem; this transport may not be important in its action for it kills germinating weeds before or soon after they emerge. The herbicidal activity of chlorthiamid, a thioamide, depends upon its decomposition to dichlobenil.

Bromoxynil and ioxynil are also of intermediate lipophilicity (log K_{ow} 2.7 and 3.1, respectively) but are acidic compounds (pK_a 3.96 and 4.06, respectively). They are applied postemergence to control broadleaf weeds. These compounds move apoplastically and, to a rather limited extent, symplastically in plants.[63,64]

9.4.1.6. Thiocarbamates

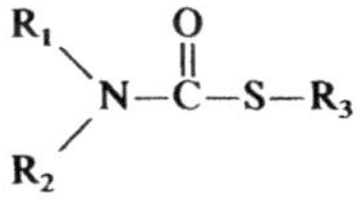

(5)

In thiocarbamates (**5**), the substituents are typically alkyl or halogenated alkyl groups containing in total 8 to 12 carbon atoms, giving log K_{ow} values in the range 2.7 to 4.8. The compounds are applied to soil, and most are mechanically incorporated immediately after application because of their high volatility. Uptake into plants can be via the air or water phase, the relative importance of these routes depending on the compound. Thiocarbamates appear to act both on roots and on the emerging shoots of weeds, especially grasses. Translocation occurs via the apoplast and is greater for the less lipophilic compounds such as molinate, but it is not clear whether this is important for the herbicidal action.

9.4.1.7. Triazines and Triazinones

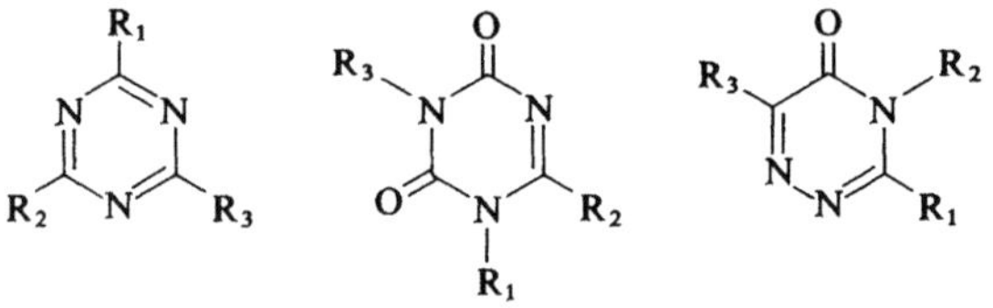

s-Triazine *s*-Triazinedione *as*-Triazinone

Applied as pre- or postemergence compounds, triazines and triazinones inhibit photosynthesis, and many of the criteria for activity are the same as for the ureas (Section 9.4.1.9). All compounds are well taken up by roots and translocated to shoots via the xylem.

In the *s*-triazine series, log K_{ow} ranges from 1.5 to 3.7. The substituent R_1 is usually chlorine, thiomethyl or methoxy, and R_2 and R_3 are various alkylamino groups. Where R_2 and R_3 are the same, the crystals of the compounds may have very high lattice energies; as a consequence, the water solubilities of such compounds, despite their intermediate lipophilicity, can be very low (e.g., 5 mg/liter for simazine). On heating, these compounds decompose before they melt, and so overly high values of log K_{ow} are estimated using Eq. (9.2) (Section 9.2.3.2) based on values of the water solubility and apparent melting point.

In the triazinones, R_1 is methyl or thiomethyl, R_2 an amino group, and R_3 a phenyl, alkyl or cycloalkyl group. Only three triazinones are available at present—metribuzin, metamitron, and the *s*-triazine-2,4-dione, hexazinone. Log K_{ow} values are generally lower than for the triazines, ranging from 0 for hexazinone to 1.7 for metribuzin.

9.4.1.8. Uracils

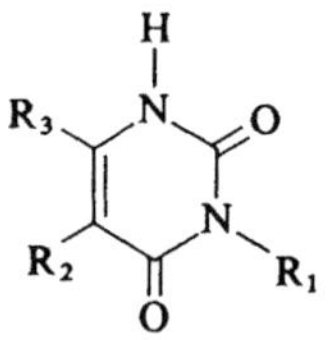

(6)

The uracils (**6**), like the ureas (Section 9.4.1.9), are applied to soil and kill weeds by inhibition of the Hill reaction in the chloroplasts of mature leaves. In the three compounds used in this class, R_1 is a large alkyl group, and R_2 and R_3 are alkyl groups or halogens. The range of lipophilicities is smaller than that of the ureas, with lenacil being rather more lipophilic than bromacil or terbacil (log $K_{ow} \approx 2$).

9.4.1.9. Ureas

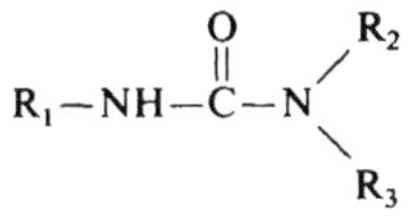

(7)

The majority of the urea herbicides (7) have R_1 as phenyl or substituted phenyl, R_2 as methyl, and R_3 as methyl or methoxy. Their mode of action involves inhibition of the Hill reaction, which occurs in the chloroplasts. Usually, they are applied preemergence to soil, and efficient uptake by roots and subsequent movement via the xylem to the mature leaves are necessary for the compounds to reach the site of biochemical action. In isolated chloroplasts, inhibition of the Hill reaction increases with increasing $\log K_{ow}$, but availability in soil and efficiency of transport to shoots decrease above $\log K_{ow} \approx 2$. The compounds thus span quite a narrow range of physicochemical properties, mostly being of intermediate lipophilicity ($\log K_{ow}$ 1.0–3.0), with only chlorobromuron, chloroxuron, and neburon being slightly more lipophilic. These properties provide a practical compromise between the conflicting requirements for availability, transport, and intrinsic activity; they also limit leaching in soil so that susceptible but more deeply rooted crop plants are not damaged. The slightly more polar compounds ethidimuron and tebuthiuron are used primarily for total vegetation control, presumably because they are too readily leached in soil to give depth protection to crop roots.

9.4.1.10. Unclassified Nonionized Compounds

Amitrole (aminotriazole) is a polar compound of $\log K_{ow}$ −0.87. It is applied to foliage and is transported symplastically throughout the plant, though its transport via phloem is modest compared to that of some of the acidic herbicides.[29] Metabolites such as 3-(3-amino-1*H*-1,2,4-triazol-1-yl)-2-aminopropionic acid also move in phloem,[29,65] although they apparently lack toxic effects.

The pyridazinone herbicides, of which norflurazon and pyrazon (chloridazon) are marketed, are applied pre- or early postemergence. The compounds are of modest to intermediate lipophilicity, and consequently they are readily taken up by the roots and translocated to the shoots, though essentially no symplastic transport has been observed.

Ethofumesate and fluridone are slightly more lipophilic; little information is available about ethofumesate whereas fluridone[66] is transported in the apoplast. Dinoseb and related phenols show limited movement in the apoplast, probably because they act primarily as contact herbicides causing rapid damage to plant membranes and hence to the transport system in plants.

Buthidazole and methazole possess heterocyclic rings containing a cyclized urea function, and, like the ureas, they act by inhibiting the Hill reaction in chloroplasts. It is believed that methazole is metabolized to a less lipophilic phenylurea that is the active compound,[67] for methazole itself is a poor inhibitor of the Hill reaction and also is too lipophilic to be well transported to leaves from roots.[68]

The other nonionized herbicides listed in Table 9.1 are rather lipophilic, having estimated values of log K_{ow} greater than 3.8. When applied to plant roots via nutrient solution, oxadiazon is translocated[69] to some extent via xylem to plant shoots, but transport of such a lipophilic compound from soil is unlikely to be very efficient. Butamifos[70] and bensulide[5] are reported not to be translocated. Pyridate is not itself translocated but is hydrolyzed in the plant to the more polar and herbicidally active 4-pyridazinol analogue, and this metabolite is moved largely in the xylem.[71] Information on the behavior of the other compounds in this group is not presently available. It seems probable that their uptake from soil application is mostly directly at the site of action in the roots or germinating shoots, as these compounds are too lipophilic to be transported well in plants.

9.4.2. Cations (Quaternized Amines)

Bipyridiniums and Difenzoquat

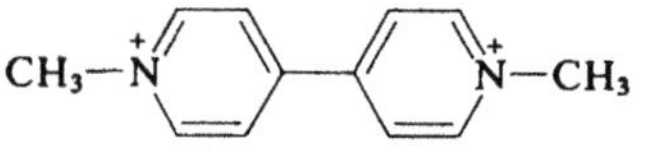

Paraquat

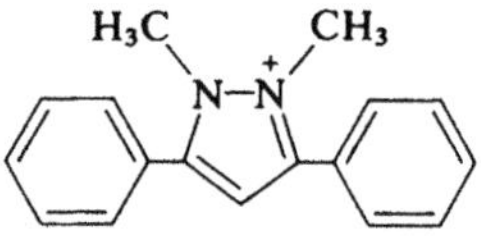

Difenzoquat

Paraquat, diquat, and morfamquat are bipyridinium dications. They are applied to foliage as the halide or sulfate salts, and, in the light, rapidly cause severe wilting and necrosis of leaves due to membrane damage. In the dark, limited movement occurs from treated to untreated leaves, and this happens even if the petiole of the treated leaf has been steam-girdled to prevent movement via phloem.[72] It has been suggested that such movement occurs by reverse flow in the xylem. Difenzoquat is a monocation, whose mode of action is uncertain though different to that of the bipyridiniums.

Simple quaternized ammonium monocations are able to move moderately well in phloem,[23,46,73] and it is probable that these herbicides also have the potential to move symplastically. However, symplastic and indeed apoplastic movement of the bipyridiniums is severely restricted by their rapid phytotoxicity and is unlikely to be of any consequence for their herbicidal action. Difenzoquat has been observed to move in phloem to a small extent, but most of the movement was acropetal.[74,75]

9.4.3. Acids

9.4.3.1. Acidic Amides (Other Than Sulfonylureas)

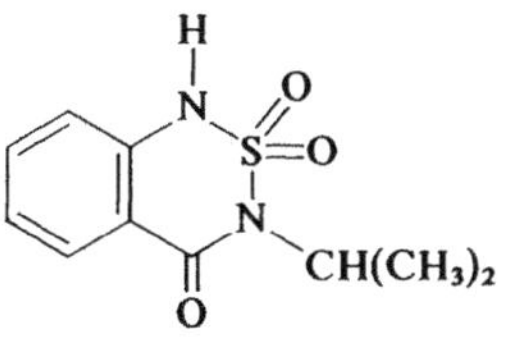

Bentazon

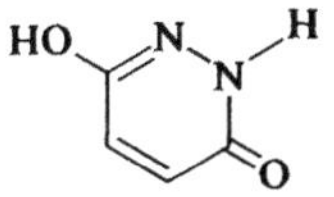

Maleic hydrazide

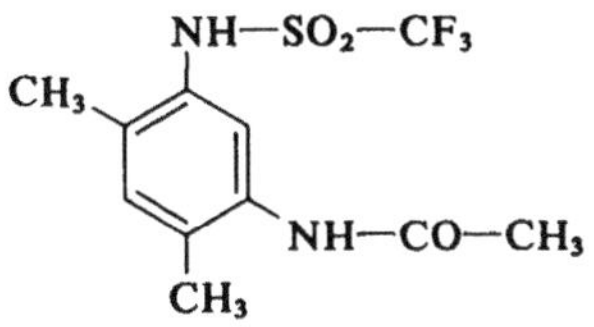

Mefluidide

Amides are not usually ionized at physiological pH, but the presence of strong electron-withdrawing groups (e.g., $-SO_2R$, $-CF_3$) attached to the nitrogen brings the pK_a down into this region. Asulam (pK_a 4.8, log $K_{ow} < 0$) is a very polar compound which is well transported in phloem, and this symplastic movement is important in preventing regrowth from roots and rhizomes of treated weeds such as bracken (*Pteridium aquilinum*).[76] Maleic hydrazide (pK_a 5.65, log K_{ow} -0.87) is a plant growth regulator with similar physicochemical properties to those of asulam and is again phloem mobile.

Bentazon, mefluidide, and perfluidone are of intermediate lipophilicity with p$K_a \approx 3.5$, 4.6, and 2.5, respectively. Mefluidide[77] is phloem translocated, but the contact herbicide bentazon is not moved symplastically[78,79]; perfluidone moves in the xylem following uptake by roots,[78,80] but experiments testing its symplastic transport have not been reported.

9.4.3.2. Aliphatic Acids

Dalapon (CH_3CCl_2COOH) and TCA (CCl_3COOH) are the main aliphatic acids used. They are quite strong acids of intermediate lipophilicity, having pK_a values of 1.8 and 0.65 respectively. Both are rapidly taken up by roots and shoots, inhibiting growth at the root tips and apical meristem. While dalapon is well translocated in the apoplast and symplast, TCA is transported more in the apoplast than in the symplast. Thus, dalapon is usually applied to foliage whereas TCA is applied to soil.

9.4.3.3. Arsenic-Containing Acids

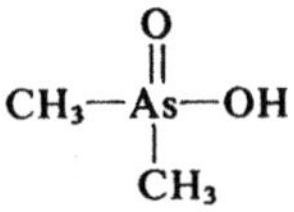

Dimethylarsinic acid

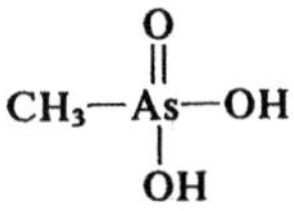

Methylarsonic acid

Dimethylarsinic acid (cacodylic acid) is a weak acid of pK_a 6.3, while methylarsonic acid (MAA) is a weak dibasic acid (pK_a 4.1 and 9.1). Being very polar, both compounds are transported in the apoplast and in the symplast, though symplastic transport in the case of dimethylarsinic acid is weak perhaps due to rapid phytotoxic action in the treated leaves.

9.4.3.4. Aryloxyphenoxypropanoic Acids

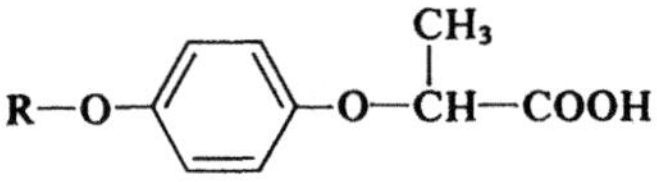

(8)

Aryloxyphenoxypropanoic acids (**8**) are rather lipophilic (log K_{ow} 3.4–4.4), with R being a substituted phenyl or other aromatic group, and have p$K_a \approx 3.5$. The compounds are applied as esters which rapidly hydrolyze in soil and plants to yield the active free acids. Application is postemergence to control grass weeds in broadleaf crops, and the compounds are also sufficiently active in soil to give prolonged control of grass weeds. The main mode of action involves inhibition of fatty acid biosynthesis (see Chapter 3). The free acids are translocated in the phloem from foliage to roots. While this phloem translocation is poor, as might be expected for such lipophilic compounds as fluazifop,[81] the intrinsic activity of these compounds is so high that rhizomatous grass weeds are very efficiently controlled.

9.4.3.5. Benzoic Acids

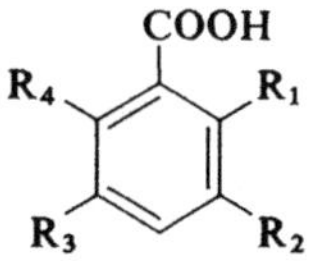

(9)

Chloramben, dicamba, and 2,3,6-TBA are trisubstituted benzoic acids of intermediate lipophilicity (log K_{ow} 2.2–4.0) and pK_a 2.3 to 3.5. They are active from applications both to soil and to foliage. Benzoic acids are well translocated in phloem, though long-distance transport of the more lipophilic compounds may be slow. One of the most important phytotoxic symptoms appears to be growth inhibition of developing buds and apices caused by the accumulation of herbicide via phloem transport.

Naptalam is an unrelated benzoic acid, being a half-amide formed by the reaction of phthalic anhydride and 1-naphthylamine; it has a pK_a of ~4 and log K_{ow} of ~3. It is applied pre- or postemergence, and its mode of action appears to be similar to that of some of the other amides. In addition, it inhibits the cell-to-cell or polar transport of auxin. It moves in the apoplast and is expected to be symplastic to some extent though information on this is sparse.[5]

Acifluorfen is also a benzoic acid but shares a common mode of action with the diphenyl ethers and has been discussed together with them in Section 9.4.1.4.

9.4.3.6. Hydroxycyclohexenones

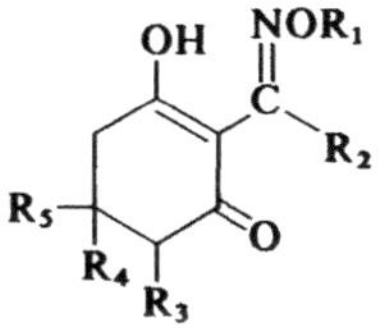

(10)

The substituents on hydroxycyclohexenones (**10**) are thioether, alkyl, or halogenated alkyl groups, with alloxydim alone possessing an R_3 group of carboxymethyl. These substituents contain 10 carbon atoms for all the compounds, and so these weak enol acids (pK_a ≈ 4.0) are rather lipophilic (log K_{ow} ≈4.2; somewhat less for alloxydim). The compounds are applied postemergence to control grasses in broadleaf crops though they also have some activity when applied to soil. Following uptake by foliage, the compounds are moved throughout the plant; while transport to the roots via the phloem is not very effective, it is sufficient to control rhizomatous grasses.[82-84]

9.4.3.7. Imidazolinones

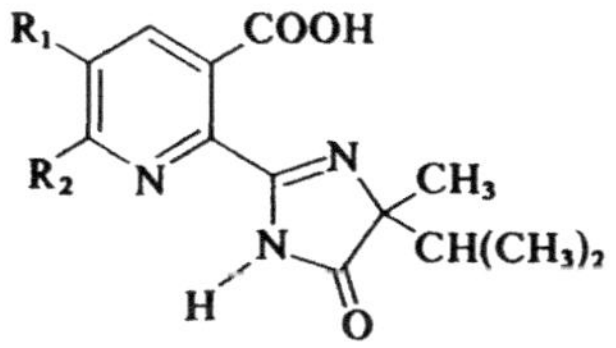

(11)

The imidazolinones (**11**) are derivatives of nicotinic acid and have an acid and a base (protonation on the imidazolinone ring) function, both of pK_a 1.8 to 3.9, and so an appreciable proportion of the molecules can be in the zwitterionic form at physiological pH. Log K_{ow} values[85] estimated for the neutral form of the molecules are 0.16 for imazapyr ($R_1 = R_2 = H$), 1.05 for imazethapyr (R_1 = Et, R_2 = H), and 1.44 for imazaquin (R_1 and R_2 fused to give a quinoline ring system).

The compounds are applied pre- or postemergence, and all are mobile in phloem,[85-89] especially imazapyr which is well translocated to roots following foliar application. In roots and apical meristems, cell division is prevented due to the inhibition of acetolactate synthase limiting production of valine, leucine, and isoleucine.

9.4.3.8. Phenoxyacetic and Other Aryloxyalkanoic Acids

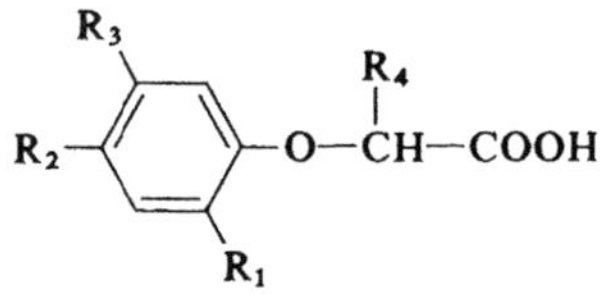

(**12**)

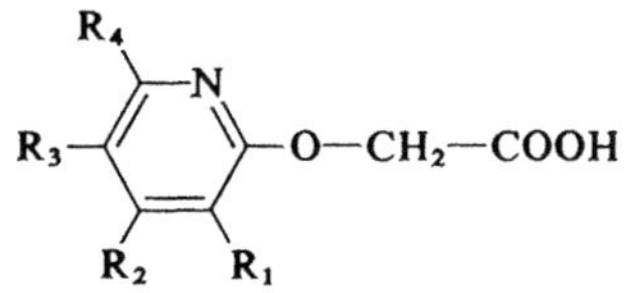

(**13**)

The most frequently used aryloxyalkanoic acids (2,4-D, MCPA, mecoprop, and 2,4,5-T) are phenoxyacetic acid derivatives (**12**) with methyl or chlorine substituents on the benzene ring, giving log K_{ow} values in the range of 2.9 to 3.6 and pK_a values of around 3.0. They are auxin mimics and are active both pre- and postemergence against broadleaf weeds. The compounds are moved very efficiently in phloem over short distances but are strongly held by ion trappping in the tissue adjacent to the vascular system, and this limits long-distance symplastic transport (e.g., from leaves to roots). Both 2,4-dichlorophenoxybutanoic acid (2,4-DB) and 2-(2,4-dichlorophenoxy)ethyl hydrogen sulfate (2,4-DES) are precursors of 2,4-D, and transport of these precursor compounds appears not to be important.

Compounds such as fluroxypyr and triclopyr (p$K_a \approx 2.7$) have the phenyl group replaced by a pyridine ring (**13**) and, as a consequence, are slightly less lipophilic than the corresponding phenoxyacetic acids. This may assist movement in phloem over longer distances, but information on their transport behavior is limited.[90,91]

9.4.3.9. Phosphorus-Containing Acids

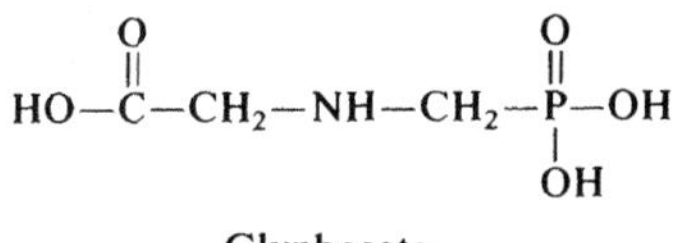

Glyphosate

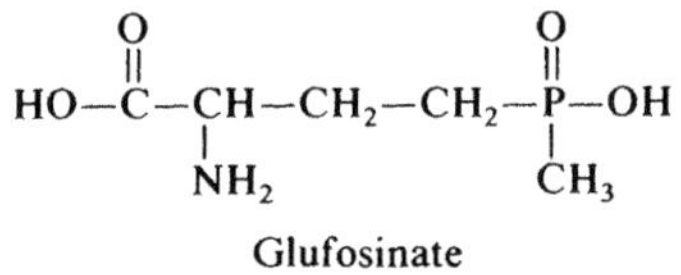

Glufosinate

Glyphosate[92] is the best known of the herbicidal phosphinic and phosphonic acids. It is a very polar compound and has three acid functions ($pK_a < 2$, 2.3 and 5.6) together with a basic amino group (pK_a 10.2). When applied to the foliage of plants, it is rapidly translocated throughout the plant and causes death by inhibiting the production of aromatic amino acids. The efficient phloem transport to rhizomes and roots makes it very effective even against well-established weeds and woody plants, and regrowth is slight. The complexity of the possible ionizations has prevented a quantitative physicochemical explanation of its phloem movement.[51]

The other three compounds in this class are also very polar, but unlike glyphosate, they have only a single acid function (of low pK_a) on the phosphorus. They are all applied to foliage. In addition to the phosphinic acid group, glufosinate has an amino acid function, and present evidence indicates that phloem transport occurs but to a limited extent.[93] Glufosinate acts by inhibiting glutamine synthase. Bilanophos is a dipeptide derivative (alanylalanine) of glufosinate produced by *Streptomyces hygroscopicus.* Application of radiolabeled bilanophos to leaves resulted in symplastic export of radiolabel, though its chemical nature was not identified[94,95]; the toxic principle is glufosinate produced by hydrolysis of bilanophos in the plant. The mode of action of fosamine is unknown, but it is well transported in phloem, and this transported material inhibits regrowth from roots and rhizomes.[96,97]

9.4.3.10. Pyridinecarboxylic Acids (Other Than Imidazolinones)

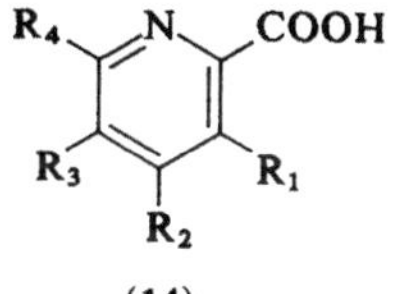

(**14**)

Picloram (**14**; $R_2 = NH_2$; $R_1, R_3, R_4 = Cl$) and clopyralid (**14**; $R_1, R_4 = Cl$; $R_2, R_3 = H$) are applied as postemergence sprays to control various broadleaf weeds. The compounds are of intermediate lipophilicity ($\log K_{ow}$ 1.0 and 1.5, respectively) and are quite strong acids ($pK_a \approx 2$ and 2.33), respectively). They are phloem mobile,[90,91] clopyralid being especially so, and appear to act as auxin mimics although picloram may also have other effects.

9.4.3.11. Sulfonylureas

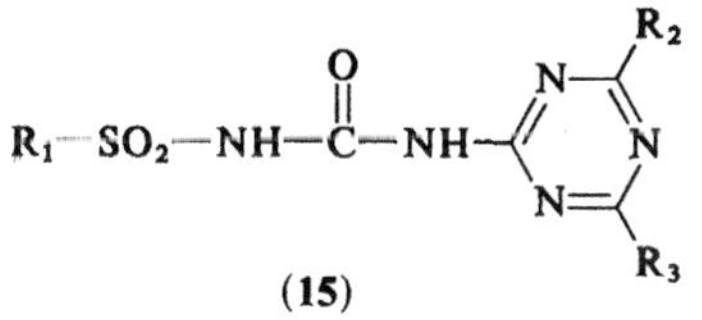

(**15**)

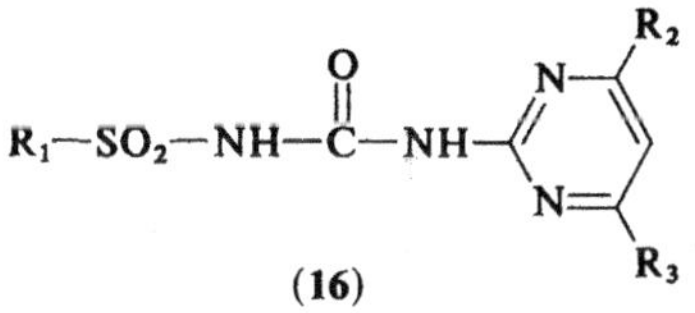

(**16**)

Sulfonylureas have structure **15** or **16**; R_1 is a benzene or thiophene ring (benzyl group in bensulfuron) containing a 2-substituent of chlorine or carboxymethyl, and R_2 and R_3 are chlorine, methoxy or methyl groups. It should be noted that the common names of these herbicides refer, where applicable, to the compound in which the 2-substituent is a free carboxyl group (i.e., bensulfuron, chlorimuron, metsulfuron, thiameturon) although the compounds are applied as the methyl esters (metsulfuron-methyl, etc.) which are themselves the active compounds.

The sulfonylureas are compounds of intermediate lipophilicity (log K_{ow} estimated in the range 1.5 to 2.5), and all possess the acidic sulfonylurea function (pK_a 3.3 to 5.2).[98] They are applied pre- or postemergence and kill plants by slowing cell division due to inhibition of the enzyme acetolactate synthase. The compounds move in both the xylem and phloem,[99-102] and their phytotoxic action is believed to be due both to direct uptake at the sites of action (e.g., root tips) and to phloem transport to the growing points.

9.4.3.12. Unclassified Carboxylic Acids

Endothall is an alicyclic dicarboxylic acid that appears to act primarily as a contact herbicide, though some movement in the xylem and phloem has been reported.

Benazolin and fenac (chlorfenac) have log K_{ow} 2 to 3 and behave similarly to the phenoxyalkanoic and benzoic acids; both are moderately mobile in the symplast.

Benzoylprop and flamprop differ by only a single substituent and have p$K_a \approx 3.5$ and log $K_{ow} \approx 3$. They are well transported in the xylem, but phloem transport is very weak. The poor symplastic transport of flamprop in castor bean observed by Rigitano *et al.*[43] could not be readily explained by these authors. However, the poor accumulation of flamprop in barley roots[13] could indicate that the anion moves through membranes with unexpected ease, and this would limit retention in the phloem.

9.5. CONCLUSIONS

The pathways of uptake and translocation of herbicides by plants appear to be largely controlled by the physicochemical properties of the herbicides. Bromilow and co-workers[12,51] have produced a simple diagram (Figure 9.8) indicating the properties required by nonionized compounds and weak acids for the various types of transport. It should be noted that the divisions between the categories are somewhat arbitrary and that even

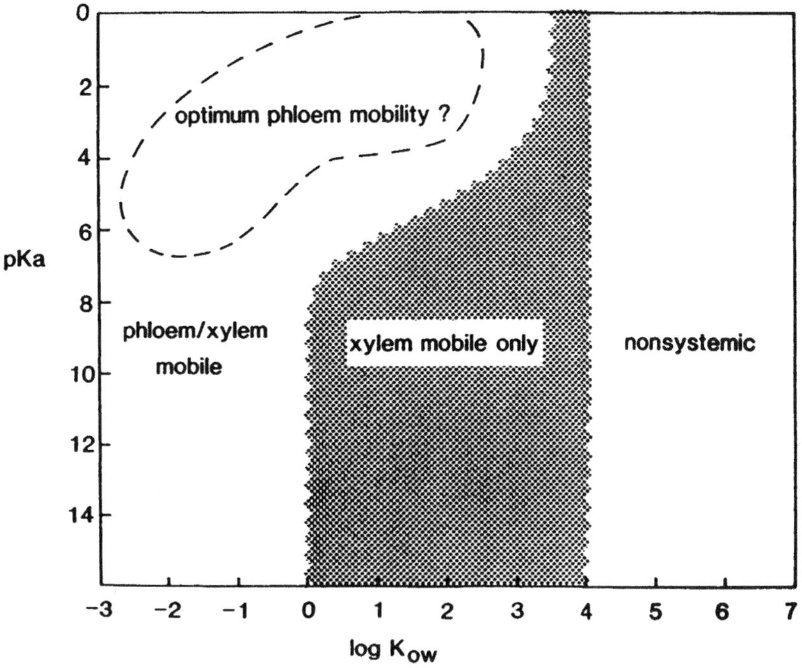

Figure 9.8. Physicochemical properties required of nonionized compounds and weak acids for various types of systemic behavior. (Reproduced from Ref. 12.)

weakly translocated molecules may be useful if they have a high intrinsic activity (e.g., aryloxyphenoxypropanoic acids).

All compounds with log K_{ow} less than 4 exhibit at least some movement in the xylem, and nonionized compounds (and probably cations also) of log K_{ow} less than 0 together with weak acids additionally move in phloem. While the accumulation of weak acids in plant compartments such as phloem is an energy-dependent process, there is no good evidence that specific carriers are used in the long-distance transport of current herbicides.

Kleier,[103] using similar concepts, has produced a computer model that predicts phloem transport of xenobiotics based upon their physicochemical properties. Topp *et al.*[104], Ryan *et al.*[105] and Boersma *et al.*[106] have attempted to assess the uptake by plants from soil of nonionized organic chemicals based upon their physicochemical properties and persistence. From our present understanding of the transport in plants of herbicides and other pesticides, it should be possible to predict approximately the behavior of new or putative herbicidal molecules.[12,51] With due account taken of the complicating factors such as availability in soil, penetration across leaf cuticles, and metabolism, this approach should be applicable to compounds used on a wide range of crop and weed species in a variety of situations.

REFERENCES

1. K. Esau, *Plant Anatomy*, 2nd ed., John Wiley and Sons, New York (1965).
2. M. H. Zimmermann and J. A. Milburn, *Encyclopedia of Plant Physiology, New Series, Volume 1. Transport in Plants I. Phloem Transport*, Springer-Verlag, Berlin (1975).
3. E. Munch, *Die Stoffbewegungen in der Pflanze*, Gustav Fischer, Jena (1930).
4. C. R. Worthing, *The Pesticide Manual*, 8th ed., The British Crop Protection Council, London (1987).
5. F. M. Ashton and A. S. Crafts, *Mode of Action of Herbicides*, John Wiley and Sons, New York (1981).
6. T. Fujita, J. Iwasa, and C. H. Hansch, *J. Am. Chem. Soc. 86*, 5175 (1964).
7. W. J. Dunn III, J. H. Block, and R. S. Pearlman (eds.), *Partition Coefficient. Determination and Estimation*, Pergamon Press, New York (1986).
8. C. H. Hansch and A. Leo, *Substituent Constants for Correlation Analysis in Chemistry and Biology*, John Wiley and Sons, New York (1979).
9. G. G. Briggs, *J. Agric. Food Chem. 29*, 1050 (1981).
10. A. Albert and E. P. Serjeant, *Ionization Constants of Acids and Bases. A Laboratory Manual*, Methuen & Co., London (1962).
11. D. D. Perrin, B. Dempsey, and E. P. Serjeant, *pK_a Prediction for Organic Acids and Bases*, Chapman and Hall, London (1981).
12. R. H. Bromilow and K. Chamberlain, in: *Mechanisms and Regulation of Transport Processes* (R. K. Atkin and D. R. Clifford, eds.), British Plant Growth Regulator Group Monograph 18, p. 113 (1989).
13. G. G. Briggs, R. L. O. Rigitano, and R. H. Bromilow, *Pestic. Sci. 19*, 101 (1987).
14. J. A. Raven, *New Phytol. 74*, 163 (1975).
15. C. R. Merritt, *Ann. Appl. Biol. 101*, 517 (1982).
16. C. R. Merritt, *Ann. Appl. Biol. 101*, 527 (1982).
17. J. A. Gougler and D. R. Geiger, *Weed Sci. 32*, 546 (1984).
18. D. R. Geiger, S. W. Kapitan, and M. A. Tucci, *Plant Physiol. 82*, 468 (1986).
19. J. C. Servaites, M. A. Tucci, and D. R. Geiger, *Plant Physiol. 85*, 370 (1987).
20. W. H. Vanden Born, H. D. Bestman, and M. D. Devine, *Proc. EWRS Symp. Factors Affecting Herbicidal Activity and Selectivity*, p. 69 (1988).
21. R. H. Bromilow, K. Chamberlain, and G. G. Briggs, *Aspects Appl. Biol. 11*, 29 (1986).
22. F. C. Hsu, R. L. Marxmiller, and A. Y. S. Yang, *Plant Physiol. 93*, 1573 (1990).
23. F. C. Hsu, D. A. Kleier, and W. R. Melander, *Plant Physiol. 86*, 811 (1988).
24. J. Groussol, S. Delrot, P. Caruhel, and J. L. Bonnemain, *Physiol. Vég. 24*, 123 (1986).
25. F. T. Lichtner, in: *Phloem Transport* (J. Cronshaw, W. J. Lucas, and R. T. Giaquinta, eds.), Vol. 1, p. 601, Alan R. Liss, New York (1986).
26. P. M. L. Tammes and J. van Die, *Acta Bot. Neerl. 13*, 76 (1964).
27. J. A. Milburn, *Pestic. Sci. 3*, 653 (1972).
28. K. Chamberlain, M. M. Burrell, D. N. Butcher, and J. C. White, *Pestic. Sci. 15*, 1 (1984).
29. R. H. Bromilow, R. L. O. Rigitano, G. G. Briggs, and K. Chamberlain *Pestic. Sci. 19*, 85 (1987).
30. W. F. Spencer and M. M. Cliath, *J. Agric. Food Chem. 22*, 987 (1974).
31. A. W. Taylor and D. E. Glotfelty, in: *Environmental Chemistry of Herbicides* (R. Grover, ed.), Vol. 1, p. 89, CRC Press, Boca Raton, Florida (1988).
32. M. G. T. Shone and A. V. Wood, *J. Exp. Bot. 25*, 390 (1974).
33. M. G. T. Shone, B. O. Bartlett, and A. V. Wood, *J. Exp. Bot. 25*, 401 (1974).
34. G. G. Briggs, R. H. Bromilow, and A. A. Evans, *Pestic. Sci. 13*, 495 (1982).
35. P. H. Nicholls and A. A. Evans, *Proceedings of the British Crop Protection Conference—Weeds*, Vol. 1, p. 333 (1985).

36. G. Ridout and R. H. Guy, in: *Pesticide Formulations, Innovations and Developments* (B. Cross and H. B. Scher, eds.), ACS Symposium Series, No. 371, p. 112, American Chemical Society, Washington, D.C. (1988).
37. K. Chamberlain, G. G. Briggs, R. H. Bromilow, A. A. Evans, and Chen Q. Fang, *Aspects Appl. Biol. 14,* 293 (1987).
38. M. T. Tyree, C. A. Peterson, and L. V. Edgington, *Plant Physiol. 63,* 367 (1979).
39. C. A. Peterson, P. P. Q. de Wildt, and L. V. Edgington, *Pestic. Biochem. Physiol. 8,* 1 (1978).
40. G. G. Briggs, R. H. Bromilow, R. Edmondson, and M. Johnston, in: *Herbicides and Fungicides—Factors Affecting Their Activity* (N. R. McFarlane, ed.), Chemical Society Special Publication, No. 29, p. 129 (1976).
41. C. E. Crisp, in: *Pesticide Chemistry: Proc. 2nd IUPAC Congress* (A. S. Tahori, ed.), Vol. 1, p. 211, Gordon and Breach, New York (1972).
42. C. E. Crisp and M. Look, in: *Advances in Pesticide Science: Proc. 4th IUPAC Congress* (H. Geissbühler, ed.), Vol. 3, p. 430, Pergamon Press, Oxford (1979).
43. R. L. O. Rigitano, R. H. Bromilow, G. G. Briggs, and K. Chamberlain, *Pestic. Sci. 19,* 113 (1987).
44. A. S. Crafts and C. E. Crisp, *Phloem Transport in Plants,* W. H. Freeman & Co., San Francisco (1971).
45. R. A. Christ, in: *Advances in Pesticide Science: Proc. 4th IUPAC Congress* (H. Geissbühler, ed.), Vol. 3, p. 420, Pergamon Press, Oxford (1979).
46. C. E. Price, in: *Advances in Pesticide Science: Proc. 4th IUPAC Congress* (H. Geissbühler, ed.), Vol. 3, p. 401, Pergamon Press, Oxford (1979).
47. L. V. Edgington, *Annu. Rev. Phytopathol. 19,* 107 (1981).
48. F. Jacob and S. Neumann, in: *Pesticide Chemistry—Human Welfare and the Environment: Proc. 5th IUPAC Congress* (J. Miyamoto and P. C. Kearney, eds.), Vol. 1, p. 357, Pergamon Press, Oxford (1983).
49. M. C. Shephard, in: *Fungicides for Crop Protection* (I. M. Smith, ed.), British Crop Protection Council Monograph 31, p. 99 (1985).
50. R. T. Giaquinta, in: *Bioregulators for Pest Control* (P. A. Hedin, ed.), p. 7, American Chemical Society, Washington, D.C. (1985).
51. R. H. Bromilow, K. Chamberlain, and A. A. Evans, *Weed Sci. 38,* 305 (1990).
52. M. Uchida, *Pestic. Biochem. Physiol. 14,* 249 (1980).
53. G. G. Briggs, R. H. Bromilow, A. A. Evans, and M. Williams, *Pestic. Sci. 14,* 492 (1983).
54. E. Barak, A. Dinoor, and B. Jacoby, *Proc. Br. Crop. Prot. Conf. Pests Dis.,* Vol. 1, p. 129 (1981).
55. G. Voss and H. Geissbühler, *Proc. 8th Br. Weed Control Conf.,* p. 266 (1966).
56. P. C. Kearney and D. D. Kaufman (eds.), *Herbicides. Chemistry, Degradation and Mode of Action,* 2nd ed., Marcel Dekker, New York (1975).
57. G. Matolcsy, M. Nádasy, and V. Andriska, *Pesticide Chemistry. Studies in Environmental Science 32,* Elsevier, Amsterdam (1988).
58. D. E. Hathway, *Molecular Mechanisms of Herbicide Selectivity,* Oxford University Press, Oxford (1989).
59. D. E. Vanstone and E. H. Stobbe, *Weed Sci. 26,* 389 (1978).
60. J. P. Walter, E. F. Eastin, and M. G. Merkle, *Weed Res. 10,* 165 (1970).
61. J. M. Higgins, T. Whitwell, F. T. Corbin, G. E. Carter, Jr., and H. S. Hill, Jr, *Weed Sci. 36,* 141 (1988).
62. B. J. Hook and S. Glenn, *Weed Sci. 32,* 691 (1984).
63. E. Grimm, S. Neumann, and B. Krug, *Biochem. Physiol. Pflanzen 182,* 323 (1987).
64. G. E. Sanders and K. E. Pallett, *Pestic. Biochem. Physiol. 28,* 163 (1987).
65. P. Massini, *Acta Bot. Neerl. 12,* 64 (1963).
66. D. F. Berard, D. P. Rainey, and C. C. Lin, *Weed Sci. 26,* 252 (1978).
67. D. W. Jones and C. L. Foy, *Pestic. Biochem. Physiol. 2,* 8 (1972).

68. D. W. Jones and C. L. Foy, *Weed Sci. 20,* 116 (1972).
69. N. R. Achhireddy, R. C. Kirkwood, and W. W. Fletcher, *Weed Sci. 32,* 727 (1984).
70. K. Wakabayashi, *Jpn. Pestic. Inf. 41,* 8 (1982).
71. P. Gaillardan, J. C. Guichaoua, J. Gasquez, and R. Scalla, *Weed Res. 29,* 45 (1989).
72. B. C. Baldwin, *Nature (London) 198,* 872 (1963).
73. C. E. Price, S. G. Boatman, and B. J. Boddy, *J. Exp. Bot. 26,* 521 (1975).
74. M. P. Sharma, W. H. Vanden Born, H. A. Fiesen, and D. K. McBeath, *Weed Sci. 24,* 379 (1976).
75. K. E. Pallett and J. C. Caseley, *Pestic. Biochem. Physiol. 14,* 144 (1980).
76. R. C. Kirkwood, *Proc. EWRS Symp. Factors Affecting Herbicidal Activity and Selectivity,* p. 245 (1988).
77. J. R. Bloomberg and L. M. Wax, *Weed Sci. 26,* 434 (1978).
78. E. W. Stoller, L. M. Wax, and R. L. Matthieson, *Weed Sci. 23,* 215 (1975).
79. A. Mine and S. Matsunaka, *Pestic. Biochem. Physiol. 5,* 444 (1975).
80. D. G. Davis and K. E. Dusbabek, *Weed Sci. 23,* 81 (1975).
81. P. Hendley, J. W. Dicks, T. J. Monaco, S. M. Slyfield, O. J. Tummon, and J. C. Barrett, *Weed Sci. 33,* 11 (1985).
82. J. R. Campbell and D. Penner, *Weed Res. 27,* 179 (1987).
83. I. Iwataki and Y. Hirono, in: *Advances in Pesticide Science: 4th IUPAC Congress* (H. Geissbühler, ed.), Vol. 2, p. 235, Pergamon Press, Oxford (1979).
84. K. Ishihara, H. Shiotani, Y. Soeda, and S. Ono, *J. Pestic. Sci. 13,* 231 (1988).
85. R. H. Bromilow and K. Chamberlain, unpublished results (1989).
86. D. L. Shaner and P. A. Robson, *Weed Sci., 33,* 469 (1985).
87. J. W. Wilcut, G. R. Wehtje, M. G. Patterson, and T. A. Cole, *Weed Sci. 36,* 5 (1988).
88. U. B. Nandihalli and L. E. Bendixen, *Weed Sci. 36,* 313 (1988).
89. D. L. Shaner, *Trop. Pest Manage. 34,* 388 (1988).
90. G. E. Sanders, L. M. Thompson, and K. E. Pallett, *Proceedings of the British Crop Protection Conference—Weeds,* Vol. 1, p. 419 (1985).
91. R. M. Gorrell, S. W. Bingham, and C. L. Foy, *Weed Sci. 36,* 447 (1988).
92. E. Grossbard and D. Atkinson (eds.), *The Herbicide Glyphosate,* Butterworths, London (1985).
93. P. Haas and F. Müller, *Proceedings of the British Crop Protection Conference—Weeds,* Vol. 3, p. 1075 (1987).
94. K. Tachibana, in: *Pesticide Science and Biotechnology: Proc. 6th IUPAC Congress* (R. Greenhalgh and T. R. Roberts, eds.), Vol. 1, p. 145, Blackwells, Oxford (1987).
95. K. Tachibana and K. Kaneko, *J. Pestic. Sci. 11,* 297 (1986).
96. R. K. Mann, W. W. Witt, and C. E. Rieck, *Weed Sci. 34,* 830 (1986).
97. R. G. Richardson, *Weed Res. 20,* 159 (1980).
98. E. M. Beyer, H. M. Brown, and M. J. Duffy, *Proceedings of the British Crop Protection Conference—Weeds,* Vol. 2, p. 531 (1987).
99. K. N. Reddy and L. E. Bendixen, *Weed Sci. 36,* 707 (1988).
100. J. H. Baird, J. W. Wilcut, G. R. Wehtje, R. Dickens, and S. Sharpe, *Weed Sci. 37,* 42 (1989).
101. J. W. Wilcut, G. R. Wehtje, M. G. Patterson, T. A. Cole, and T. V. Hicks, *Weed Sci. 37,* 175 (1989).
102. A. R. Leys and F. W. Slife, *Weed Sci. 36,* 1 (1988).
103. D. A. Kleier, *Plant Physiol. 86,* 803 (1988).
104. E. Topp, I. Scheunert, A. Attar, and F. Korte, *Ecotoxicol. Environ. Saf. 11,* 219 (1986).
105. J. A. Ryan, R. M. Bell, J. M. Davidson, and G. A. O'Connor, *Chemosphere 17,* 2299 (1988).
106. L. Boersma, F. T. Lindstrom, C. McFarlane, and E. L. McCoy, *Soil Sci. 146,* 403 (1988).

10

Herbicide Metabolism as a Basis for Selectivity

W. JOHN OWEN

10.1. INTRODUCTION

There are a number of factors which can contribute to herbicide selectivity, including soil placement, rates of absorption and subsequent translocation, localization (both within the plant and at the subcellular level), and transformation to products of modified phytotoxicity. In addition, the recent work on the elucidation of the modes of action of the aryloxyphenoxypropionate and cyclohexanedione groups of herbicides has highlighted the importance of species differences in sensitivity of the target site, in this case the enzyme acetyl-CoA carboxylase. The monocot and dicot enzymes studied to date show a remarkable difference in sensitivity to these herbicides which correlates very well with the high level of resistance among dicots. Biotypes of *Chenopodium album* and *Amaranthus hybridus* L. resistant to atrazine as a result of a mutation in the 32-kDa protein component of photosystem II would constitute a further example.

Herbicide selectivity often results from a complex interaction of a number of the factors indicated above though there are many examples where a single dominant factor is responsible. There are a number of well-documented cases in which this is true of herbicide uptake and movement and of differential metabolism to less phytotoxic products. The purpose of the present chapter is to elaborate on the latter of these phenomena in

W. JOHN OWEN • DowElanco Ltd., Letcombe Laboratory, Letcombe Regis, Wantage, Oxon OX12 9JT, United Kingdom.

such a way as to indicate the wide range of herbicide detoxification mechanisms encountered in plants in the context, whenever possible, of our knowledge of the metabolic fate of herbicides of current commercial interest.

In addition to meeting requirements for registration purposes, information on the transformation of herbicides in plants can also contribute to our overall understanding of their modes of action. The data obtained may also be of use in the design of herbicide progenitors which require metabolic transformation for activation in target susceptible species. The genetic engineering of herbicide-resistant crop plants carrying herbicide-detoxifying genes also requires a detailed knowledge of metabolism and of the enzymes which mediate the key detoxification steps.

10.2. EVIDENCE FOR THE ROLE OF VARIOUS DETOXIFICATION MECHANISMS IN HERBICIDE SELECTIVITY

10.2.1. Hydrolysis

10.2.1.1. Carboxylic Esters

Hydrolysis of carboxylic ester groups of various xenobiotics is of widespread occurrence in plants. In general, the substrate specificity of plant carboxylesterases appears quite broad, as indicated by the ability of wheat to cleave the methyl ester of diclofop (**1**, Figure 10.1), the ethoxybutyl ester of the pyridinyloxyacetic acid herbicide triclopyr,[1] and the 2-ethylhexyl ester of a phthalate plasticizer.[2] Aryloxyphenoxypropionates such as diclofop, herbicidal auxins such as 2,4-D and triclopyr, and other herbicides containing carboxylic acid functions are generally applied to plants as ester formulations. These are considered to be herbicidally inactive but are rapidly activated within the plant by conversion to the corresponding free acids (Figure 10.1). The rapidity with which deesterification occurs in many plant species renders it unlikely that ester hydrolysis contributes appreciably to the selectivity of existing acidic herbicides. In the case of the aryloxyphenoxypropionates, previous assumptions concerning the requirement for ester hydrolysis for activity have recently been validated in studies with haloxyfop. This compound was more active (>100-fold) than the corresponding methyl ester in inhibiting acetyl-CoA carboxylase, the recently discovered target site for this group of herbicides.[3] More detailed information will be required in order to assess whether significant species differences in specificity exist with respect to other ester functions; if so, these may be exploitable in the design of selective herbicides.

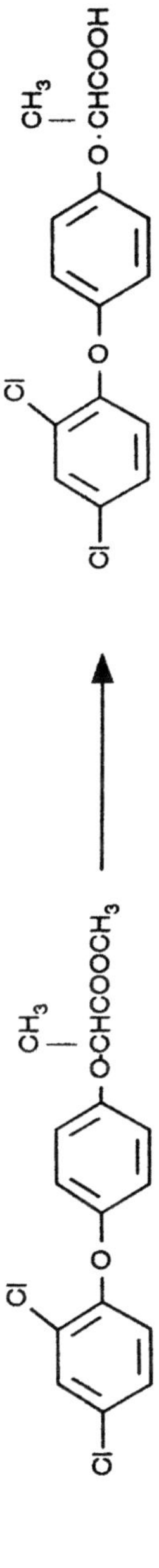

(1)

Figure 10.1. Hydrolytic cleavage of diclofop-methyl.

10.2.1.2. Amides

The amide linkage is similar to the ester linkage and is hydrolyzed in plants by amidases. There has been speculation, however, that some carboxylamidases may be carboxylesterases with extended specificity; partially purified carboxylesterases from mammalian sources are known to utilize amide substrates.[4]

It has been known for some time that amino acid amide conjugates of 2,4-D and the related 2,4,5-T are prominent metabolites of these phenoxyalkanoic acid herbicides in some plant species, notably legumes such as soybean.[5-10] The predominant amino acid conjugates of 2,4-D occurring in such species are the 2,4-D-aspartate and 2,4-D-glutamate conjugates. Recent work in our own laboratory has also led to the identification of aspartate and glutamate amide conjugates of triclopyr as the major products formed from this pyridinyloxyacetic acid derivative in soybean suspension cultures and intact plants of *Stellaria media*.[1,11] In contrast, glycine, alanine, and valine conjugates were the dominant metabolites of the naturally occurring auxin 3-indoleacetic acid (IAA) in *Parthenocissus tricuspidata*.[12] Studies of the fate of amino acid amide conjugates of 2,4-D in soybean callus tissue have indicated that such metabolites are readily hydrolyzed in plant tissue.[6,13] This probably accounts for their observed biological activity.[14,15] Conversion of 2,4-D to amide conjugates does not, in itself, appear to constitute an effective detoxification mechanism unless formation of the amide conjugates occurs in a subcellular compartment from which the specific amidases are excluded. In the case of IAA, however, conjugate formation probably is related to translocation and the regulation of localized concentrations of active auxin.

Plant species may differ significantly in their ability to cleave amide conjugates. While this has not been extensively investigated, examination of the herbicidal properties of 2,4-D–amino acid conjugates using green beans, pea, sunflower, and soybean showed a similar response.[15] Application of the dibenzyl ester of triclopyr-aspartate to young plants of *C. album* resulted in hydrolysis of the amide bond at a faster rate than in the case of *S. media*[16]; this may be of significance with respect to the greater susceptibility of the former species to triclopyr.

10.2.1.3. Arylamides

Available evidence suggests that arylamide bonds generally are quite resistant to hydrolytic attack in plant tissues. A notable exception, however, is the substituted propionanilide propanil (**2**), which is rapidly hydrolyzed in resistant rice plants but not in susceptible weeds such as *Echinochloa crus-galli* (Figure 10.2). Hydrolysis of propanil is mediated by a specific

Figure 10.2. Amide hydrolysis of propanil.

arylacylamidase enzyme, the activity of which is some 60-fold greater in rice than in *E. crus-galli.*[17,18] Similar enzyme activities have been recorded in extracts from 20 other species including tulip,[19] *Taraxacum officinale,*[20] and lettuce,[21] the last mentioned species being the best source of the enzyme.

Rice is also capable of hydrolyzing the arylamide bond of naproanilide, a related naphthoxypropionanilide derivative, suggesting that the specificity of the rice arylacylamidase extends beyond propanil.[22] A product of this cleavage was the phytotoxic naphthoxypropionic acid (NOP), implying that, in this case, amide hydrolysis resulted in bioactivation. Subsequent rapid detoxification of NOP by ring hydroxylation and glycosidation appeared to account for the tolerance of rice to naproanilide. While amide hydrolysis was also rapid in the susceptible weed species *Sagittaria pygmaea,* further metabolism produced mainly methyl and glucose esters rather than hydroxylated products. Since the methyl ester was reported to be phytotoxic and the glucose ester represented a potential source of NOP,[22] this difference in metabolite profile between the two species was considered to offer a plausible explanation for the herbicidal selectivity of naproanilide.

10.2.1.4. Nitrile Groups

Nitrile hydrolysis, in contrast, seems to occur readily in plants and proceeds via the corresponding amide to the carboxylic acid function, which may subsequently be decarboxylated. This metabolic route has been firmly established for bromoxynil in wheat[23] and for cyanazine in wheat, potato, and maize.[24,25] Since both herbicides are subjected to metabolism by a number of competing pathways, however, reliable conclusions regarding the relative importance of cyano group hydrolysis to detoxification could not be reached. In a comparative study of the metabolism of cyanazine in plants, nitrile hydrolysis was not a major route in the tolerant weed *Amaranthus retroflexus* L., in which a glutathione conjugate of cyanazine was the dominant product.[26] This latter species may lack the ability to hydrolyze nitrile functions, or this metabolic route may be unable to compete with a very efficient glutathione conjugation mechanism. Currently, there is little evidence to suggest that nitrile hydrolysis contributes significantly to herbicide selectivity. A notable exception, however, is the successful generation of transgenic tobacco plants transformed with a nitrilase gene from the soil bacterium *Klebsiella ozaenae*; this gene confers resistance to bromoxynil.[27]

10.2.1.5. Chlorine Hydrolysis

In the case of chloro-*s*-triazine herbicides, exemplified by simazine, atrazine, and cyanazine, a competing detoxification mechanism involves a rapid nucleophilic displacement of the chloro group substituted at the

2-position of the triazine ring. Such reactions are nonenzymatic but are catalyzed by the cyclic hydroxamate 2,4-dihydroxy-3-keto-7-methoxy-1,4-benzoxazine (benzoxazinone), which occurs naturally in the roots of tolerant species such as maize.[28,29] The products of hydrolytic chlorine displacement are hydroxytriazines which are produced in insignificant amounts in species which do not contain benzoxazinones such as cotton, pea, sorghum, and Johnson grass (*Sorghum halepense*).[29] As will be discussed below, the metabolite profile of atrazine in maize, and other species of panicoid grasses, results from the concerted action of three modes of metabolic attack. *N*-Dealkylation, C-2 chlorine hydrolysis, and nucleophilic displacement of C-2 chlorine by glutathione may all contribute to the resistance of some Panicoideae to atrazine.[30] Rapid conversion to hydroxytriazine alone, however, confers tolerance to this chloro-*s*-triazine in *Coix lacryma-jobi*, in which glutathione conjugation plays a very minor role in detoxification. In contrast, the ability of susceptible panicoid grasses such as *E. crus-galli* and *Setaria glauca* to carry out any of the above biotransformations was very limited.[30]

10.2.2. Reductive Deamination

Reductive reactions are not generally of common occurrence in plants and usually are insufficiently rapid to play a significant role in pesticide detoxification and selectivity. A reaction which has been recorded, however, is the reduction of aromatic nitro groups to the corresponding anilines, and the observed reduction of botran (2,6-dichloro-4-nitroaniline) by axenic callus cultures of soybean[31] demonstrated the ability of plant tissue *per se* to mediate such biotransformations. While such reactions appear not to occur extensively in higher plants, the reductive deamination (Figure 10.3) of the substituted triazinone metribuzin (**3**) to **4** represents a major detoxification mechanism which may account for the tolerance of soybean cultivars to metribuzin.[32] Thus, deamination in leaves of tolerant "Bragg"

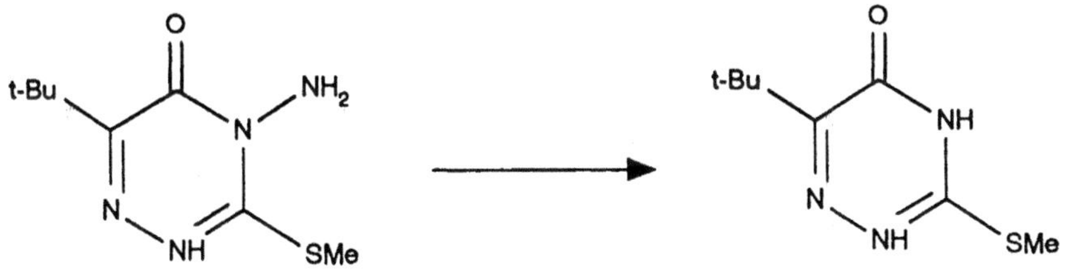

(3) (4)

Figure 10.3. Reductive deamination of metribuzin.

and "Tracy M" was considerably more rapid than that observed in susceptible "Tracy" and "Coker 156." Subsequently, the deaminase enzyme mediating this reaction was characterized by Fedtke and Schmidt, who found that the enzyme functioned anaerobically, requiring a reduced flavin cofactor, and was located in the peroxisomal fraction.[32] The rates of photosynthetic CO_2 fixation in metribuzin-treated leaves of the various cultivars were correlated with deaminase activities in leaf extracts, suggesting that this mode of metabolic attack contributed in a major way to cultivar tolerance. A similar enzyme system has been implicated in the detoxification of the related triazinone metamitron in tolerant sugar beet.[33] In this study also, rates of deamination correlated with increasing tolerance among sugar beet and five weed species. The reaction product was less inhibitory in the Hill reaction of isolated chloroplasts, the target site of the triazinones, than was metamitron itself.[33]

10.2.3. Oxidative Reactions

10.2.3.1. Alkyl Hydroxylation

Generally, oxidation of alkyl groups is considered to be mediated by that class of microsomal mixed-function oxidases (*mfo*s) having an obligate requirement for cytochrome P450 as the hydroxylating species. Consequently, oxidative reactions involved in herbicide detoxification in plants can frequently be inhibited by specific inhibitors of P450-containing hydroxylases, such as the suicide substrate 1-aminobenzotriazole (ABT), or by carbon monoxide. In a number of studies, such inhibition has been interpreted as providing indirect evidence for involvement of a species of P450 in oxidative herbicidal metabolism.[34–36]

For both *N*- and *O*-substituted alkyl groups, initial hydroxylation results in the formation of alcohol intermediates that are generally unstable and, unless rapidly glycosylated, decompose to liberate formaldehyde in the case of a methyl group. These processes are referred to as *N*- and *O*-dealkylations. Further oxidation of *C*-alkyl-derived alcoholic functions to corresponding carboxylic acids is generally slow and is probably of limited relevance to selectivity. In the case of carboxylic acids, rates of glucose ester formation may also be sufficiently rapid to further reduce the opportunity for decarboxylation.

C-Alkyl hydroxylations are of very common occurrence in plants; notable species differences exist, however, which are of significance with respect to selectivity. The tolerance of certain broadleaf species to chlorsulfuron (**5**) results from hydroxylation of the 4-methyl substituent on the triazine ring to produce **6**, followed by glycosylation (Figure 10.4),[37] rather than 5′ ring hydroxylation, as in the case of graminaceous crops. The fact that **6** retained herbicidal activity implies that for this metabolic route the

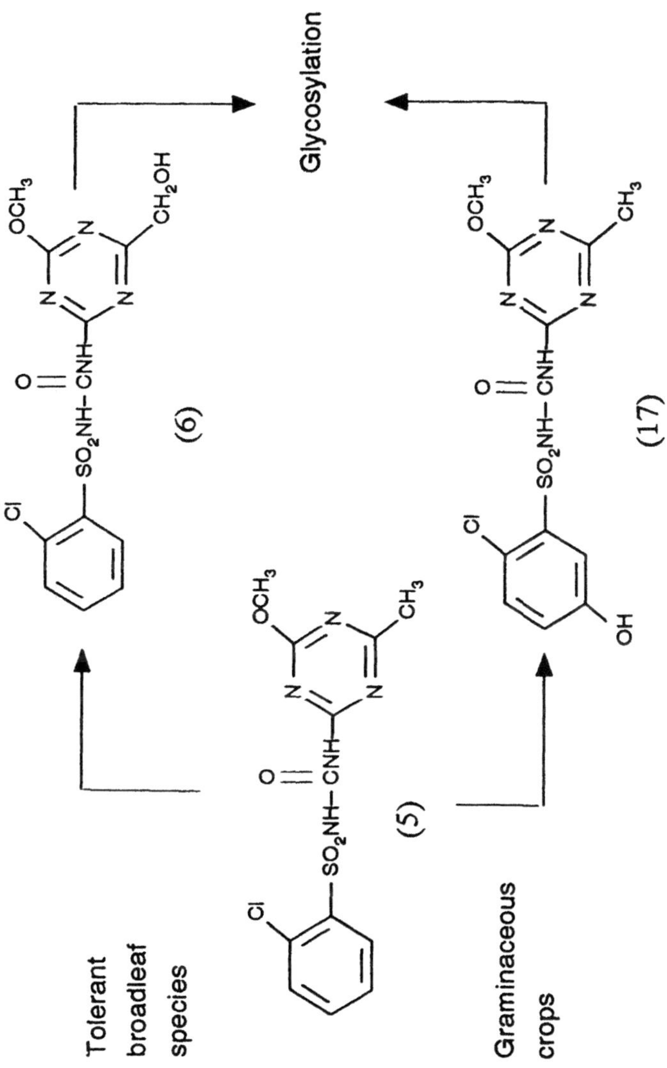

Figure 10.4. Chlorsulfuron metabolism in plants.

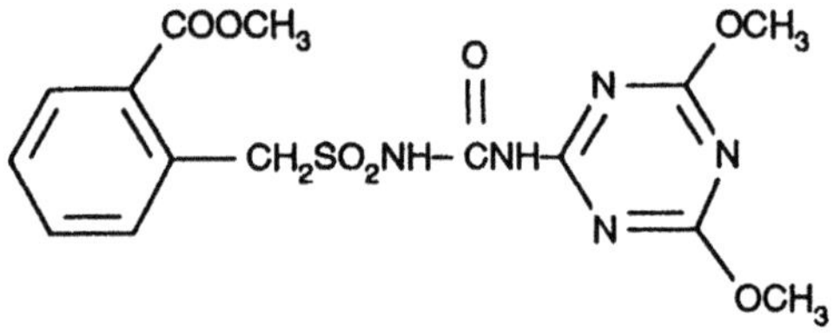

(7)

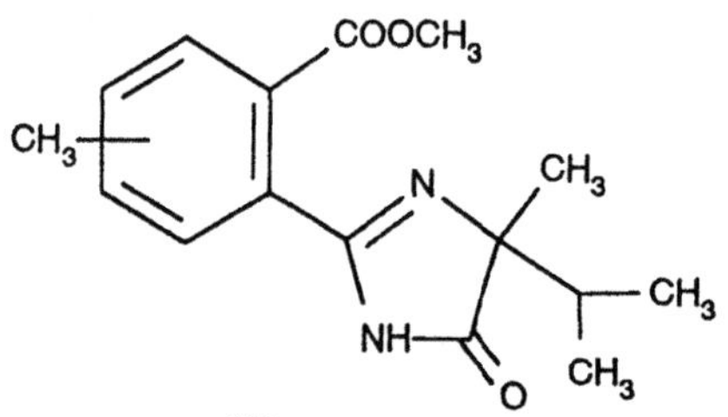

(8)

Figure 10.5a. Structures for bensulfuron (7) and Assert (8).

subsequent glycosylation step may be critical for detoxification.[38] A particularly interesting feature of the sulfonylurea herbicides is that comparatively subtle modifications to the substitution patterns around the parent ring systems can markedly alter selectivity properties. Thus, whereas chlorsulfuron is phytotoxic to rice, the related compound bensulfuron-methyl (**7**) is used for selective weed control in this crop.[39] Tolerance of rice to **7** apparently correlated with an ability to *O*-demethylate one of the pyrimidine methoxy substituents to the corresponding hydroxy derivative.[40]

Acetolactate synthase (ALS) is the enzyme target site of the sulfonylureas. In common with the imidazolinone aryl carboxylates, these herbicides inhibit valine and isoleucine biosynthesis. The imidazolinones are exemplified by Assert (**8**), which consists of a mixture of *m*- and *p*-isomers. The selectivity of ALS inhibitors invariably can be accounted for by differential metabolism or uptake or related phenomena rather than by any significant inherent difference in the properties of the ALS enzymes of crop and weed species. Assert is another example of a herbicide activated in plants by deesterification to the phytotoxic acid, and susceptible species such as *Avena fatua* (wild oat) are unable to metabolize the molecule further. Facile ring-methyl hydroxylation to the *m*- and *p*-benzyl alcohols, however, followed by glycosylation confers tolerance to maize and wheat.[41]

Turning to the substituted phenylureas, such as monuron (**9**) and diuron (**10**), *N*-dealkylation features prominently in their metabolism in plants (Figure 10.5b). Whereas monodemethylated derivatives possess residual phytotoxicity, *N*-didealkylation generally results in complete detoxification, as indicated by failure of such products to inhibit the Hill reaction of

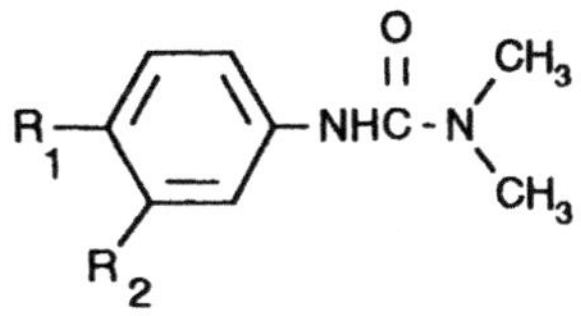

(9) : R_1 = Cl, R_2 = H
(10) : R_1 = Cl, R_2 = Cl
(11) : R_1 = CH_3, R_2 = Cl
(12) : R_1 = Cl, R_2 = CH_3

Figure 10.5b. Structures of some substituted phenylurea herbicides.

isolated chloroplasts. Plant species differ widely in their abilities to mediate either one or both of these *N*-dimethylation steps. This has been correlated with degree of susceptibility to various dimethylphenylureas in a number of crop/weed selectivity situations.[42] With the notable exception of cotton, however, the ability of crop species to detoxify monuron and diuron is generally inadequate for these phenylmethylureas to be used selectively in agriculture.[34,35,43]

Subsequently, selectivity was introduced into the substituted phenylureas by replacement of the *para* chlorine with alkyl groups such as methyl and isopropyl in the case of chlortoluron (**11**) and isoproturon or an alkoxy function as in metoxuron. Substitution of the *m*-chloro group of diuron with methyl to form the chlortoluron analogue CGA 43057 (**12**), however, did not result in any significant improvement in selectivity. This served to highlight the importance of positional effects in influencing the ability of plants to metabolize a particular substituent. The presence of alkyl or alkoxy functions in the *para* position of the phenylurea ring, however, provides an additional mode of oxidative attack, and extensive ring-methyl hydroxylation was considered to be the basis of the tolerance of both wheat and barley to chlortoluron.[43,44] Though hydroxylation of the ring methyl also occurred in cotton, *N*-didealkylation was very prominent, as with diuron, and probably accounted for tolerance. On the other hand, susceptible species such as *Lolium perenne*, *Avena fatua*, and *Alopecurus myosuroides* (blackgrass) metabolized chlortoluron to only a very limited extent and never beyond *N*-monodealkylation.[43] The inability of wheat and barley to oxidize the *meta*-located ring-methyl substituent of the experimental herbicide CGA 43057 correlated with the lack of selectivity of this compound for these species.[43] Not unexpectedly, the tolerance of cotton to this latter methylphenylurea was also explained by facile *N*-didealkylation.

The margin of selectivity of the alkyl- and alkoxyphenylureas between crop and weed species is often modest. In the case of some wheat and barley varieties, it is inadequate for selective control of *A. myosuroides* and other graminaceous weed species. Studies on the fate of chlortoluron in leaf disks of susceptible and tolerant wheat and barley cultivars have indicated a metabolic basis for the differing responses to the herbicide. Thus, detoxification by ring-methyl oxidation represented only a minor route in leaf disks of susceptible cultivars, in which metabolism was generally slow and restricted to *N*-monodealkylation,[45] a profile resembling that of the susceptible grass weeds. In contrast, in leaf disks of tolerant cultivars, *N*-dealkylation was relatively unimportant compared to hydroxylation of the ring-methyl substituent, as was recorded previously for whole wheat plants. This distinction between cultivars was less apparent in a similar study by Cabanne *et al.* using intact plants, albeit of different cultivars.[46] Wheat varieties, however, show parallel differences in response to the related phenylurea metoxuron, and in this case a difference in relative contributions to metabolism of *N*-demethylation versus *O*-demethylation of the ring methoxy substituent to give a nonphytotoxic phenol was the major factor that correlated with varietal reaction.[47]

Differential metabolism appears to be an important basis of selectivity also in the phenoxyalkanoic acids. *O*-Dealkylation leading to detoxification of 2,4-D by side-chain removal, as indicated by release of glyoxylic acid from side-chain-labeled 2,4-D, has been implicated as being important to selectivity in certain plant species. In the case of the phenoxyacetic acids, however, side-chain removal might also be achieved by *O*-demethylation of the dichloroanisole resulting from an initial decarboxylation. In many plants such as susceptible black currant (*Ribes nigrum* L.), these pathways occur to only a minor extent, but they constitute a major degradative mechanism in red currant (*Ribes sativum* Syme), where they confer tolerance.[48]

N-Dealkylation also plays a role in the metabolism and selectivity of *s*-triazine herbicides such as atrazine,[49,50] simazine,[51] simetryn,[52] and terbutryn.[53] In species in which this pathway was prominent, the major products identified were monodealkylated triazines, which retained phytotoxicity. However, in a study of atrazine metabolism in *C. album*, *C. strictum*, and *Amaranthus ponelli*, Khan *et al.* identified the bound residue from resistant biotypes as *N*-desethylatrazine and suggested that sequestration of phytotoxic metabolites contributed to tolerance to the herbicide.[54]

A recent study of the comparative aspects of the metabolism of atrazine and terbutryn in cell suspension cultures of potato indicated that plant species do not necessarily always metabolize a particular moiety in an identical manner. Whereas oxidative attack at the *N*-ethyl group occurred in the case of atrazine at C-1, the carbon adjacent to the substituted nitrogen, resulting in *N*-dealkylation, hydroxylation of terbutryn occurred predominantly at the C-2 position to form the hydroxyethyl derivative as the major metabolite.[53] Consequently, the susceptibility of a particular site to oxidative attack may be influenced by the nature and position of other substituent groups within the molecule and how they may modify relevant properties.

10.2.3.2. Aryl Hydroxylation

As a consequence of the aromatic nature of most herbicides, it is not surprising that ring hydroxylations feature prominently in higher plant detoxification pathways.[55–57] Species differ, however, in their ability to aryl hydroxylate herbicide substrates, presumably because of variation in the specificities or levels of expression of the *mfo* enzymes responsible for these oxidative reactions. For example, bentazon (**13**, Figure 10.6) has a 200-fold selectivity margin between rice and the sedge weed *Cyperus serotinus* attributable to a rapid conversion in the crop species to 6-hydroxybentazon (**14**), followed by glycosidation to give **15**.[58] Metabolism in *C. serotinus* does not occur to a significant extent as is the case in other susceptible species such

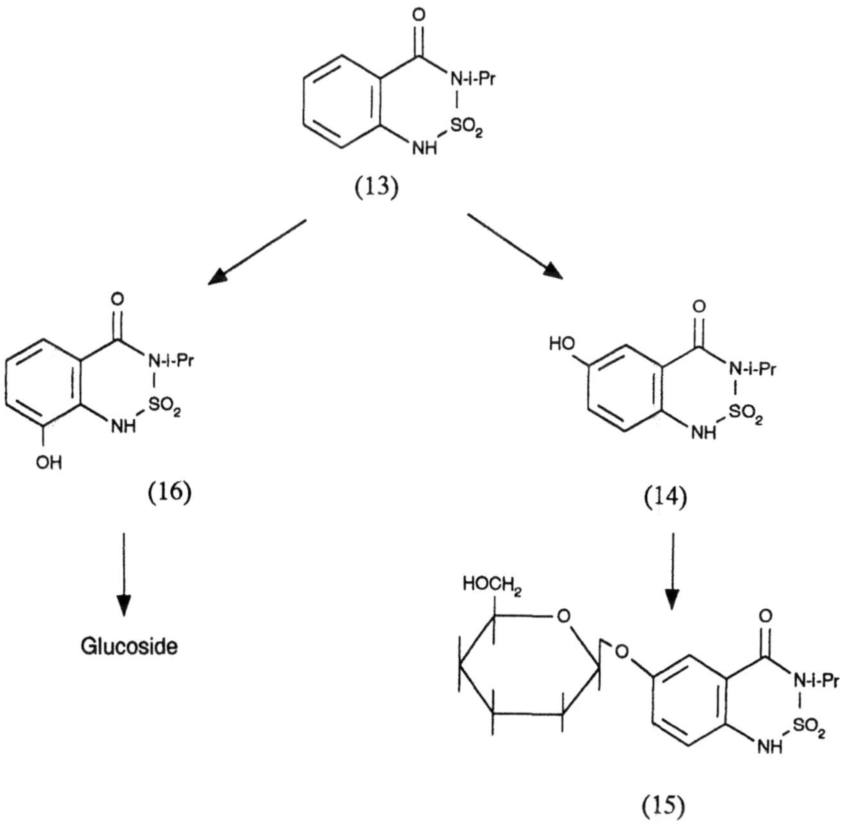

Figure 10.6. Pathways of bentazon metabolism in plants.

as *Cirsium arvense*, *Eleocharis kuroguwai*, *Sagittaria pygmaea*, and *Solanum nigrum.*[58-60] Rapid formation of water-soluble glycosidic conjugates as a consequence of initial ring hydroxylation also occurred in tolerant maize, navy bean (*Phaseolus vulgaris*), *E. crus-galli*,[58,61] and wheat.[62] In rice, wheat, and a number of other species examined, hydroxylation of bentazon occurs exclusively at the 6-position of the aromatic ring, whereas in soybean,[63,64] 8-hydroxybentazon (**16**) is also a prominent metabolite. This suggests either a broader specificity for the soybean P450-linked *mfo* mediating aryl hydroxylation or the presence of an additional isozyme. Differences in tolerance to bentazon also occur between various cultivars of soybean[65,66] and have been attributed to the relative rates of ring hydroxylation at both the 6- and 8-positions.[67,68]

Despite the remarkably high biological activity of the more recently introduced sulfonylurea herbicides, their margin of selectivity can be very large; that for chlorsulfuron (**5**, Figure 10.4) approaches some 4000-fold between cereals and sensitive weeds.[69] Selectivity of chlorsulfuron between tolerant (wheat, barley, *A. fatua*, and *Poa annua*) and susceptible (cotton, soybean, mustard, and sugar beet) species has been attributed to a striking difference in ability of young plants to oxidize chlorsulfuron to the 5′-hydroxy derivative (**17**, Figure 10.4), which accumulated in tolerant species as the *O*-glucoside.[69] Differences in uptake or translocation, however, appeared to be independent of species response and were not considered important to selectivity. The observation that **17** was also a potent inhibitor of ALS,[70] the target enzyme of the sulfonylureas, suggests that for this route of chlorsulfuron metabolism also the subsequent glycosidation (Figure 10.4) constitutes an important detoxification step. Differential metabolism of chlorsulfuron has also been implicated as a mechanism of selectivity between wheat and barley (tolerant), *Viola bicolor* (moderately susceptible), and sugar beet and *Matricaria chamomilla* (susceptible),[71] though in this case the products were not characterized.

An early example of herbicide metabolism by aromatic hydroxylation was demonstrated in the case of 2,4-D (**18**, Figure 10.7). The work of Guroff and co-workers provided evidence for a phenomenon which has become known as the "NIH" shift.[72] Hydroxylation of 2,4-D occurred preferentially at the 4-position, displacing the *para*-chlorine to give a mixture of 4-OH-2,5-D (**19**) and 4-OH-2,3-D (**20**); product **19** usually, though not always, predominates. In some species, elimination of the *para*-chlorine may also occur concomitantly with hydroxylation as evidenced by the identification of 4-OH-2-chlorophenoxyacetic acid and 2-OH-4-chlorophenoxyacetic acid.[73,74]

Unsubstituted positions on the ring are also hydroxylated to varying degrees to give 3-OH-2,4-D,[73] 5-OH-2,4-D,[75] and 6-OH-2,4-D.[74] While 3-OH-2,4-D retains auxin activity, the major products of hydroxylation of

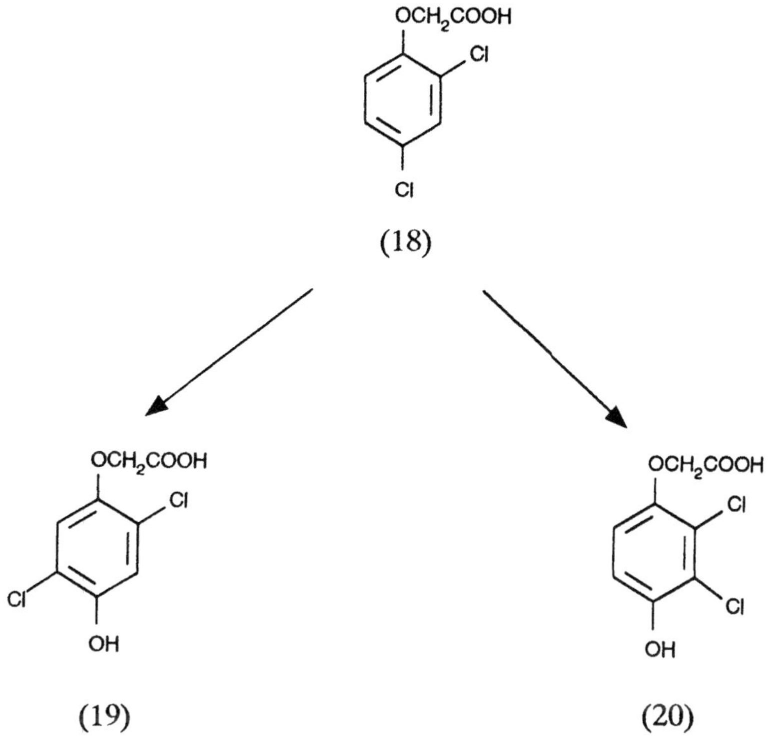

Figure 10.7. Ring hydroxylation of 2,4-D involving the "NIH" shift.

2,4-D, **19** and **20**, are both inactive, and consequently their formation constitutes an effective detoxification mechanism.[5,76] From these and other data, it appears that hydroxylation via the "NIH" shift may be restricted to the *para* position. Interestingly, the "NIH" shift has not been reported to occur in the case of the related phenoxyalkanoic acid 2-methyl-4-chlorophenoxyacetic acid (MCPA), which is hydroxylated predominantly at the ring methyl substituent.[77-79] Species tolerant to 2,4-D by virtue of their ability to catalyze the "NIH" shift include barley, oat, and wheat.[74,75] The pattern of hydroxylated metabolites in a cell suspension culture of hexaploid wheat reflected that obtained in intact plants.[10] Other species utilizing hydroxylation of 2,4-D as an important detoxification mechanism include cucumber and leguminous plants such as soybean.[8,75] The major metabolites formed in soybean, however, were amino amide conjugates, notably the 2,4-D-glutamic acid conjugate, although the proportion of phenolic derivatives increased with time.[5,6] A plausible interpretation of these data would be that the amide conjugates, rather than 2,4-D itself, may be the preferred substrates for aromatic hydroxylation. In contrast to other cereals, rice

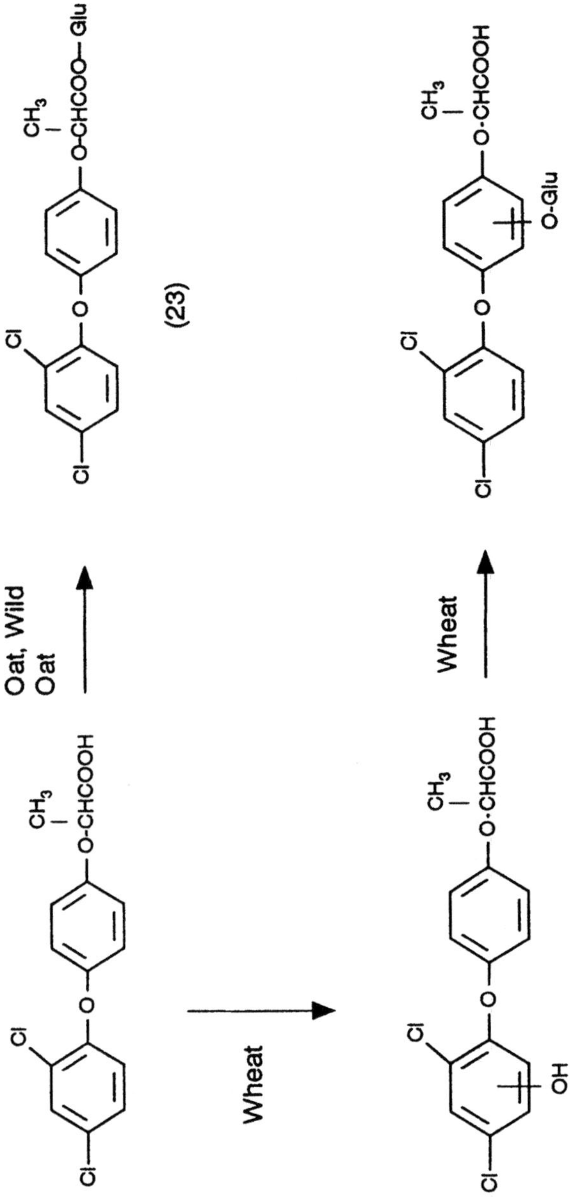

Figure 10.8. Fate of diclofop in plants.

callus was a poor hydroxylating tissue and accumulated a glycosyl ester of 2,4-D,[12] which was also the major initial product in cucumber.[8]

Considerable attention has been given to the role of ring hydroxylation in the selectivity of diclofop (Figure 10.8) following its generation from its inactive methyl ester (**1**, Figure 10.1) by the action of plant carboxylesterases. In resistant cereals, notably wheat, diclofop is rapidly inactivated by hydroxylation of the dichlorophenyl ring to give **21** followed by glycosidation, which results in the accumulation of **22**.[80–82] In contrast, aryl hydroxylation is of very minor importance in susceptible oat and wild oat (*A. fatua*), in which diclofop is primarily converted to the neutral glycosyl ester (**23**).[81] Polar conjugates of diclofop have also been identified as the major metabolites of diclofop-methyl in other susceptible species such as *E. crus-galli* and maize.[83,84] Though glucose esters of acidic herbicides might be expected to accumulate in cell vacuoles, as for glycosidic conjugates, the susceptibility of oat and *A. fatua* to diclofop-methyl implies that in this case compartmentalization of metabolites is inadequate to confer tolerance. Subsequent liberation of the phytotoxic acid from **23** presumably induces the susceptible response. This interpretation is supported by the observation that the acetylated derivative of **23** was as inhibitory as diclofop-methyl to growth of oat roots.[82] Metabolism of diclofop-methyl in cell suspension and callus cultures of oat and *A. fatua*, respectively, also resulted predominantly in the formation of **23**,[85] whereas ring hydroxylation was the major route in suspension cultures of diploid wheat (*Triticum monococcum*).[86] Earlier reports suggested that the aryl hydroxylation of diclofop in wheat occurred at all three unsubstituted sites on the dichlorophenyl ring, with the 5′-hydroxy derivative predominating.[80,81] Recent reexamination of hydroxylated products, however, using more stringent spectroscopic methodologies, has clearly indicated the presence of 4′-hydroxydichlorophenyl derivatives.[87] Consequently, these data represent the first demonstration of the operation of the "NIH" shift in the oxidative metabolism of a pesticide other than 2,4-D.

10.2.3.3. Manipulation of Oxidative Metabolism

Comparatively little progress has been made with studies of the enzymology of oxidative metabolism of pesticides, primarily because of the somewhat cryptic nature of the plant microsomal P450-linked *mfo*s which are generally believed to catalyze these reactions. As a result, conversion of a herbicide substrate to its oxygenated product is usually either impossible to detect or is extremely slow in freshly prepared microsomes; the necessary substrate specificity and kinetic studies have eluded investigators of xenobiotic metabolism. Thus, in previous work in our own laboratory, it was not possible to detect any activity of the *O*-demethylase which presumably converts metoxuron to the corresponding nonphytotoxic phenol, thereby conferring tolerance to wheat cultivars.[47] However, we were able to demonstrate an *N*-demethylase active in the conversion of metoxuron to the monomethyl derivative, which was enriched in the microsomal fraction of wheat. In whole plants of wheat, this latter route did not constitute an important metabolic pathway. Not surprisingly, the *N*-demethylase from both tolerant and susceptible cultivars showed little difference in specific activity, which was on the order of 6.8 nmol/30 min per mg of protein.[47] This value compared favorably with that of 2.16 nmol/30 min per mg of protein quoted by Frear *et al.* for a monuron *N*-demethylase in microsomes prepared from etiolated cotton hypocotyls.[35,88] Frear *et al.* also provided evidence implicating involvement of P450 in the oxidative *N*-demethylation of monuron. Later, ring hydroxylation of diclofop was demonstrated in wheat microsomal preparations, and a P450 binding spectrum was recorded.[89] Recently, microsomes were isolated from soybean suspension-cultured cells which retained ability to hydroxylate bentazon.[90] Evidence such as that reported above for the involvement of cytochrome P450 in xenobiotic metabolism in plant tissues is rare. However, other documented examples include a *p*-chloro-*N*-methylaniline *N*-demethylase from castor bean endosperm[91] and avocado,[92–94] an aldrin epoxidase from pear roots,[95] a *p*-nitroanisole *O*-demethylase from avocado mesocarp,[92,93] and a 2,4-D aryl hydroxylase from cucumber leaves.[96] Sweetser has speculated on the involvement of an inducible cytochrome P450-dependent *mfo* in the hydroxylation of chlorsulfuron in wheat.[38] This was based on the incorporation of a single atom of ^{18}O from molecular oxygen into a hydroxylated metabolite and the sensitivity of chlorsulfuron metabolism to inhibitors of protein biosynthesis.

While progress is being made in this area of plant metabolism, difficulties associated with the isolation of plant microsomal preparations active in the utilization of other herbicide substrates, and with the purification of the active *mfo* enzymes themselves, are still to be resolved. Consequently, only slow progress has been made on the identification of specific *mfo* polypeptides suitable as candiates for gene transfer studies. Such studies are aimed at manipulating crop plants for herbicide resistance based on enhanced oxidative detoxification.

On the basis of the above, it might be expected that compounds which function as inhibitors of P450-linked *mfo*s offer potential for use as synergists for herbicides initially detoxified in weed species via microsomal hydroxylation. This appears to be the case. Thus, the enzyme-activated inhibitor (suicide substrate) ABT blocked the oxidation of chlortoluron and isoproturon in wheat[97] and enhanced the toxicity of both herbicides.[98] ABT was found to preferentially inhibit the ring-methyl oxidation versus *N*-demethylation of chlortoluron, but such selectivity was not observed in a study of the influence of this inhibitor on the fate of chlortoluron in

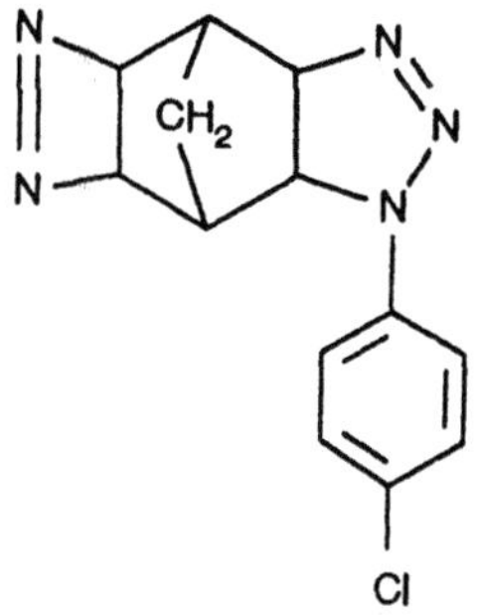

Figure 10.9. Structure of the plant growth regulator, tetcyclasis.

(24)

suspension-cultured cells of maize and cotton.[36] In a related study using potato tuber slices, the methyl hydroxylation of MCPA was similarly susceptible to ABT.[79]

In our experience, the most potent inhibition of oxidative metabolism has been achieved using the plant growth regulator tetcyclasis (**24**). Metabolism of chlortoluron was inhibited by 80% and 94% in cell suspension cultures of maize and cotton, respectively, following a 24-h pretreatment with tetcyclasis at a concentration of 10 μM (Table 10.1), ABT and 2,4-dichlorophenoxypropyne (2,4-DP) being considerably less potent; both ring-methyl oxidation and *N*-demethylation were similarly susceptible.[36]

Table 10.1. Effect of Monooxygenase Inhibitors on the Metabolism of Chlortoluron in Maize and Cotton Suspension Cultures[a,b]

		Inhibition of metabolism (%) at inhibitor concentration of:				
Species	Inhibitor	10^{-8} *M*	10^{-7} *M*	10^{-6} *M*	10^{-5} *M*	10^{-4} *M*
Maize (Black Mexican Sweet)	1-ABT	ND[c]	ND	ND	13	42
	2,4-DP	ND	ND	ND	0	53
	Tetcyclasis	28	72	79	80	81
Cotton (Delta Pine var. 61)	1-ABT	ND	ND	ND	17	49
	2,4-DP	ND	ND	ND	9	75
	Tetcyclasis	0	46	86	94	93

[a] Reproduced from *Progress in Pesticide Biochemistry and Toxicology*, Vol. 6, with permission.
[b] Cells were treated five days after subculture with inhibitors for 24 h, followed by incubation with [^{14}C-*phenyl*]chlortoluron. Maize cells had been subcultured with a 20% inoculum and were allowed to metabolize chlortoluron for 16 h (1-ABT and 2,4-DP experiments) or 6 h (tetcyclasis experiments). For 1-ABT and 2,4-DP experiments in cotton, cells were subcultured with a 10% inoculum and allowed to metabolize chlortoluron for 8 h; in tetcyclasis experiments with cotton, a 20% subculture inoculum was used with a 2-h period for chlortoluron incubation.
[c] ND: Not determined.

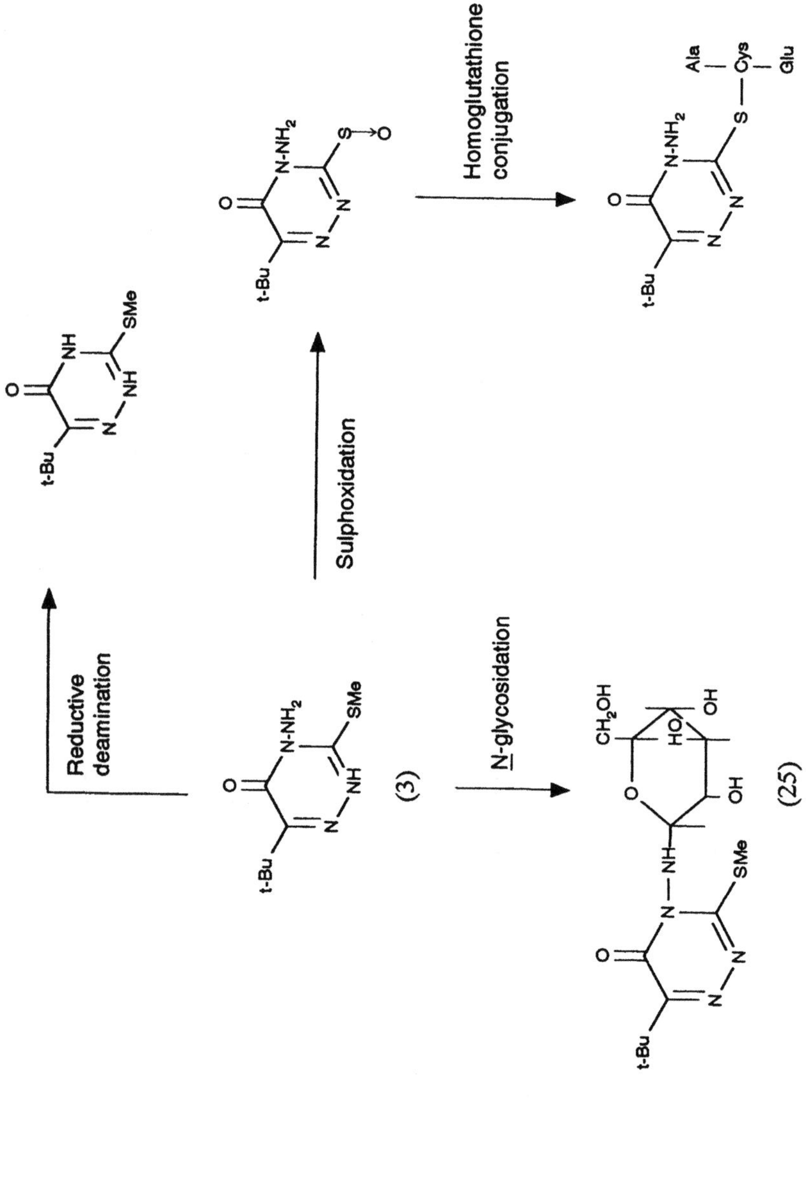

Figure 10.10. Routes of metribuzin metabolism in plants.

Though these observations point to both *C*- and *N*-alkyl hydroxylation of chlortoluron being mediated by the same species of *mfo*, an alternative explanation would be provided by a general lack of specificity in the action of tetcyclasis. This latter possibility received support from the recorded ability of tetcyclasis to interfere with *mfo*s involved in kaurene oxidation[99] and in the biosynthesis of plant sterols and xanthophylls.[100,101]

Studies reported in this section would appear to confirm that a real potential exists for modifying the ability of plants to detoxify herbicides, and thus change their selectivity profiles, by limiting the activities of their oxidative enzyme machinery. Further developments along these lines, however, will need to be underpinned by a greater fundamental knowledge of the biochemistry of microsomal hydroxylation reactions, in both crop and weed species.

10.2.3.4. Sulfoxidation

Oxidizable sulfur moieties are of common occurrence in pesticide molecules and have been shown to be sulfoxidized by numerous plant species[102-105]; the precise mechanism of such oxidations in plants, however, is not clear. In mammalian systems, sulfide substrates are oxidized primarily by microsomal cytochrome P450-dependent monooxygenases (*mfo*s),[106] though an FAD-dependent *S*-oxygenase has also been identified in microsomal preparations from pig liver.[107] Other sulfur oxidations have been shown to be mediated by horseradish peroxidase, such as that of the heterocyclic sulfide chlorpromazine; the carbamate insecticide mesurol, an aromatic-alkyl sulfide, however, was not a substrate.[55] Resulting products are not only more water soluble but tend to be more chemically reactive, pesticidally active, and short-lived in the environment than the precursor sulfides. Thus, the thiocarbamate herbicide EPTC is activated by oxidation to the corresponding sulfoxide.[108] For these reasons, plant sulfoxidation is a very significant factor in the biochemistry, physiology, and environmental chemistry of pesticides.

In addition to its role in the bioactivation of various pesticides, where species differences can account for selectivity,[108] sulfoxidation of alkyl sulfide groups is often an important initial step in detoxification. Thus, the sulfoxides generated by the oxidation of the methyl sulfide ring substituents of terbutryn and metribuzin (**3**, Figure 10.10) in turn are substrates for glutathione *S*-transferases; these catalyze their conversion to nonphytotoxic peptide conjugates.[109] Earlier studies by the Fargo group on the fate of 2-methylthio-*s*-triazines in plants indicated that this class of herbicide could only be detoxified by this latter route subsequent to sulfoxidation; the parent compounds, however, apparently inhibited the glutathione *S*-transferase activity of maize.[110] More recent evidence has implicated the sulfone derivative of the thiocarbamate EPTC, rather than the sulfoxide, as the true carbamoylating agent.[111]

Further metabolism of a prominent glutathione conjugate of the fungicide pentachloronitrobenzene (PCNB) in suspension-cultured cells of peanut resulted in the formation of phenyl methyl sulfides.[112-114] Such metabolites arose as a result of the concerted action of C—S lyase and *S*-adenosyl-L-[methyl]-methionine-dependent methyltransferase enzymes on the cysteine conjugate derived from the product of the glutathione *S*-transferase reaction. In the same work, it was noted that phenyl methyl sulfides became oxidized *in vivo* to the sulfoxides, which might subsequently serve as substrates for a further cycle of glutathione conjugation. The ability to mediate enzymatic sulfoxidation is also retained by suspension cultures as evidenced by the demonstrated ability of cells of cotton, carrot, and tobacco to oxidize *p*-chlorophenyl methyl sulfide (PCPMS), a model alkyl sulfide.[115] Rates of PCPMS sulfoxidation to the sulfoxide were greatest in cotton and poorest in tobacco, indicating the existence of exploitable species differences with respect to this mode of oxidative attack also.

10.2.4. Glycosidation

In the foregoing section, reference was made on a number of occasions to the production of glycosidic conjugates, generally *O*-β-D-glucosides, in plants. Formation of such phase 2 metabolites almost invariably follows the generation of hydroxylated derivatives though the extent of conjugation may vary between both species and plant tissues. The additional ability of plants to synthesize the glucose esters of acidic herbicides has also been referred to in the context of diclofop (Figure 10.8). Such esters, however, are inherently unstable, and their formation, even if extensive, probably does not constitute an effective detoxification mechanism. In general, glycosidation may contribute to overall detoxification of a particular herbicide by virtue of the enhanced water solubility of the products, which facilitates their disposal in the vacuole. However, in the case of chlorsulfuron metabolism in tolerant plant species, glycosidation resulted in the conversion of phytotoxic hydroxylated metabolites to inactive glycosides.[38] In suspension cultures of soybean and wheat, however, an *O*-glycosyltransferase was directly responsible for detoxification of pentachlorophenol.[116]

Though less common, formation of *N*-glycosides also occurs in plants, and direct *N*-glucosidation of chloramben represents a significant detoxification pathway in tolerant but not in susceptible species.[117,118] Susceptibility was associated with the conversion of chloramben to its *O*-glucose ester as the prominent metabolic route. Subsequently, the *N*-glucosyltransferase responsible for detoxification of chloramben in tolerant soybean and tomato was isolated and partially characterized.[119,120] Reference has already been made to the roles of reductive deamination and sulfoxidation followed by glutathione conjugation in the detoxification of metribuzin (**3**, Figures 10.3 and 10.10) in various plant species. However, *N*-glycoside formation (**25**, Figure 10.10) offers an additional metabolic route for detoxification of this triazinone and is particularly prominent in conferring metribuzin tolerance to some soybean cultivars such as "Bragg."[121] In contrast, in susceptible "Coker 102" and "Semmes," *N*-glycosidation occurred to only a minor extent. Other workers have recorded significant differences in metribuzin metabolism between tetraploid and diploid soybean lines,[122] metabolism being three- to fourfold more rapid in tetraploids than in diploids. In this study also, *N*-glucosidation was considered to be the major detoxification route involved. This distinction was mirrored in corresponding suspension cultures derived from tetraploid and diploid lines, which provided further evidence in support of the relevance of plant cell cultures for use in herbicide metabolism and selectivity studies.

Formation of glycoside conjugates does not necessarily represent the termination of a particular herbicide detoxification sequence in plants, and the glycoside conjugates are frequently subjected to further modification. Thus, **25** was acylated to the *N*-malonylglucoside in tomato plants,[120] as was the β-D-glucosyl conjugate of pentachlorophenol in suspension cultures of soybean and wheat.[116]

10.2.5. Glutathione Conjugation

Previous reference has been made to the established involvement of glutathione conjugation in the detoxification of the chloro-*s*-triazines and the sulfoxide metabolites of 2-methylthiotriazines and the triazinone metribuzin. A more complete list of herbicides and their derivatives which are now known to serve as substrates for the glutathione *S*-transferases of various plant species is presented in Figure 10.11. Glutathione conjugation products generally represent the dominant metabolites of chlorotriazines, such as atrazine, in tolerant maize and related panicoid species. Maize is particularly tolerant of atrazine, a response which results from the combined operation of glutathione conjugation, C-2 chlorine hydrolysis, and, to a lesser extent, *N*-dealkylation. On the other hand, facile glutathione *S*-transferase activity alone confers atrazine tolerance to sorghum.[123] Not all maize cultivars are atrazine tolerant, however, and the susceptibility of line GT112 has been attributed to low glutathione *S*-transferase activity.[124] Similarly, activity of the enzyme was deficient in the susceptible panicoid grasses *E. crus-galli* and *S. glauca*.[49]

Other classes of herbicide for which glutathione conjugation constitutes a major detoxification mechanism include the nitrodiphenyl ethers acifluorfen,[125] fluorodifen,[126–128] and fomesafen,[129] the chloroacetanilides metolachlor,[130–132] alachlor,[133] and propachlor,[134,135] the oxirane derivative tridiphane,[136] and the sulfonylurea chlorimuron-ethyl (Classic).[70] The last

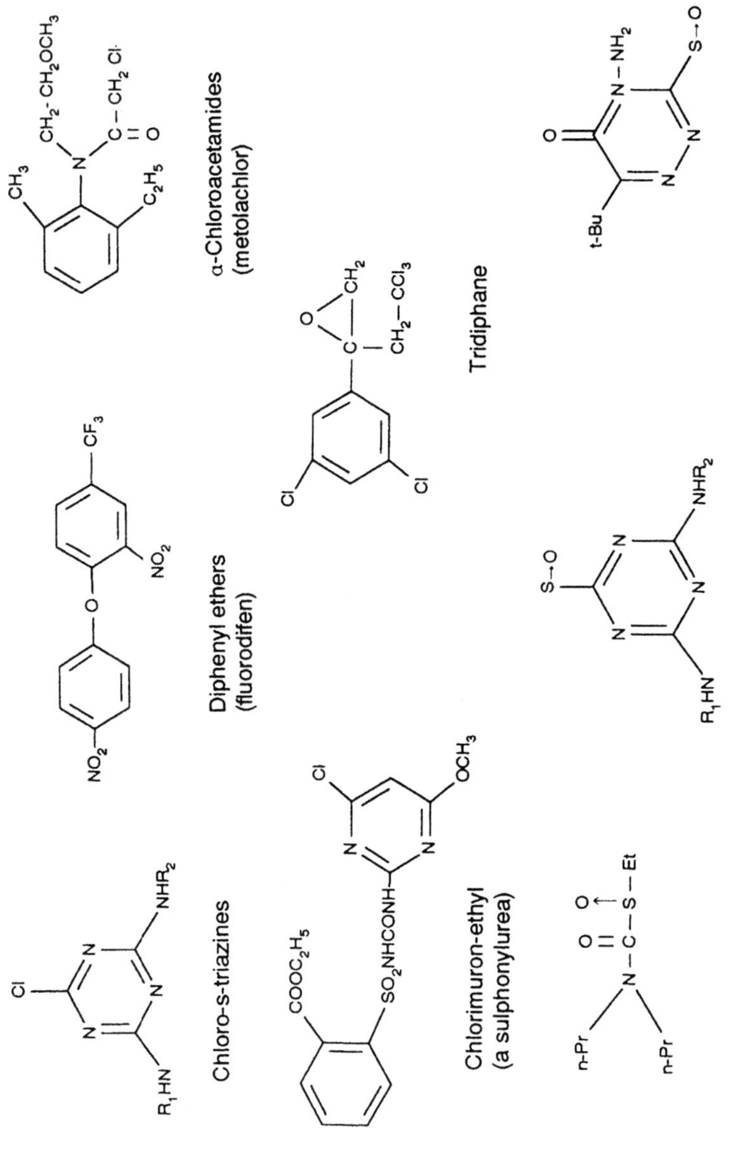

Figure 10.11. Herbicides and their metabolites that serve as substrates for plant glutathione *S*-transferases.

mentioned compound has been developed for broadleaf weed control in tolerant soybean, which rapidly detoxifies Classic by nucleophilic displacement of the pyrimidine chlorine by homoglutathione. In the nitrodiphenyl ethers, the action of glutathione *S*-transferases results in cleavage of the aromatic ether bond. Consequently, fluorodifen was cleaved to produce *S*-(2-nitro-4-trifluoromethylphenyl)glutathione and *p*-nitrophenol,[137] while in soybean acifluorfen yielded 2-chloro-4-trifluoromethylphenol and *S*-(3-carboxy-4-nitrophenyl)homoglutathione.[125] Extractable glutathione *S*-transferase activity utilizing fluorodifen as substrate was higher in resistant species such as cotton, maize, pea, peanut, okra, and soybean than in susceptible tomato, cucumber, and marrow.[126] Thus, the relative distribution and levels of expression of the transferase apparently account for fluorodifen selectivity. Surprisingly, the glutathione *S*-transferase of pea was found, on further characterization, to be remarkably specific for fluorodifen and did not utilize even structurally related herbicides such as nitrofen.[126] The fact that in other plant species nitrofen is degraded by ether cleavage suggests the existence of significant species differences in the specificity of glutathione *S*-transferases toward nitrodiphenyl ethers[138]; the enzyme from pea was also unable to utilize atrazine as substrate. This serves to illustrate the difficulties of predicting the detoxification potential of a plant species with other substrates undergoing similar modes of metabolism.

Studies on the metabolism of herbicides by glutathione peptide conjugation have led to the recognition of the existence of isoenzymic forms with differing specificities for various classes of herbicide substrates. Isoenzyme multiplicity was suggested by the difference in response to atrazine and EPTC between maize (tolerant to both herbicides) and *Panicum miliaceum* (atrazine-tolerant only) despite the fact that both herbicides undergo similar routes of detoxification by glutathione conjugation.[139] More recent studies using maize cell suspension cultures suggested that the isoenzymes responsible for the detoxification of the chloroacetanilide metolachlor were distinct from the enzyme which conjugated atrazine.[131] Subsequent analysis of whole protein extracts of maize leaves by chromatofocusing confirmed these findings and distinguished three peaks of glutathione *S*-transferase activity differing in their activity toward the two herbicides.[140] Apparently, this isoenzyme multiplicity results from multiple genes, which may also give rise to various molecular weight forms of the enzyme.[141]

In this context, it should be mentioned that a number of herbicide safeners, or antidotes, seem to exert their effect by elevating endogenous levels of glutathione and glutathione *S*-transferase activity in treated plants[131,133] (Chapter 6). In the case of dichlormid (R-25788) applied to maize, the latter effect was attributed to a direct influence of the safener on the level of transcription of the genes encoding the metolachlor-specific, but not the atrazine-specific, isoenzymes of glutathione *S*-transferase.[142]

As was the case with glycosides, tripeptide conjugates frequently do not represent terminal metabolites but are subjected to further modification. Thus, in maize, *S*-(metolachlor)glutathione was rapidly converted, by the concerted action of carboxypeptidase and γ-glutamyl transpeptidase enzymes, to the corresponding cysteine conjugate,[130,143] which was subsequently oxidized to the thiolactic acid conjugate. Similar pathways have been recorded for the fate of glutathione and homoglutathione conjugates of other herbicides though malonylation of the cysteine conjugates may be of more common occurrence than conversion to thiolactic derivatives.

While the role of glutathione *S*-transferases in the biotransformation of chloroacetanilides is firmly established, their contribution to selectivity of these herbicides is somewhat uncertain. Thus, the observation that propachlor served as a substrate for the transferases from 10 agriculturally important plant species implies that certain types of glutathione *S*-transferase activity may be widely distributed among higher plants. Conversely, formation of a metolachlor–glutathione conjugate was found to be some eightfold faster in suspension-cultured cells of maize compared to rice. This suggests that rates of conjugation alone could account for the differing sensitivities to metolachlor between these species.[143]

Reference has already been made to the fact that, in addition to some parent herbicides, the sulfoxide metabolites of dithiocarbamates (e.g., EPTC), 2-methylthio-*s*-triazines (e.g., terbutryn), and 3-methylthio-1,2,4-triazine-5-ones (e.g., metribuzin; Figure 10.10) are also substrates for plant glutathione *S*-transferases; species differences in capacity for their conjugation can contribute to selectivity. Detoxification of metribuzin in plants occurs by three competing pathways (Figure 10.10), but a recent report suggested that homoglutathione conjugation following sulfoxidation may be the most important route in soybean.[109] No comparative studies correlating metabolism with plant tolerance, however, were recorded.

Studies with other soybean cultivars referred to in previous sections of this chapter implicated reductive deamination or *N*-glucosidation as the dominant metabolic routes conferring tolerance. Though all of these studies on the fate of metribuzin in soybean agree on a major role for differential metabolism in determining cultivar response, it would appear that tolerant cultivars differ with respect to the relative importance of the three detoxification routes.

10.3. CONCLUDING REMARKS

The literature relating to the fate of herbicides in plants is vast, and in the course of the present chapter it has only been possible to draw upon examples so as to encompass a representative spectrum of the various modes of metabolic attack most frequently encountered. Hopefully, the investigations cited will serve to underline the importance of differential metabolism in determining selectivity of many herbicides between both species and crop cultivars. However, metabolism is only one of a number of factors which may potentially play a role in determining the response of a particular crop or weed species to a herbicide. While it might constitute a major determinant in one crop/weed situation, it may be of limited importance in others, where uptake and translocation could dominate. Cell suspension cultures allow the study of the ability of the plant cells *per se* to mediate particular biotransformations in the absence of problems associated with plant translocation processes. Studies reported here further emphasize their usefulness in investigations of selective herbicide metabolism. In turn, a thorough understanding of the metabolic fate of herbicides in plants might lead to the development of new herbicides which exploit differential metabolism to achieve selectivity or of herbicide progenitors which are activated by metabolism in target weed species. Finally, though the biorational design of herbicide synergists based on inhibition of oxidative metabolism is a very interesting and challenging concept, progress in this area must await further unraveling of the fundamental biochemistry of the target microsomal *mfo* enzymes.

REFERENCES

1. P. Lewer and W. J. Owen, *Proc. Br. Crop Prot. Conf.—Weeds*, Vol. 1, p. 353 (1987).
2. H.-W. Krell and H. Sandermann, *J. Agric. Food Chem.* *34*, 194 (1986).
3. J. Secor, C. Cséke, and W. J. Owen, in: *Biocatalysis in Agricultural Biotechnology* (J. R. Whitaker and P. E. Sonnet, eds.), ACS Symposium Series, No. 389, p. 265, American Chemical Society, Washington, D.C. (1989).
4. K. A. Hassal, *The Chemistry of Pesticides: Their Metabolism, Mode of Action and Uses in Crop Protection*, MacMillan, London (1982).
5. C.-S. Feung, R. H. Hamilton, and F. H. Witham, *J. Agric. Food Chem.* *19*, 475 (1971).
6. C.-S. Feung, R. H. Hamilton, and R. O. Mumma, *J. Agric. Food Chem.* *21*, 637 (1973).
7. D. I. Chkanikov, A. M. Makeev, N. N. Pavlova, and V. P. Dubovoi, *Sov. Plant Physiol.* *19*, 364 (1972).
8. D. I. Chkanikov, N. N. Pavlova, A. M. Makeev, T. A. Nazarova, and A. Yu. Makoveichuk, *Sov. Plant Physiol.* *24*, 457 (1977).
9. M. Arjmand, R. H. Hamilton, and R. O. Mumma, *J. Agric. Food Chem.* *26*, 1125 (1978).
10. D. Scheel and H. Sandermann, *Planta* *152*, 248 (1981).
11. P. Lewer and W. J. Owen, *Pestic. Biochem. Physiol.* *33*, 249 (1989).
12. C.-S. Feung, R. H. Hamilton, and R. O. Mumma, *J. Agric. Food Chem.* *24*, 1013 (1976).
13. W. J. Owen, R. H. Hamilton, and R. O. Mumma, unpublished results (1984).
14. C.-S. Feung, R. O. Mumma, and R. H. Hamilton, *J. Agric. Food Chem.* *22*, 307 (1974).
15. C.-S. Feung, R. H. Hamilton, and R. O. Mumma, *J. Agric. Food Chem.* *25*, 898 (1977).
16. P. Lewer and W. J. Owen, *Pestic. Biochem. Physiol.* *36*, 187 (1990).
17. D. S. Frear and G. G. Still, *Phytochemistry* *7*, 913 (1968).
18. R. E. Hoagland, *Plant Cell Physiol.* *19*, 1019 (1978).
19. R. E. Hoagland and G. Graf, *Phytochemistry* *11*, 521 (1972).
20. R. E. Hoagland, *Phytochemistry* *14*, 383 (1975).
21. R. E. Hoagland, G. Graf, and E. D. Handel, *Weed Res.* *14*, 371 (1974).
22. M. Oyamada, T. Tanaka, Y. Takasawa, and T. Takematsu, *J. Pestic. Sci.* *11*, 197 (1986).

23. J. L. Buckland, R. F. Collins, and E. M. Pullin, *Pestic. Sci. 4*, 149 (1973).
24. K. I. Beynon, G. Stoydin, and A. N. Wright, *Pestic. Sci. 3*, 293 (1972).
25. K. I. Beynon, G. Stoydin, and A. N. Wright, *Pestic. Sci. 3*, 379 (1972).
26. A. Waring, V. T. Edwards, and W. J. Owen, unpublished results (1986).
27. D. M. Stalker, K. E. McBride, and L. D. Malyj, *Science 242*, 419 (1988).
28. R. H. Hamilton and D. E. Moreland, *Science 135*, 373 (1962).
29. R. H. Hamilton, *J. Agric. Food Chem. 12*, 14 (1964).
30. K. I. N. Jensen, G. R. Stephenson, and L. A. Hunt, *Weed Sci. 25*, 212 (1977).
31. G. L. Lamoureux, personal communication.
32. C. Fedtke and R. R. Schmidt, in: *Pesticide Chemistry: Human Welfare and the Environment* (S. Matsunaka, D. H. Hutson, and S. D. Murphy, eds.), Vol. 3, p. 177, Pergamon Press, Oxford (1983).
33. C. Fedtke and R. R. Schmidt, *Z. Naturforsch., C 34*, 948 (1979).
34. D. S. Frear, *Science 162*, 674 (1968).
35. D. S. Frear, H. R. Swanson, and F. S. Tanaka, *Phytochemistry 8*, 2157 (1969).
36. D. J. Cole and W. J. Owen, *Plant Sci. 50*, 13 (1987).
37. J. M. Hutchison, R. Shapiro, and P. B. Sweetser, *Pestic. Biochem. Physiol. 22*, 243 (1984).
38. P. B. Sweetser, *Proc. Br. Crop Prot. Conf.—Weeds*, Vol. 3, p. 1147 (1985).
39. T. B. Ray, *Proc. Br. Crop Prot. Conf.—Weeds*, Vol. 1, p. 131 (1985).
40. H. Brown, personal communication.
41. M. A. Brown, T. Y. Chiu, and P. Miller, *Pestic. Biochem. Physiol. 27*, 24 (1987).
42. H. Geissbuhler, in: *Degradation of Herbicides* (P. C. Kearney and D. D. Kaufman, eds.), p. 79, Marcel Dekker, New York (1969).
43. P. J. Ryan, D. Gross, W. J. Owen, and T. L. Laanio, *Pestic. Biochem. Physiol. 16*, 213 (1981).
44. P. J. Ryan and W. J. Owen, *Proc. Br. Crop Prot. Conf.—Weeds*, Vol. 1, p. 317 (1982).
45. P. J. Ryan and W. J. Owen, *Aspects Appl. Biol. 3*, 63 (1983).
46. F. Cabanne, P. Gaillardon, and R. Scalla, *Pestic. Biochem. Physiol. 23*, 212 (1985).
47. R. Emami-Saravi, Ph.D. thesis, University of London (1979).
48. L. C. Luckwill and C. P. Lloyd-Jones, *Ann. Appl. Biol. 48*, 613 (1960).
49. K. I. N. Jensen, J. D. Bandeen, and V. S. Machado, *Can. J. Plant Sci. 57*, 1169 (1977).
50. J. Gressel, R. H. Shimabukuro, and M. E. Duysen, *Pestic. Biochem. Physiol. 19*, 361 (1983).
51. O. A. Akinyemiju, D. I. Dickmann, and R. A. Leavitt, *Weed Sci. 31*, 775 (1983).
52. H. Matsumoto and K. Ishizuka, *Weed Res. (Japan) 25*, 185 (1980).
53. R. Edwards and W. J. Owen, *Pestic. Biochem. Physiol. 34*, 246 (1989).
54. S. U. Khan, S. I. Warwick, and P. B. Marriage, *Weed Res. 25*, 33 (1985).
55. G. L. Lamoureux and D. S. Frear, in: *Xenobiotic Metabolism—In Vitro Methods* (G. D. Paulson, D. S. Frear, and E. P. Marks, eds.), ACS Symposium Series, No. 97, p. 72, American Chemical Society, Washington, D.C. (1979).
56. D. J. Cole, in: *Progress in Pesticide Biochemistry and Toxicology* (D. H. Hutson and T. R. Roberts, eds.), Vol. 3, p. 199, Wiley, Chichester (1983).
57. D. E. Hathway, *Biol. Rev. 61*, 435 (1986).
58. A. Mine, M. Miyakado, and S. Matsunaka, *Pestic. Biochem. Physiol. 5*, 566 (1975).
59. M. D. Mahoney and D. Penner, *Weed Sci. 23*, 272 (1975).
60. D. Penner, *Weed Res. 15*, 259 (1975).
61. M. D. Mahoney and D. Penner, *Weed Sci. 23*, 265 (1975).
62. G. Retzlaff and R. Hamm, *Weed Res. 16*, 263 (1976).
63. S. Otto, P. Beutel, N. Drescher, and R. Huber, in: *IUPAC: Advances in Pesticide Science* (H. Geissbuhler, ed.), Part 3, p. 551, Pergamon Press, Oxford (1979).
64. T. M. Sterling and N. E. Balke, *Weed Sci. 36*, 558 (1988).
65. L. M. Wax, R. L. Bernard, and R. M. Hayes, *Weed Sci. 22*, 35 (1974).
66. R. M. Hayes and L. M. Wax, *Weed Sci. 23*, 516 (1975).
67. T. M. Sterling and N. E. Balke, *Pestic. Biochem. Physiol. 34*, 39 (1989).
68. J. A. Connelly, M. D. Johnson, J. W. Gronwald, and D. L. Wyse, *Weed Sci. 36*, 417 (1988).

69. P. B. Sweetser, G. S. Schow, and J. M. Hutchison, *Pestic. Biochem. Physiol. 17*, 18 (1982).
70. H. M. Brown and S. M. Neighbors, *Pestic. Biochem. Physiol. 29*, 112 (1987).
71. F. Muller, B. H. Kang, and F. T. Maruska, *Meded. Fac. Landbouwwetensch., Rijksuniv. Gent 49*, 1091 (1984).
72. G. Guroff, J. W. Daly, D. M. Jerina, J. Renson, B. Witkop, and S. Udenfriend, *Science 157*, 1524 (1967).
73. C.-S. Feung, R. H. Hamilton, and R. O. Mumma, *J. Agric. Food Chem. 23*, 373 (1975).
74. D. W. Bristol, A. M. Ghanuni, and A. E. Oleson, *J. Agric. Food Chem. 25*, 1309 (1977).
75. C.-S. Feung, S. L. Loerch, R. H. Hamilton, and R. O. Mumma, *J. Agric. Food Chem. 26*, 1064 (1978).
76. R. H. Hamilton, J. Hurter, J. K. Hall, and C. D. Ercegovich, *J. Agric. Food Chem. 19*, 480 (1971).
77. D. J. Cole and B. C. Loughman, *J. Exp. Bot. 34*, 1299 (1983).
78. D. J. Cole and B. C. Loughman, *Plant Cell Rep. 3*, 5 (1984).
79. D. J. Cole and B. C. Loughman, *Physiol. Veg. 23*, 879 (1985).
80. S. G. Gorbach, K. Kuenzler, and J. Asshauer, *J. Agric. Food Chem. 25*, 507 (1977).
81. R. H. Shimabukuro, W. C. Walsh, and R. A. Hoerauf, *J. Agric. Food Chem. 27*, 615 (1979).
82. A. Jacobson, R. H. Shimabukuro, and C. McMichael, *Pestic. Biochem. Physiol. 24*, 61 (1985).
83. P. F. Boldt and A. R. Putnam, *Weed Sci. 29*, 237 (1981).
84. K. Gorecka, R. H. Shimabukuro, and W. C. Walsh, *Physiol. Plant. 53*, 55 (1981).
85. J. A. Dusky, D. G. Davis, and R. H. Shimabukuro, *Physiol. Plant. 54*, 490 (1982).
86. J. A. Dusky, D. G. Davis, and R. H. Shimabukuro, *Physiol. Plant. 49*, 151 (1980).
87. F. S. Tanaka, R. G. Wien, B. L. Hoffer, and R. H. Shimabukuro, Abstracts, American Chemical Society Meeting and Third Chemical Congress of North America, 166, Toronto, Canada, June, 1988, American Chemical Society, Washington, D.C.
88. D. S. Frear, H. R. Swanson, and F. S. Tanaka, in: *Recent Advances in Phytochemistry* (V. C. Runeckles and T. C. Tso, eds.), Vol. 5, p. 225, Academic Press, London (1972).
89. J. J. McFadden, D. S. Frear, and E. R. Mansager, *Pestic. Biochem. Physiol. 34*, 92 (1989).
90. N. Balke, personal communication (1989).
91. O. Young and H. Beevers, *Phytochemistry 15*, 379 (1976).
92. F. J. McPherson, A. Markham, J. W. Bridges, G. C. Hartman, and D. V. Parke, *Biochem. Soc. Trans. 3*, 281 (1975).
93. F. J. McPherson, A. Markham, J. W. Bridges, G. C. Hartman, and D. V. Parke, *Biochem. Soc. Trans. 3*, 283 (1975).
94. D. R. Dohn and R. L. Krieger, *Arch. Biochem. Biophys. 231*, 416 (1984).
95. J. W. Earl and I. R. Kennedy, *Phytochemistry 14*, 1507 (1975).
96. A. M. Makeev, A. Yu. Makoveichuk, and D. I. Chkanikov, *Dokl. Bot. Sci. 223*, 36 (1977).
97. F. Cabanne, P. Gaillardon, R. Scalla, and F. Durst, *Proc. Br. Crop Prot. Conf.—Weeds*, Vol. 3, 1163 (1985).
98. P. Gaillardon, F. Cabanne, R. Scalla, and F. Durst, *Weed Res. 25*, 397 (1985).
99. J. C. Graebe, in: *Plant Growth Substances* (P. F. Wareing, ed.), p. 71, Academic Press, London (1982).
100. K. Grossman, W. Rademacher, and J. Jung, *J. Plant Growth Regul. 2*, 19 (1983).
101. G. Sandmann and P. M. Bramley, *Biochim. Biophys. Acta 843*, 73 (1985).
102. R. L. Metcalf, T. R. Fukuto, C. Collins, K. Borck, H. T. Burk, H. T. Reynolds, and M. F. Osman, *J. Agric. Food Chem. 14*, 579 (1966).
103. P. W. Muller, P. H. Payot, and J. R. Geigy, in: *Isotopes in Weed Research*, International Atomic Energy Agency, Vienna, pp. 61–68 (1966).
104. M. C. Bowman, M. Beroza, and J. A. Harding, *J. Agric. Food Chem. 17*, 138 (1969).
105. N. R. Andrawes, W. P. Bagley, and R. A. Herett, *J. Agric. Food Chem. 19*, 731 (1971).
106. T. Nagatsugawa and M. A. Morelli, in: *Insect Biochemistry and Physiology* (C. F. Wilkinson, ed.), Plenum, New York (1976).

107. L. L. Poulsen, R. M. Hyslop, and D. M. Ziegler, *Arch. Biochem. Biophys. 198,* 78 (1979).
108. J. E. Casida, R. A. Gray, and H. Tilles, *Science 184,* 573 (1974).
109. D. S. Frear, H. R. Swanson, and E. R. Mansager, *Pestic. Biochem. Physiol. 23,* 56 (1985).
110. D. S. Frear and H. R. Swanson, *Phytochemistry 9,* 2123 (1970).
111. L. Horvath and A. Pulay, *Pestic. Biochem. Physiol. 14,* 265 (1980).
112. G. L. Lamoureux and D. G. Rusness, *Pestic. Biochem. Physiol. 14,* 50 (1980).
113. D. G. Rusness and G. L. Lamoureux, *J. Agric. Food Chem. 28,* 1070 (1980).
114. G. L. Lamoureux, J.-H. Gouot, D. G. Davis, and D. G. Rusness, *J. Agric. Food Chem. 29,* 996 (1981).
115. L. C. Blair, F. W. Slife, A. Felsot, and M. J. Plewa, *Pestic. Biochem. Physiol. 21,* 291 (1984).
116. R. Schmitt, J. Kaul, T. van der Trenck, E. Schaller, and H. Sandermann, *Pestic. Biochem. Physiol. 24,* 77 (1985).
117. E. W. Stoller and L. M. Wax, *Weed Sci. 16,* 283 (1968).
118. D. S. Frear, H. R. Swanson, E. R. Mansager, and R. G. Wien, *J. Agric. Food Chem. 26,* 1347 (1978).
119. D. S. Frear, *Phytochemistry 7,* 381 (1968).
120. D. S. Frear, E. R. Mansager, H. R. Swanson, and F. S. Tanaka, *Pestic. Biochem. Physiol. 19,* 270 (1983).
121. A. E. Smith and R. E. Wilkinson, *Physiol. Plant. 32,* 253 (1974).
122. P. Abusteit, F. J. Corbin, D. P. Schmitt, J. W. Burton, A. D. Worsham, and L. Thompson, *Weed Sci. 33,* 618 (1985).
123. G. L. Lamoureux, L. E. Stafford, R. H. Shimabukuro, and R. G. Zaylskie, *J. Agric. Food Chem. 21,* 1020 (1973).
124. R. H. Shimabukuro, D. S. Frear, H. R. Swanson, and W. C. Walsh, *Plant Physiol. 47,* 10 (1971).
125. D. S. Frear, H. R. Swanson, and E. R. Mansager, *Pestic. Biochem. Physiol. 20,* 299 (1983).
126. D. S. Frear and H. R. Swanson, *Pestic. Biochem. Physiol. 3,* 473 (1973).
127. R. H. Shimabukuro, G. L. Lamoureux, H. R. Swanson, W. C. Walsh, L. E. Stafford, and D. S. Frear, *Pestic. Biochem. Physiol. 3,* 483 (1973).
128. H. Diesperger and H. Sandermann, *Planta 146,* 643 (1979).
129. J. D. H. L. Evans, B. D. Cavell, and R. R. Hignett, *Proc. Br. Crop Prot. Conf.—Weeds,* Vol. 1, p. 345 (1987).
130. P. Blattmann, D. Gross, H. P. Kriemler, and K. Ramsteiner, Presented at the 6th IUPAC International Congress on Pesticide Chemistry, Ottawa, August, 1986.
131. R. Edwards and W. J. Owen, *Planta 169,* 208 (1986).
132. E. P. Fuerst and J. W. Gronwald, *Weed Sci. 34,* 354 (1986).
133. T. J. Mozer, D. C. Tiemeier, and E. G. Jaworski, *Biochemistry 22,* 1068 (1983).
134. G. L. Lamoureux, L. E. Stafford, and F. S. Tanaka, *J. Agric. Food Chem. 19,* 346 (1971).
135. R. R. S. Burkholder, M.S. thesis, North Dakota State University, Fargo (1978).
136. G. L. Lamoureux and D. G. Rusness, *Pestic. Biochem. Physiol. 26,* 323 (1986).
137. G. L. Lamoureux and D. G. Rusness, *Proceedings of the 5th IUPAC Congress in Pesticide Chemistry* (J. Miyamoto and P. C. Kearney, eds.), Vol. 3, p. 295, Pergamon Press, Oxford (1983).
138. D. Hawton and E. H. Strobbe, *Weed Sci. 19,* 555 (1971).
139. G. Ezra and G. R. Stephenson, *Pestic. Biochem. Physiol. 24,* 207 (1985).
140. R. Edwards and W. J. Owen, *Biochem. Soc. Trans. 15,* 1184 (1987).
141. R. E. Moore, M. S. Davies, K. M. O'Connell, E. I. Harding, R. C. Wiegand, and D. C. Tiemeier, *Nucleic Acids Res. 14,* 7227 (1986).
142. R. Edwards and W. J. Owen, *Planta 175,* 99 (1988).
143. W. J. Owen and B. Donzel, unpublished results (1982).

Chemical Names of Compounds Mentioned in the Text

acetochlor	2-chloro-*N*-(ethoxymethyl)-*N*-(2-ethyl-6-methylphenyl)acetamide
acifluorfen	5-[2-chloro-4-(trifluoromethyl)phenoxy]-2-nitrobenzoic acid
aclonifen	2-chloro-6-nitro-3-phenoxybenzenamine
alachlor	2-chloro-*N*-(2,6-diethylphenyl)-*N*-(methoxymethyl)acetamide
aldicarb sulfone (Aldoxycarb)	2-methyl-2-(methylsulfonyl)propanol *O*-[(methylamino)carbonyl] oxime
aldrin	1,2,3,4,10,10-hexachloro-1,4,4α,5,8,8α-hexahydro-1,4:5,8-dimethanonaphthalene
alloxydim	methyl 2,2-dimethyl-4,6-dioxo-5-[1-[(2-propenyloxy)amino]butylidene]cyclohexanecarboxylate
ametryn	*N*′-ethyl-*N*′-(1-methylethyl)-6-(methylthio)-1,3,5-triazine-2,4-diamine
amiprophos-methyl	*O*-methyl *O*-(4-methyl-2-nitrophenyl) (1-methylethyl)phosphoramidothioate
amitrole	1*H*-1,2,4-triazole-3-amine
AMO 1618	2′-isopropyl-4′-(trimethylammonium chloride)-5′-methylphenyl-1-piperidinecarboxylate
ancymidol	α-cyclopropyl-α-(4-methoxyphenyl)-5-pyrimidinemethanol
anilofos	*S*-[2-[(4-chlorophenyl)(1-methylethyl)amino]-2-oxoethyl]*O*,*O*-dimethyl phosphorodithioate

Assert (AC-222,293)	isomeric mixture of methyl 6-[4-isopropyl-4-methyl-5-oxo-2-imidazolin-2-yl]-*m*-toluate and methyl 2-[4-isopropyl-4-methyl-5-oxo-2-imidazolin-2-yl]-*p*-toluate
asulam	methyl [(4-aminophenyl)sulfonyl]carbamate
atraton	*N*-ethyl-6-methoxy-*N'*-(1-methylethyl)1-3,5-triazine-2,4-diamine
atrazine	6-chloro-*N*-ethyl-*N'*-(1-methylethyl)-1,3,5-triazine-2,4-diamine
aziprotryne	4-azido-*N*-(1-methylethyl)-6-(methylthio)-1,3,5-triazin-2-amine
barban	4-chloro-2-butynyl 3-chlorophenylcarbamate
benazolin	4-chloro-2-oxo-3(2*H*)-benzothiazoleacetic acid
benefin	*N*-butyl-*N*-ethyl-2,6-dinitro-4-(trifluoromethyl)-benzenamine
bensulfuron	2-[[[[[(4,6-dimethoxy-2-pyrimidinyl)amino]-carbonyl]amino]sulfonyl]methyl]benzoic acid
bensulide	*O,O*-bis(1-methylethyl) *S*-[2-[(phenylsulfonyl)-amino]ethyl] phosphorodithioate
bentazon	3-(1-methylethyl)-(1*H*)-2,1,3-benzothiadiazin-4(3*H*)-one 2,2-dioxide
benzadox	[(benzoylamino)oxy]acetic acid
benzofluor	*N*-[4-(ethylthio)-2-(trifluoromethyl)phenyl]-methanesulfonamide
benzoylprop	*N*-benzoyl-N-(3,4-dichlorophenyl)-DL-alanine
benzthiazuron	*N*-2-benzothiazolyl-*N'*-methylurea
bialaphos	4-(hydroxymethylphosphinyl)-L-2-aminobutanoyl-L-alanyl-L-alanine
bifenox	methyl 5-(2,4-dichlorophenoxy)-2-nitrobenzoate
bromacil	5-bromo-6-methyl-3-(1-methylpropyl)-2,4(1*H*,3*H*)pyrimidinedione
bromofenoxim	3,5-dibromo-4-hydroxybenzaldehyde *O*-(2,4-dinitrophenyl)oxime
bromoxynil	3,5-dibromo-4-hydroxybenzonitrile
butachlor	*N*-(butoxymethyl)-2-chloro-*N*-(2,6-diethyl-phenyl)acetamide
butamifos	*O*-ethyl *O*-(5-methyl-2-nitrophenyl) (1-methyl-propyl)phosphoramidothioate
buthidazole	3-[5-(1,1-dimethylethyl)-1,3,4-thiadiazol-2-yl]-4-hydroxy-1-methyl-2-imidazolidinone
butralin	4-(1,1-dimethylethyl)-*N*-(1-methylpropyl)-2,6-dinitrobenzenamine

buturon	*N'*-(4-chlorophenyl)-*N*-methyl-*N*-(1-methyl-2-propynyl)urea
butylate	*S*-ethylbis(2-methylpropyl)carbamothioate
cacodylic acid	dimethylarsinic acid
carbetamide	(*R*)-*N*-ethyl-2-[[(phenylamino)carbonyl]oxy]-propanamide
CDAA	2-chloro-*N,N*-di-2-propenylacetamide
CDEC (Sulfallate)	2-chloro-2 propenyl diethylcarbamodithioate
CGA-133205	*O*-[1,3-dioxolan-2-yl-methyl]-2,2,2-trifluoro-4'-chloroacetophenone-oxime
chlomethoxyfen	4-(2,4-dichlorophenoxy)-2-methoxy-1-nitrobenzene
chloramben	3-amino-2,5-dichlorobenzoic acid
chloramphenicol	D(−)-*threo*-2,2-dichloro-*N*-[β-hydroxy-α-(hydroxymethyl)-4-nitrophenethyl]acetamide
chlorbromuron	*N'*-(4-bromo-3-chlorophenyl)-*N*-methoxy-*N*-methylurea
chlorbufam	1-methyl-2-propynyl (3-chlorophenyl)carbamate
chlordane	1,2,4,5,6,7,8,8-octachloro-2,3,3a,4,7,7a-hexahydro-4,7-methano-1*H*-indene
chlorfenprop-methyl	methyl α,4-dichlorobenzenepropanoate
chlorfenvinphos	2-chloro-1-(2,4-dichlorophenyl)ethenyl diethyl phosphate
chlorimuron	2-[[[[(4-chloro-6-methoxy-2-pyrimidinyl)amino]-carbonyl]amino]sulfonyl]benzoic acid
chlormequat chloride	2-chloro-*N,N,N*-trimethylethanaminium ion
chlornitrofen	4-nitrophenyl 2,4,6-trichlorophenyl ester
chloroxuron	*N'*-[4-(4-chlorophenoxy)phenyl]-*N,N'*-dimethyl-urea
chlorphonium chloride	tributyl[(2,4-dichlorophenyl)methyl]phosphonium
chlorpropham	1-methylethyl 3-chlorophenylcarbamate
chlorsulfuron	2-chloro-*N*-[[(4-methoxy-6-methyl-1,3,5-triazin-2-yl)amino]carbonyl]benzenesulfonamide
chlorthal-dimethyl	dimethyl 2,3,5,6-tetrachlorobenzene-1,4-dicarboxylate
chlorthiamid	2,6-dichlorobenzenecarbothioamide
chlortoluron	*N'*-(3-chloro-4-methylphenyl)-*N,N*-dimethylurea
cinmethylin	*exo*-1-methyl-4-(1-methylethyl)-2-[(2-methyl-phenyl)methoxy]-7-oxabicyclo[2.2.1]heptane
clethodim	(*E,E*)-(±)-2-[1-[[(3-chloro-2-propenyl)oxy]imino]propyl]-5-[2-(ethylthio) propyl]-3-hydroxy-2-cyclohexen-1-one

clomazone	2-[(2-chlorophenyl)methyl]-4,4-dimethyl-3-isoxazolidinone
cloproxydim	(*E,E,*)-2-[1-[[3-chloro-2-propenyl)oxy]imino]butyl]-5-[2-(ethylthio)propyl]-3-hydroxy-2-cyclohexen-1-one
clopyralid	3,6-dichloro-2-pyridinecarboxylic acid
cyanazine	2-[[4-chloro-6-(ethylamino)-1,3,5-triazin-2-yl]amino]-2-methylpropanenitrile
cycloate	*S*-ethyl cyclohexylethylcarbamothioate
cycloxydim	2-[1-(ethoxyimino)butyl]-3-hydroxy-5-(tetrahydro-2*H*-thiopyran-3-yl)-2-cyclohexen-1-one
cyometrinil	(*Z*)-α[cyanomethoxyimino]benzeneacetonitrile
2,4-D	(2,4-dichlorophenoxy)acetic acid
2,6-D	(2,6-dichlorophenoxy)acetic acid
dalapon	2,2-dichloropropanoic acid
daminozide	butanedioic acid mono(2,2-dimethylhydrazide)
2,4-DB	4-(2,4-dichlorophenoxy)butanoic acid
DCPA	dimethyl 2,3,5,6-tetrachloro-1,4-benzenedicarboxylate
2,4-DES	2-(2,4-dichlorophenoxy)ethyl hydrogen sulfate
desmedipham	ethyl [3-[[(phenylamino)carbonyl]oxy]phenyl]-carbamate
desmetryn	*N*-methyl-*N'*-(1-methylethyl)-6-(methylthio)-1,3,5-triazine-2,4-diamine
diallate	*S*-(2,3-dichloro-2-propenyl) bis(1-methylethyl)-carbamothioate
dicamba	3,6-dichloro-2-methoxybenzoic acid
dichlobenil	2,6-dichlorobenzonitrile
dichlormate	(3,4-dichlorobenzene)methyl methylcarbamate
dichlormid	2,2-dichloro-*N,N*-di-2-propenylacetamide
dichlorprop	(±)-2-(2,4-dichlorophenoxy)propanoic acid
diclofop	(±)-2-[4-(2,4-dichlorophenoxy)phenyxy]propanoic acid
diethatyl	*N*-(chloroacetyl)-*N*-(2,6-diethylphenyl)glycine
dietholate	*O,O*-diethyl *O*-phenyl phosphorothioate
difenoxuron	*N'*-[4-(4-methoxyphenoxy)phenyl]-*N,N*-dimethyl-urea
difenzoquat	1,2-dimethyl-3,5-diphenyl-1*H*-pyrazolium ion
diflufenican	*N*-(2,4-difluorophenyl)-2-[3-(trifluoromethyl)-phenoxy]-3-pyridinecarboxamide
difunone	5-(dimethylaminomethylene)-2-oxo-4-phenyl-2,5-dihydrofurane-carbonitrile

dimefuron	*N'*-[3-chloro-4-[5-(1,1-dimethylethyl)-2-oxo-1,3,4-oxadiazol-3(2*H*)-yl]phenyl]-*N,N*-dimethylurea
dimethachlor	2-chloro-*N*-(2,6-dimethylphenyl)-*N*-(2-methoxyethyl)acetamide
dimethametryn	*N*-(1,2-dimethylpropyl)-*N'*-ethyl-6-(methylthio)-1,3,5-triazine-2,4-diamine
dinitramine	N^3,N^3-diethyl-2,4-dinitro-6-(trifluoromethyl)-1,3-benzenediamine
dinoseb	2-(1-methylpropyl)-4,6-dinitrophenol
dinoterb	2-(1,1-dimethylethyl)-4,6-dinitrophenol
diphenamid	*N,N*-dimethyl-α-phenylbenzeneacetamide
dipropetryn	6-(ethylthio)-*N,N'*-bis(1-methylethyl)-1,3,5-triazine-2,4-diamine
diquat	6,7-dihydrodipyrido[1,2-*a*:2',1'-*c*]pyrazinediium ion
diuron	*N'*-(3,4-dichlorophenyl)-*N,N*-dimethylurea
DNOC	2-methyl-4,6-dinitrophenol
DPX-F5384	see bensulfuron-methyl
DPX-F6025	see chlorimuron-ethyl
DPX-M6316	methyl 3-[[[(4-methoxy-6-methyl-1,3,5-triazin-2-yl)amino]carbonyl]aminosulfonyl]-2-thiophenecarboxylate
eglinazine	*N*-[4-chloro-6-(ethylamino)-1,3,5-triazin-2-yl]glycine
endothal	7-oxabicyclo[2.2.1]heptane-2,3-dicarboxylic acid
EPTC	*S*-ethyl dipropylcarbamothioate
ethalfluralin	*N*-ethyl-*N*-(2-methyl-2-propenyl)-2,6-dinitro-4-(trifluoromethyl)benzenamine
ethidimuron	*N*-[5-(ethylsulfonyl)-1,3,4-thiadiazol-2-yl]-*N,N'*-dimethylurea
ethofumesate	(±)-2-ethoxy-2,3-dihydro-3,3-dimethyl-5-benzofuranyl methanesulfonate
fenac	2,3,6-trichlorobenzeneacetic acid
fenoprop	(±)-2-(2,4,5-trichlorophenoxy)propanoic acid
fenoxaprop	(±)-2-[4-[(6-chloro-2-benzoxazolyl)-oxy]phenoxy]propanoic acid
fenuron	*N,N*-dimethyl-*N'*-phenylurea
flamprop	*N*-benzoyl-*N*-(3-chloro-4-fluorophenyl)-DL-alanine
fluazifop	(±)-2-[4-[[5-(trifluoromethyl)-2-pyridinyl]oxy]phenoxy]propanoic acid
fluchloralin	*N*-(2-chloroethyl)-2,6-dinitro-*N*-propyl-4-(trifluoromethyl)benzenamine

flumetralin	2-chloro-*N*-[2,6-dinitro-4-(trifluoromethyl)-phenyl]-*N*-ethyl-6-fluorobenzenemethanamine
fluometuron	*N*,*N*-dimethyl-*N*′-[3-(trifluoromethyl)phenyl]urea
fluorochloridone	3-chloro-4-(chloromethyl)-1-[3-(trifluoromethyl)-phenyl]-2-pyrrolidinone
fluorodifen	2-nitro-1-(4-nitrophenoxy)-4-trifluoromethylbenzene
fluoroglycofen	carboxymethyl 5-[2-chloro-4-(trifluoromethyl)-phenoxy]-2-nitrobenzoate
flurazole	phenylmethyl 2-chloro-4-(trifluoromethyl)-5-thiazolecarboxylate
fluridone	1-methyl-3-phenyl-5-[3-(trifluoromethyl)phenyl]-4-(1*H*)-pyridinone
fluroxypyr	[(4-amino-3,5-dichloro-6-fluoro-2-pyridinyl)oxy]acetic acid
fomesafen	5-[2-chloro-4-(trifluoromethyl)phenoxy]-*N*-(methyl-sulfonyl)-2-nitrobenzamide
fosamine	ethyl hydrogen (aminocarbonyl)phosphonate
glufosinate	(±)-2-amino-4-(hydroxymethylphosphinyl)butanoic acid
glyphosate	*N*-(phosphonomethyl)glycine
haloxydine	3,5-dichloro-2,6-difluoro-4-pyridinol
haloxyfop	2-[4-[[3-chloro-5-(trifluoromethyl)-2-pyridinyl]oxy]phenoxy]propanoic acid
hexaflurate	potassium hexafluoroarsenate
hexazinone	3-cyclohexyl-6-(dimethylamino)-1-methyl-1,3,5-triazine-2,4(1*H*,3*H*)-dione
HOE 70542	ethyl 1-(2,4-dichlorophenyl)-5-trichloromethyl-1*H*-1,2,4-triazole-3-carboxylate
IAAA	3-indoleacetic acid
imazapyr	(±)-2-[4,5-dihydro-4-methyl-4-(1-methylethyl)-5-oxo-1*H*-imidazol-2-yl]-3-pyridinecarboxylic acid
imazaquin	2-[4,5-dihydro-4-methyl-4-(1-methylethyl)-5-oxo-1*H*-imidazol-2yl]-3-quinolinecarboxylic acid
imazethapyr	(±)-2-[4,5-dihydro-4-methyl-4-(1-methylethyl)-5-oxo-1*H*-imidazol-2-yl]-5-ethyl-3-pyridine-carboxylic acid
ioxynil	4-hydroxy-3,5-diiodobenzonitrile
isopropalin	4-(1-methylethyl)-2,6-dinitro-*N*,*N*-dipropyl-benzenamine
isoproturon	*N*,*N*-dimethyl-*N*′-[4-(1-methylethyl)phenyl]urea

isouron	*N'*-[5-(1,1-dimethylethyl)-3-isoxazolyl]-*N,N*-dimethylurea
isoxaben	*N*-[3-(1-ethyl-1-methylpropyl)-5-isoxazolyl]-2,6-dimethoxybenzamide
karbutilate	3-[[(dimethylamino)carbonyl]amino]phenyl (1,1-dimethylethyl)carbamate
lactofen	(±)-2-ethoxy-1-methyl-2-oxoethyl 5-[2-chloro-4-(trifluoromethyl)phenoxy]-2-nitrobenzoate
lenacil	3-cyclohexyl-6,7-dihydro-1*H*-cyclopentapyrimidine-2,4(3*H*,5*H*)-dione
linuron	*N'*-(3,4-dichlorophenyl)-*N*-methoxy-*N*-methylurea
MAA	methylarsonic acid
MCPA	(4-chloro-2-methylphenoxy)acetic acid
MCPB	4-(4-chloro-2-methylphenoxy)butanoic acid
mecoprop	(±)-2-(4-chloro-2-methylphenoxy)propanoic acid
mefenacet	2-(2-benzothiazolyloxy)-*N*-methyl-*N*-phenylacetamide
mefluidide	*N*-[2,4-dimethyl-5-[[(trifluoromethyl)-sulfonyl]amino]phenyl]acetamide
mephenate	4-chlorophenyl methylcarbamate
metamitron	4-amino-4,5-dihydro-3-methyl-6-phenyl-1,2,4-triazin-5-one
metazachlor	*N*-(2,6-dimethylphenyl)-*N*-(1-pyrazol-1-ylmethyl)acetamide (9cl)
metflurazon	4-chloro-5-(dimethylamino)-2-[(3-trifluoromethyl)phenyl]-3(2*H*)-pyridazinone
methabenzthiazuron	*N*-2-benzothiazolyl-*N,N'*-dimethylurea
methazole	2-(3,4-dichlorophenyl)-4-methyl-1,2,4-oxadiazolidine-3,5-dione
methoprotryne	*N*-(3-methoxypropyl)-*N'*-(1-methylethyl)-6-(methylthio)-1,3,5-triazine-2,4-diamine
methoxyphenone	4 methoxy-3,3'-dimethylbenzophenone
metobromuron	*N'*-(4-bromophenyl)-*N*-methoxy-*N*-methylurea
metolachlor	2-chloro-*N*-(2-ethyl-6-methylphenyl)-*N*-(2-methoxy-1-methylethyl)acetamide
metoxuron	*N'*-(3-chloro-4-methoxyphenyl)-*N,N*-dimethylurea
metribuzin	4-amino-6-(1,1-dimethylethyl)-3-(methylthio)-1,2,4-triazin-5(4*H*)-one
metsulfuron	2-[[[[(4-methoxy-6-methyl-1,3,5-triazin-2-yl)amino]carbonyl]amino]sulfonyl]benzoic acid

MH	1,2-dihydro-3,6-pyridazinedione
molinate	*S*-ethyl hexahydro-1*H*-azepine-1-carbothioate
monalide	*N*-(4-chlorophenyl)-2,2-dimethylpentanamide
monochloroacetic acid	
monolinuron	*N*′-(4-chlorophenyl)-*N*-methoxy-*N*-methylurea
monuron	*N*′-(4-chlorophenyl)-*N*,*N*-dimethylurea
morfamquat	1,1′-bis[2-(3,5-dimethyl-4-morpholinyl)-2-oxoethyl]-4,4′-bipyridinium ion
NAA	1-naphthaleneacetic acid
naphthalic anhydride	naphthalene-1,8-dicarboxylic acid anhydride
napropamide	*N*,*N*-diethyl-2-(1-naphthalenyloxy)propanamide
naptalam	2-[(1-naphthalenylamino)carbonyl]benzoic acid
neburon	*N*-butyl-*N*′-(3,4-dichlorophenyl)-*N*-methylurea
nitralin	4-(methylsulfonyl)-2,6-dinitro-*N*,*N*-dipropyl-benzenamine
nitrofen	2,4-dichloro-1-(4-nitrophenoxy)benzene
nitrofluorfen	2-chloro-1-(4-nitrophenoxy)-4-(trifluoromethyl)-benzene
norea	3a*α*,4*α*,5*α*,7*α*,7a*α*-*N*,*N*-dimethyl-*N*′-(octahydro-4,7-methano-1*H*-inden-5-yl)urea
norflurazon	4-chloro-5-(methylamino)-2-[3-(trifluoromethyl)phenyl]-3(2*H*)-pyridazione
orbencarb	*S*-[2-(chlorophenyl)methyl] diethylcarbamothioate
oryzalin	4-(dipropylamino)-3,5-dinitrobenzenesulfonamide
oxabetrinil	*α*-[(1,3-dioxolan-2-yl)methoxyimino]benzene-acetonitrile
oxadiazon	3-[2,4-dichloro-5-(1-methylethoxy)phenyl]-5-(1,1-dimethylethyl)-1,3,4-oxadiazol-2(3*H*)-one
oxyfluorfen	2-chloro-1-(3-ethoxy-4-nitrophenoxy)-4-(trifluoromethyl)benzene
paclobutrazol	(*R**,*R**)-(±)-*β*-[(4-chlorophenyl)methyl]-*α*-(1,1-dimethylethyl)-1*H*-1,2,4-triazole-1-ethanol
paraquat	1,1′-dimethyl-4,4′-bipyridinium ion
pebulate	*S*-propyl butylethylcarbamothioate
pendimethalin	*N*-(1-ethylpropyl)-3,4-dimethyl-2,6-dinitro-benzenamine
pentanochlor	*N*-(3-chloro-4-methylphenyl)-2-methylpentanamide

perfluidone	1,1,1-trifluoro-*N*-[2-methyl-4-(phenylsulfonyl)phenyl]methanesulfonamide
phenisopham	3-[[(1-methylethoxy)carbonyl]amino]phenyl ethylphenylcarbamate
phenmedipham	3-[(methoxycarbonyl)amino]phenyl (3-methylphenyl)carbamate
phenophylate	*N*(2,4-dichlorophenoxycarbonyl)pyrrolidine
phosphinothricin (glufosinate)	(±)-2-amino-4-(hydroxymethylphosphinyl)butanoic acid (9 Cl)
picloram	4-amino-3,5,6-trichloro-2-pyridinecarboxylic acid
piperophos	*S*-[2-(2-methyl-1-piperidinyl)-2-oxoethyl] *O,O*-dipropylphosphorodithioate
pretilachlor	2-chloro-*N*-(2,6-diethylphenyl)-*N*-(2-propoxyethyl)acetamide
prodiamine	2,4-dinitro-N^3,N^3-dipropyl-6-(trifluoromethyl)-1,3-benzenediamine
profluralin	*N*-(cyclopropylmethyl)-2,6-dinitro-*N*-propyl-4-(trifluoromethyl)benzenamine
proglinazine	*N*-[4-chloro-6-[(1-methylethyl)amino]-1,3,5-triazin-2-yl]glycine
prometon	6-methoxy-*N,N′*-bis(1-methylethyl)-1,3,5-triazine-2,4-diamine
prometryn	*N,N′*-bis(1-methylethyl)-6-(methylthio)-1,3,5-triazine-2,4-diamine
pronamide	3,5-dichloro-*N*-(1,1-dimethyl-2-propynyl)benzamide
propachlor	2-chloro-*N*-(1-methylethyl)-*N*-phenylacetamide
propanil	*N*-(3,4-dichlorophenyl)propanamide
propazine	6-chloro-*N,N′*-bis(1-methylethyl)-1,3,5-triazine-2,4-diamine
propham	1-methylethyl phenylcarbamate
pyrazon	5-amino-4-chloro-2-phenyl-3(2*H*)-pyridazinone
pyrichlor	2,3,5-trichloro-4-pyridinol
pyridate	*O*-(6-chloro-3-phenyl-4-pyridazinyl) *S*-octyl carbonothioate
pyrozolate (DTP)	1,3-dimethyl-4-(2,4-dichlorobenzoyl)-5-hydroxypyrazole
quinchlorac	3,7-dichloro-8-quinolinecarboxylic acid
quinmerac	7-chloro-3-methyl-8-quinolinecarboxylic acid
quizalofop	(±)-2-[4-[(6-chloro-2-quinoxalinyl)oxy]phenoxy]propanoic acid
secbumeton	*N*-ethyl-6-methoxy-*N′*-(1-methylpropyl)-1,3,5-triazine-2,4-diamine

sethoxydim	2-[1-(ethoxyimino)butyl]-5-[2-(ethylthio)propyl]-3-hydroxy-2-cyclohexen-1-one
siduron	*N*-(2-methylcyclohexyl)-*N*′-phenylurea
simazine	6-chloro-*N*,*N*′-diethyl-1,3,5-triazine-2,4-diamine
simetryn	*N*,*N*′-diethyl-6-(methylthio)-1,3,5-triazine-2,4-diamine
sulfometuron	2-[[[[(4,6-dimethyl-2-pyrimidinyl)amino]-carbonyl]amino]sulfonyl]benzoic acid
2,4,5-T	(2,4,5-trichlorophenoxy)acetic acid
2,4,6-T	(2,4,6-trichlorophenoxy)acetic acid
2,3,6-TBA	2,3,6-trichlorobenzoic acid
TCA	trichloroacetic acid
tebutam	2,2-dimethyl-*N*-(1-methylethyl)-*N*-(phenyl-methyl)propanamide
tebuthiuron	*N*-[5-(1,1-dimethylethyl)-1,3,4-thiadiazol-2-yl]-*N*,*N*′-dimethylurea
terbacil	5-chloro-3-(1,1-dimethylethyl)-6-methyl-2,4(1*H*,3*H*)-pyrimidinedione
terbumeton	*N*-(1,1-dimethylethyl)-*N*′-ethyl-6-methoxy-1,3,5-triazine-2,4-diamine
terbuthylazine	6-chloro-*N*-(1,1-dimethylethyl)-*N*′-ethyl-1,3,5-triazine-2,4-diamine
terbutryn	*N*-(1,1-dimethylethyl)-*N*′-ethyl-6-(methylthio)-1,3,5-triazine-2,4-diamine
tetcyclasis	(3aα,4β,4aα,6aα,7β,7aα)-1-(4-chlorophenyl)-3a,4,4a,6a,7,7a-hexahydro-4,7-methano-1*H*-[1,2]diazeto[3,4-*f*]benzotriazole
thiameturon	3-[[[[(4-methoxy-6-methyl-1,3,5-triazin-2-yl)amino]carbonyl]amino]sulfonyl]-2-thio-phenecarboxylic acid
thiazafluron	*N*,*N*′-dimethyl-*N*-[(5-trifluoromethyl)-1,3,4-thiadiazol-2-yl]urea
thiobencarb	*S*-[(4-chlorophenyl)methyl] diethylcarbamothioate
thiocarbazil	*S*-(phenylmethyl) bis(1-methylpropyl)carbamo-thioate
tralkoxydim	2-[1-(ethoxyimino)prophyl]-3-hydroxy-5-mesityl cyclohex-2-enone
triallate	*S*-(2,3,3-trichloro-2-propenyl) bis(1-methylethyl)-carbamothioate
triclopyr	[(3,5,6-trichloro-2-pyridinyl)oxy]acetic acid
tridiphane	2-(3,5-dichlorophenyl)-2-(2,2,2-trichloroethyl)-oxirane

trietazine	6-chloro-*N,N,N′*-triethyl-1,3,5-triazine-2,4-diamine
trifluralin	2,6-dinitro-*N,N*-dipropyl-4-(trifluoromethyl)-benzenamine
vernolate	*S*-propyl dipropylcarbamothioate
xylachlor	2-chloro-*N*-(2,3-dimethylphenyl)-*N*-(1-methylethyl)acetamide

Index

GPSR Compliance

The European Union's (EU) General Product Safety Regulation (GPSR) is a set of rules that requires consumer products to be safe and our obligations to ensure this.

If you have any concerns about our products, you can contact us on ProductSafety@springernature.com

In case Publisher is established outside the EU, the EU authorized representative is:

Springer Nature Customer Service Center GmbH
Europaplatz 3
69115 Heidelberg, Germany

Batch number: 10755284

Printed by Printforce, the Netherlands